Electron-Molecule Scattering and Photoionization

PHYSICS OF ATOMS AND MOLECULES

Electron-Molecule Scattering and Photoionization

Edited by
P. G. Burke
The Queen's University of Belfast
Belfast, Northern Ireland

and
J. B. West
SERC Daresbury Laboratory
Warrington, England

Plenum Press • New York and London

Library of Congress Cataloging in Publication Data

International Symposium on Electron–Molecule Scattering and Photoionization (1987: Daresbury Laboratory)
  Electron–molecule scattering and photoionization / edited by P. G. Burke and J. B. West
    p.   cm. — (Physics of atoms and molecules)
  "Proceedings of an International Symposium on Electron–Molecule Scattering and Photoionization, held July 18-19, 1987, at SERC Daresbury Laboratory, Warrington, England" — T.p. verso.
  Bibliography: p.
  Includes index.
  ISBN-13:978-1-4612-8309-6   e-ISBN-13:978-1-4613-1049-5
  DOI: 10.1007/978-1-4613-1049-5
  1. Electron–molecule scattering — congresses. 2. Photoionization — Congresses. I. Burke, P. G. II. West, J. B. (John B.) III. Title. IV. Series.
QC793.5.E6281583   1987   88-19610
539.7'2112 — dc19   CIP

Proceedings of an International Symposium on Electron–Molecule
Scattering and Photoionization, held July 18-19, 1987,
at SERC Daresbury Laboratory, Warrington, England

© 1988 Plenum Press, New York
Softcover reprint of the hardcover 1st edition 1988

A Division of Plenum Publishing Corporation
233 Spring Street, New York, N.Y. 10013

PREFACE

This volume contains the invited papers and selected contributed papers presented at the International Symposium on 'Electron-Molecule Scattering and Photoionization' held at SERC's Daresbury Laboratory, Cheshire, England from 18th to 19th July, 1987. This Symposium was a Satellite Meeting to the XVth International Conference on the Physics of Electronic and Atomic Collisions (ICPEAC) and follows a tradition of Satellite Meetings in related areas of collisions held in association with previous ICPEAC's. In order to make this volume as representative of the Symposium as possible 'Hot Topics' presented orally at the meeting together with a few papers selected by the Programme Committee from the contributed posters are included. The Editors are grateful to the authors for responding rapidly to the invitation to submit their contributions for inclusion in the volume, as indeed they are grateful to all the authors for the high quality of their contributions.

The Symposium brought together over 100 scientists from many countries and from broad interdisciplinary backgrounds to hear about current rapid advances in electron-molecule scattering and photoionization. These advances have been stimulated on the experimental side by the increasing availability of electron beams with millivolt energy resolution, by synchrotron radiation sources and by intense tunable lasers. On the theoretical side the introduction of new computational methods enables accurate predictions to be made, resulting in a new and deeper understanding of the basic physical processes involved. Demands for a detailed knowledge of these processes in many applications where photons, electrons and molecules are involved, for example in plasma physics, laser physics, atmospheric and interstellar science, isotope separation, MHD power generation, electrical discharges and radiation chemistry and physics, have stimulated research in this area.

We now turn to a discussion of the contributions on the experimental side. J.L. Dehmer describes how the high resolution of the laser is used in multiphoton ionization experiments to carefully prepare initial excited states with well-defined quantum numbers for subsequent ionization through a resonant transition, demonstrating the power and sensitivity of the technique with some data on hydrogen. Recent work on double photoionization is covered by I. Nenner, using it as a probe of electron correlation in atoms and molecules, particularly in the latter case where molecular relaxation can play a role in the process. Multiple photoionization is attracting more attention recently, as is evident from A. Yagishita's summary of work done at the Photon Factory in Japan on atoms and diatomic molecules. In this case the work shows how the branching ratios in the various multiply. charged states vary with photon energy, and examines the various processes, mainly associated with auto-ionization, which lead to the charge states seen.

A review of electron momentum spectroscopy by E. Weigold demonstrates how powerful this technique has become in recent years, having progressed beyond energy level measurements to the determination of

electron momentum distributions. Experimental mapping of wavefunctions
for molecules is described and shows how this technique has been of
considerable help in guiding and assisting development of theory. The
subject of shape resonant phenomena is taken up by R. Hall; he shows how
high sensitivity measurements capable of measuring cross sections as
small as 10^{-22} cm^2 highlight resonant phenomena, and goes on to demonstrate how the general technique of low energy electron scattering is
applied to the bonding mechanisms of molecules on surfaces. Further work
along these lines, on physisorbed O_2, is described by R.E. Palmer using
the angular distribution of the scattered electrons to determine
molecular orientation. G.C. King summerized recent threshold electron
measurements using both electron impact and photoionization: he has been
able to study individual metastable excitation functions and also to
examine Wannier ridge resonances in the case of helium near the double
ionization threshold.

In an unusual application of synchrotron radiation D. Field shows
how very high resolution electron beams of low energy can be produced and
demonstrates their application to precision spectroscopy of negative
molecular ion resonances. G. Knoth describes further experiments using
low energy electron impact, concentrating on rotational excitations; by
examining the angular distributions of the scattered electrons he deduces
that several partial waves are involved, rather than a Feshbach type
resonance. N. Böwering gives an account of experiments on HI using
multiphoton ionization to provide rotationally resolved photoelectron
spectra and circularly polarized synchrotron radiation to yield information on electron spin polarisation, thus providing a very detailed
picture of the ionization of this molecule.

On the theoretical side the scene is set by comprehensive reviews by
L.A. Collins and B.I. Schneider on recent ab-initio methods for electron-
molecule collisions with emphasis on diatomic molecules and by
F.A. Gianturco and S. Scialla, who discuss recent advances which have been
made in theory and calculation of electron scattering by polyatomic
molecules. In the area of diatomic molecules, new approaches such as the
linear algebraic equations method, the Schwinger variation method and the
R-matrix method have reached a stage where accurate ab-initio procedures
are beginning to open up this field, while for polyatomic molecules new
ab-initio exchange and polarization potentials are now being used giving
excellent agreement with experiment. A. Herzenberg then reviews
resonance collisions of electrons with molecules and solids. The
important role which resonances play in electron molecule collisions was
first established in the early 1960s, however there are still many
controversial features of such collisions such as their application to
threshold peaks similar to those observed in HF and HCl. In addition,
exciting new developments in resonance collisions in solids are discussed
and their relationship to gas phase processes reviewed. The role of
resonances in electron-molecule scattering is discussed further in the
next paper by J.P. Gauyacq, who develops an ab-initio effective range
theory for e-O_2 collisions.

The following two papers by B.C. Saha and by C.A. Weatherford and
A. Temkin discuss recent accurate low energy e-N_2 scattering calculations. The low energy cross section in this system is dominated by non-
resonant scattering in the $^2\Pi_g$ state and has provided a rich testing
ground for electron-molecule experiment and theory. These papers show
that accurate calculations for this system are being carried out. The
following paper by C. Gillan also discusses e-N_2 scattering as well as
recent calculations for e-HF scattering obtained using the R-matrix
method. It is shown that this method provides a general framework for
discussing a broad range of electron-molecule collision processes.

Turning to photoionization, the next paper by J. Tennyson presents recent calculations of the vibrationally and rotationally resolved photoelectron spectra of H_2. Surprisingly for such a simple system, there are still unresolved questions, but this paper goes further than any previous work in answering them. Finally, the last paper by E.A.G. Armour discusses the effect of a change of charge in the scattering of low energy positrons by hydrogen molecules. This introduces new processes such as positronium formation and positron annihilation which are not present in electron scattering. Nevertheless, this paper shows that very rapid advances are being made in this complementary area of scattering.

The Editors wish to conclude by acknowledging with thanks the financial support received from CRAY Research Inc., Floating Point Systems, the SERC Synchrotron Radiation Facility Committee, the SERC Collaborative Computational Project No. 2, and from Plenum Publishing Corporation. They also express their sincere thanks to Mrs. Shirley Lowndes and her staff at the Daresbury Laboratory, whose dedication, help and support contributed greatly to the success of the Symposium.

P.G. Burke and J.B. West

CONTENTS

PHOTOIONIZATION DYNAMICS OF EXCITED MOLECULAR STATES[*]

J. L. Dehmer, M. A. O'Halloran, F. S. Tomkins, P. M. Dehmer, and S. T. Pratt

Argonne National Laboratory
Argonne, Illinois 60439, U.S.A.

INTRODUCTION

Resonance Enhanced Multiphoton Ionization (REMPI) utilizes tunable dye lasers to ionize an atom or molecule by first preparing an excited state by multiphoton absorption and then ionizing that state before it can decay. This process is highly selective with respect to both the initial and resonant intermediate states of the target, and it can be extremely sensitive. In addition, the products of the REMPI process can be detected as needed by analyzing the resulting electrons, ions, fluorescence, or by additional REMPI. This points to a number of opportunities for exploring excited state physics and chemistry at the quantum-state-specific level. Here we will first give a brief overview of the large variety of experimental approaches to excited state phenomena made possible by REMPI. Then we will examine in more detail, recent studies of the three photon resonant, four photon (3+1) ionization of H_2 via the C $^1\Pi_u$ state. Strong non-Franck-Condon behavior in the photoelectron spectra of this nominally simple Rydberg state has led to the examination of a variety of dynamical mechanisms. Of these, the role of doubly excited autoionizing states now seems decisive. Progress on photoelectron studies of autoionizing states in H_2, excited in a (2+1) REMPI process via the E,F $^1\Sigma_g^+$ will also be briefly discussed.

To illustrate the potential of REMPI, we will outline several different types of experiments that can be carried out using the REMPI excitation processes shown schematically in Figure 1. In Figure 1a, two photons from a "pump" laser with frequency $h\nu_1$ are used to excite an individual rotational (not shown) and vibrational level of an excited electronic state AB^*. An independently tunable "probe" laser of frequency $h\nu_2$ is used to further excite the AB^* level to the manifold of rotational-vibrational levels of a higher excited state AB^{**}. A third photon of frequency $h\nu_1$ or $h\nu_2$ is used to ionize AB^{**}. In this case, one is interested in the $AB^* \rightarrow AB^{**}$ transition rather than the ionization step, so the continuum is represented simply by a structureless hatched area. In Figure 1b, a similar process is indicated; however, in this

---

[*]Work supported by U.S. Department of Energy, Office of Health & Environmental Research, under Contract W-31-109-Eng-38, and by Office of Naval Research.

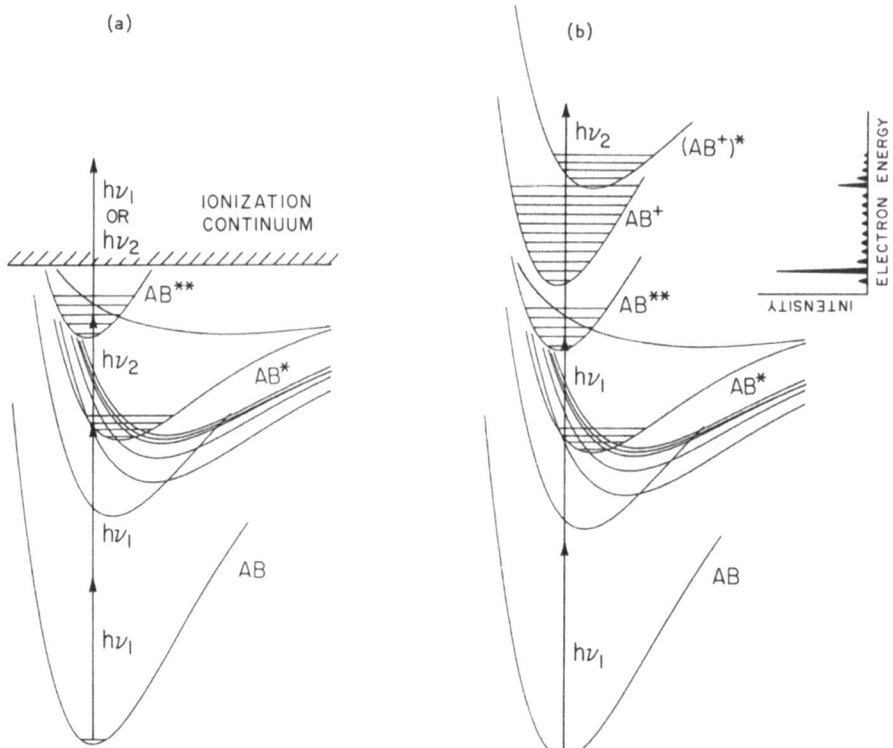

Fig. 1. Schematic potential energy diagram showing two different REMPI processes in a diatomic molecule.

case, the state AB** is produced by two photons of the pump laser, with the probe laser accessing the ionization continuum directly. Here, one is interested in the ionization step itself, and so the accessible ionic states AB$^+$ and (AB$^+$) are shown explicitly.

The resonant multicolor excitation schemes represented schematically in Figure 1 permit us to address many problems in molecular science which were either very difficult or unimaginable with conventional excitation sources. These include the following. (a) By varying hν_2 (in Figure 1a) and detecting the total (or mass selected) ion current as a function of wavelength, one performs optical-optical double resonance (OODR) spectroscopy. This generates spectroscopic information on AB** which typically lies in the vacuum ultraviolet (VUV) with single photon sources, but in the visible or ultraviolet with multiphoton sources. This produces high resolution spectroscopic information without the need of a large vacuum spectrograph. More importantly, the hν_2 transition originates from a single rotational level of AB*, which greatly simplifies the spectrum. Use of OODR techniques also enables the direct study of states that are dipole forbidden in single photon excitation. (b) If the excited state AB** is predissociated, e.g., by the repulsive curve in Figure 1, it is possible to probe in detail the mechanisms of the dissociation process by analyzing both the internal energy distribution of the photofragments and the time dependence of their formation. In addition, photodissociation often is one of the simplest and most convenient methods of producing open shell atoms, free radicals or transient species for further spectroscopic study. (c) Measurement of

2

the photoelectron energy distribution (indicated by the inset in Figure 1b) will give the relative probabilities of producing alternative ionic states and, thus will directly reflect the photoionization dynamics of individual excited quantum states. It also will be possible to determine photoelectron branching ratios at various points within an autoionizing resonance, which will be an extremely sensitive probe of the interactions between the discrete state and the various ionization continua. At present, such measurements are being performed using synchrotron radiation light sources with modest wavelength resolution ($\sim$ 0.2-0.5 Å); however, this wavelength resolution is rarely sufficient to sample different regions within a single autoionizing resonance. (d) Since the ionization step in Figure 1b is performed with a visible or UV wavelength, simple rotation of a retardation plate will produce a photoelectron angular distribution, which accesses further dynamical information and also reflects the orientation of the excited state AB^{**}, resulting from the multiphoton excitation process. (e) Preparation of an excited state AB^{**}, followed by a delayed laser probe can monitor the time evolution of intramolecular rearrangement and/or decay processes. In molecules more complicated than that indicated in Figure 1, this procedure can monitor the time evolution of vibrational energy redistribution. In this case, picosecond lasers would be required to capture the normally very fast internal rearrangement. Use of a delayed probe beam can also be used to characterize collisional effects on a prepared state. (f) Using the high degree of selectivity, and hence, sensitivity of either excitation mechanism in Figure 1, it is possible to directly probe free radicals, clusters, ions and other transient species which are formed as minor components in complex mixtures. (g) Many possible chemical uses of the general scheme in Figure 1 can also be readily seen. For instance, by suitable selection of AB^{**} in Figure 1b, it is possible to produce AB^{+} or AB^{+*} in particular vibrational and rotational quantum states in order to study the dependence of subsequent chemical transformations on varying degrees of internal energy in different electronic or nuclear modes. Also, by using the selectivity of the excitation process, it is possible to monitor the reactants and products of elementary chemical reactions at the quantum-state-specific level.

This list of possibilities is not exhaustive, but it is ample to show the great scientific potential of REMPI. In what follows, we will present specific examples of REMPI studies in hydrogen in order to illustrate the utility of multiphoton excitation in gaining new insight into molecular photoionization dynamics.

PHOTOIONIZATION OF H_2 C $^1\Pi_u$, v', J'

Background

Several years ago, we reported the photoelectron spectra obtained following three photon resonant, four photon (3+1) ionization via the C $^1\Pi_u$, v' = 0 - 4 levels of molecular hydrogen (Pratt et al., 1984). The C $^1\Pi_u$ Rydberg state corresonds to the $1s\sigma_g 2p\pi_u$ configuration, and has a potential curve similar to that of the ground state of H_2^+ (Huber and Herzberg, 1979). Thus, the vibrational overlap integrals between a particular vibrational level, v', of the C $^1\Pi_u$ state and the vibrational levels of the X $^2\Sigma_g^+$ ion, v^+, will be nearly unity for $v^+ = v'$, and nearly zero for $v^+ \neq v'$. On the basis of the Franck-Condon principle, the photoelectron spectra are therefore expected to show strong $v^+ = v'$ peaks, with very little intensity in the $v^+ \neq v'$ peaks. Qualitatively, these expectations are fulfilled, as seen in Figure 2, which shows the photoelectron spectra obtained following (3+1) ionization via the C $^1\Pi_u$, v' = 0 - 4 Q(1) transitions. These spectra were obtained along the

3

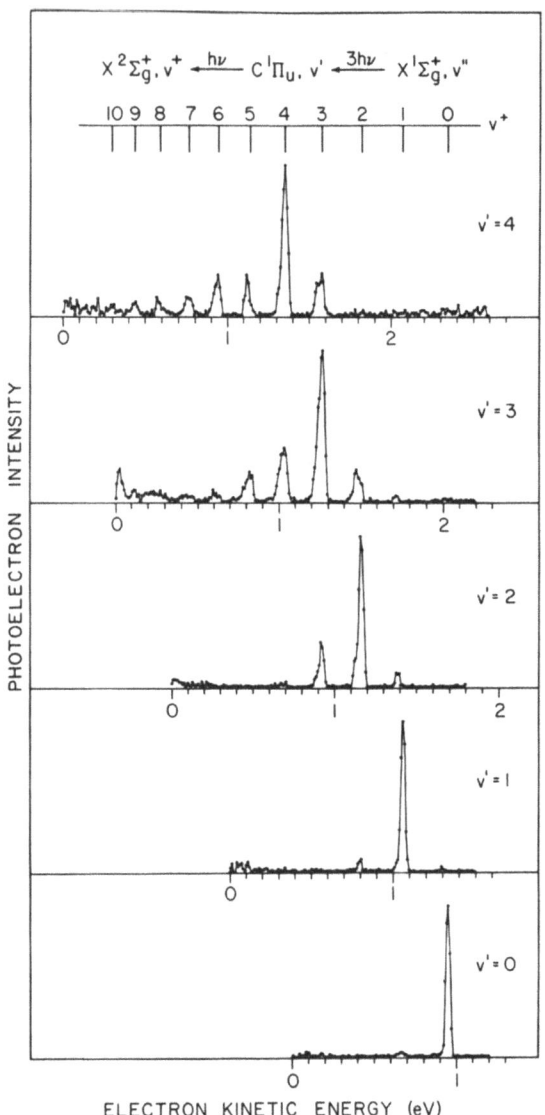

Fig. 2. Photoelectron spectra of H_2 determined along the laser polarization axis ($\theta = 0°$) at the wavelengths of the resonant three-
photon $C\ ^1\Pi_u$, $v' = 0 - 4 \leftarrow X\ ^1\Sigma_g^+$, $v'' = 0$, Q(1) transitions.

laser polarization axis ($\theta = 0°$) using an electrostatic energy analyzer. The Q(1) transitions were chosen to access the Π^- component of the C $^1\Pi_u$ state, which is unperturbed by the B $^1\Sigma_u^+$ state. In each spectrum the $v^+ = v'$ peak is the most intense and the $v^+ \neq v'$ peaks are significantly weaker, in accord with the Franck–Condon arguments. However, a quantitative comparison of the relative intensities with theoretical Franck–Condon factors reveals significant discrepancies, particularly for $v' = 3$ and 4. For example, in the $v' = 4$ spectrum the observed $v^+ = 3$, 5, and 6 peaks are too large by factors of 3, 2, and 23, respectively (Pratt et al., 1984).

The discrepancy between the experimental results and theoretical Franck–Condon factors could have a number of sources. These include (1) a kinetic energy dependence of the electronic transition matrix element, which must be taken into account even within the Franck–Condon approximation; (2) an internuclear distance (R) dependence of the same electronic matrix element which, by definition, constitutes a breakdown of the Franck–Condon approximation; and (3) a v^+-dependence of the photoelectron angular distributions. Dixit et al. (1984) have included all three effects in theoretical calculations of the $\theta = 0°$ spectra of Figure 2. However, the agreement with experiment, while improved, is still not good. Recently, we have measured angle-integrated branching ratios using two different techniques. In the first experiment (Pratt et al., 1986), the photoelectron angular distributions were determined for each spectrum and then integrated to give branching ratios. In the second experiment (O'Halloran et al., 1987) the integrated branching ratios were determined directly using a magnetic bottle electron spectrometer with 2π steradian collection efficiency (Kruit and Read, 1983). The two measurements are in good agreement. Figure 3 shows a comparison of the angle-integrated branching ratios calculated by Dixit et al. (1984) with the results of O'Halloran et al. (1987). The discrepancies are quite apparent, particularly for $v' = 3$ and 4, where the observed distributions are much broader than the theoretical predictions. This indicates that the photoionization of the C $^1\Pi_u$ state is more complicated than the direct excitation of the Rydberg electron into the X $^2\Sigma_g^+$ continuum, and suggests that another mechanism is important for higher v'. It is worth noting that the photoelectron angular distributions (Pratt et al., 1986) for the $v^+ \neq v'$ photoelectron bands are generally more isotropic than those for the $v^+ = v'$ bands, which also suggests another mechanism is responsible for the observed intensity of these bands.

Additional evidence for the complexity of the photoionization process is found in the rotational structure of the vibrational bands in the C $^1\Pi_u$, $v' = 0 - 4$ spectra. Assuming that only s and d partial waves contribute to the outgoing electron, the selection rules for photoionization from the C $^1\Pi_u$, $J' = 1$ state (Pratt et al., 1986; Dixit and McKoy, 1986) indicate that the H_2^+ ion can only be formed in the rotational levels $N^+ = 1$, or 3. This is confirmed by the observation of only $N^+ = 1$ and 3 photoelectron peaks in the C $^1\Pi_u$, $v' = 0 - 4$ spectra shown in Figure 4. These spectra were obtained at somewhat higher resolution than those of Figure 2 by using the magnetic bottle electron spectrometer. (Note that the spectra shown in Figure 4 are angle-integrated spectra.) In these spectra, the relative intensity of the $N^+ = 3$ photoelectron peak increases dramatically with increasing v', both for $v^+ = v'$ and for $v^+ \neq v'$. In addition, for a given intermediate level, v', the $N^+ = 3$ rotational peaks tend to be larger relative to the $N^+ = 1$ peaks in the $v^+ \neq v'$ bands. This definitely indicates that the photoionization mechanism is changing with increasing vibrational quantum number, v', and that this mechanism contributes to the intensity of the $v^+ \neq v'$ bands.

It is sometimes useful to describe photoionization dynamics in terms of the angular momentum transfer, j_t, which is defined as the angular

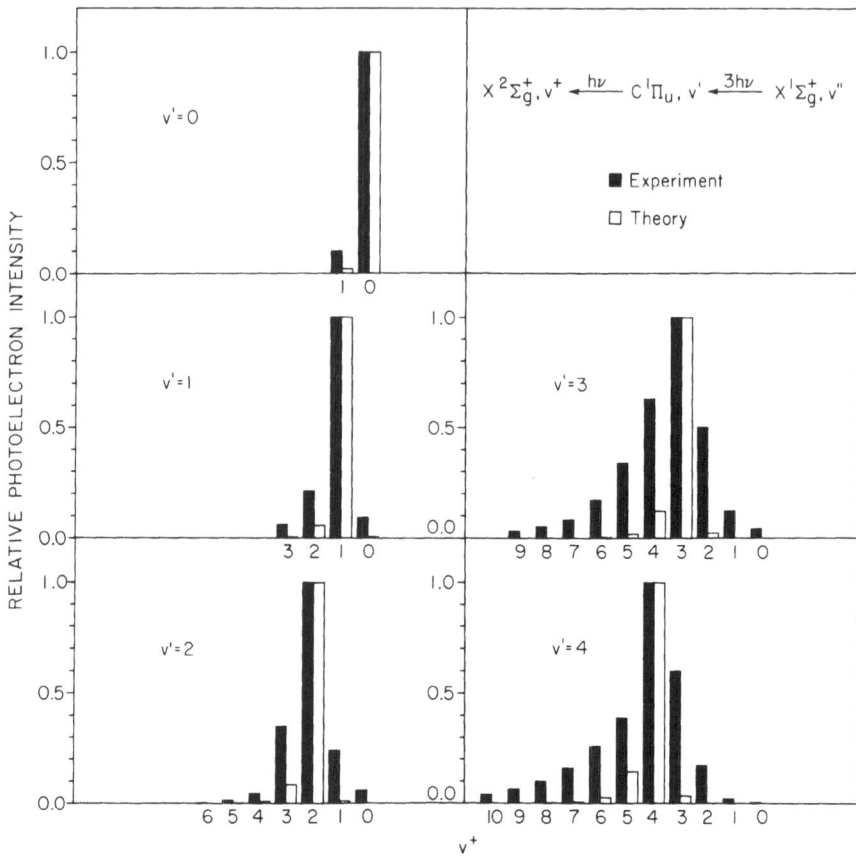

Figure 3. Vibrational branching ratios determined for three photon resonant, four photon ionization of H_2 via $C\ ^1\Pi_u$, v', Q(1) transitions. The vibrational level of the $C\ ^1\Pi_u$ state is denoted by v', and that of the ion by v^+. The calculation is that of Dixit, Lynch, and McKoy (1984).

momentum exchanged between the unobserved initial and final angular momenta (Dill, 1972, 1976; Fano and Dill, 1972). If only s and d partial waves are considered, the N^+ = 1 peaks in Figure 4 can only arise from j_t = 1 processes, while the N^+ = 3 peaks can only arise from j_t = 3 processes (O'Halloran et al., 1987). If the photoionization process is divided into two parts, corresponding to the initial excitation followed by the photoelectron escape, then only j_t = 1 processes can be created in the excitation step (Dill, 1976), and the higher value of angular momentum transfer, j_t = 3, must result from anisotropic interactions between the ion core and the escaping photoelectron. Figure 4 indicates that these anisotropic interactions, and thus the j_t = 3 processes, become increasingly important with increasing vibrational quantum number, and are relatively more important for $v^+ \neq v'$.

The Role of Doubly Excited Electronic States

The increasing intensity of j_t = 3 processes with increasing v' indicates that photoionization from these levels of the $C\ ^1\Pi_u$ state does not proceed by the direct ejection of the Rydberg electron. If, instead, the photoionization process involves excitation to an autoionizing level

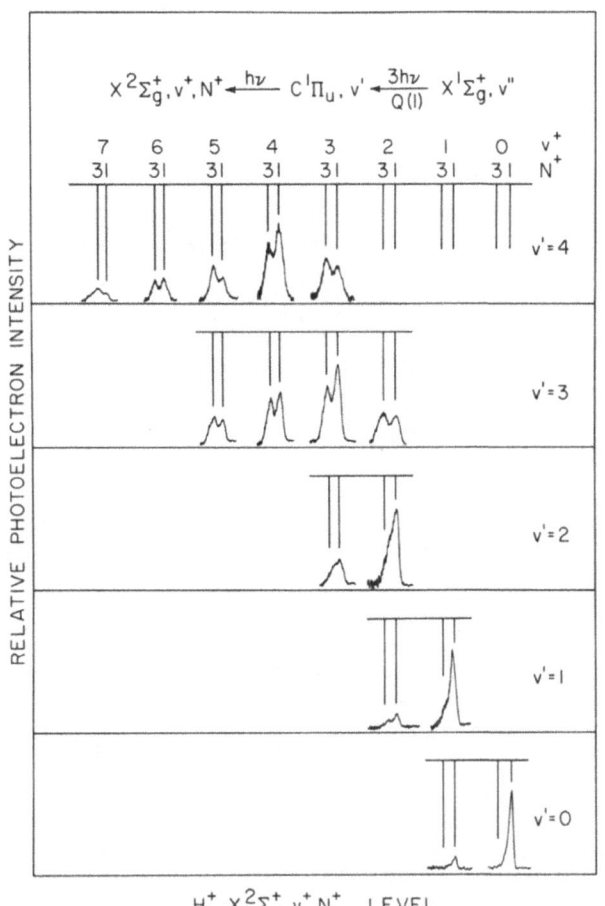

Figure 4. Photoelectron spectra determined at the wavelengths of the three photon resonant H_2 C $^1\Pi_u$, v' + X $^2\Sigma_g^+$, v" = 0 Q(1) transitions. The spectra of individual ionic vibrational bands were recorded separately, with retarding voltages chosen so as to achieve comparable energy resolution for each vibrational band. Note that the horizontal scale does not indicate energy, although the individual vibrational bands are plotted with the same energy scale. The integrated areas of the vibrational bands are set equal to the ionic vibrational branching ratios.

at the four photon energy followed by decay into the ionization continuum, a mechanism would be provided for both the non-Franck-Condon behavior and for the increasing importance of anisotropic electron-ion interactions. On the basis of crude wavelength dependent studies, performed by pumping different rotational levels within the C $^1\Pi_u$, v' bands, we have concluded that it is unlikely that sharp autoionizing resonances are responsible for the observed behavior. In addition, the $\Delta v = -1$ propensity rule for vibrational autoionization (Berry, 1966) and the energetics for rotational autoionization suggest that neither of these processes contributes to the present observations.

However, Chupka (1987) has recently suggested that doubly excited states at the four photon energy will play an important role in the (3+1) ionization via the C $^1\Pi_u$ state. In particular, he has argued that the $2p\sigma_u 2p\pi_u$ doubly excited state will have significant oscillator strength from the $1s\sigma_g 2p\pi_u$ C $^1\Pi_u$ state, and that autoionization of this doubly excited state will lead to non-Franck-Condon vibrational branching ratios. Cornaggia et al. (1987) have also suggested that doubly excited states will be more important for multiphoton ionization via the C $^1\Pi_u$ state into the g continuum than for ionization via the E,F $^1\Sigma_g^+$ state into the u continuum, for which they performed calculations. Independently, Hickman (1987a,b) has performed model calculations of the vibrational branching ratios following autoionization of the $2p\sigma_u 2p\pi_u$ doubly excited state accessed by (3+1) excitation via the C $^1\Pi_u$ state. Using the $2p\sigma_u 2p\pi_u$ potential curve of Guberman (1983), Hickman (1987a,b) has obtained very encouraging agreement with the experimental results.

In a classical or semi-classical time dependent framework (Chupka, 1987), such a process could be viewed as the production of a wave-packet on the repulsive curve of the doubly excited state. As it evolves in time, the wave-packet can be decomposed into outgoing and incoming components; the latter will be reflected and subsequently interfere with the originally outgoing component. Because the doubly excited state has a finite width for autoionization, as the wave-packet evolves it will have some probability for transitions into the $^2\Sigma_g^+$ continuum. In this model, the production of $v^+ > v'$ photoelectron bands arises from autoionization as the wave-packet evolves to dissociation products, while those with $v^+ < v'$ arise from autoionization of the incoming component propagating to smaller R.

This model also introduces the possibility that some of the molecules in the doubly excited state will not autoionize, but rather will dissociate into a ground state atom and an excited state atom. Excited states having n = 3 − 5 have been observed (Bonnie et al., 1986a,b; O'Halloran et al., 1987; Xu et al., 1987), and may result from curve crossings of the repulsive $2p\sigma_u 2p\pi_u$ state with singly excited Rydberg states at large internuclear distance. In general, the dissociation processes for the C $^1\Pi^-$ levels are much weaker than the ionization processes. However, the (3+1) spectra via high-lying vibrational levels of the B $^1\Sigma_u^+$ state in the same energy region exhibit nearly complete dissociation, and the C $^1\Pi^+$ levels, which interact with the B $^1\Sigma^+$ state, generally display significantly more dissociation than the corresponding C $^1\Pi^-$ levels. The increased dissociation for the B $^1\Sigma_u^+$ levels may arise from two sources. First, the B $^1\Sigma_u^+$ state samples a much larger range of internuclear distance than the C $^1\Pi_u$ state, and may have a significant direct photodissociation cross section. Second the $(2p\sigma_u)^2$ doubly excited state is "configurationally" allowed from the $1s\sigma_g 2p\sigma_u$ B $^1\Sigma_u^+$ state, which could produce more dissociation than the $2p\sigma_u 2p\pi_u$ state.

The relative positions of the C $^1\Pi_u$ and $2p\sigma_u 2p\pi_u$ potential curves indicate that the transition will occur from the outer turning point of the C $^1\Pi_u$ state to the inner turning point of the $2p\sigma_u 2p\pi_u$ state. Increasing the vibrational quantum number has two effects: the four photon energy is increased, and the outer turning point of the lower state is moved to larger R. For v' = 0, both the total energy and the outer turning point are too small for transitions to the doubly excited state to be very important. As v' is increased, both the overlap with the doubly excited state and the energy requirement improve, and the effects attributable to the doubly excited state become more noticeable, as is the case for v' = 3 and 4. Eventually, as v' is increased further, the energy will be too high and the overlap again will be too poor for the $2p\sigma_u 2p\pi_u$ state to play a role. In the region just above v' = 4 the qualitative model described above corresponds to excitation to somewhat larger R than the classical inner turning point, with considerable amplitude for autoionization at smaller internuclear distances. This would lead to an increase in the population of vibrational levels of H_2^+ with $v^+ < v'$.

As is seen in Figure 5, these arguments are supported by the photo-electron spectra obtained at θ = 0° following (3+1) ionization via the C $^1\Pi_u$, v' = 5, 6 Q(1) transitions. In both spectra, the v^+ = v' peak is the largest, with much smaller $v^+ \neq$ v' peaks.

Comparison with Figure 2 reveals the distribution of $v^+ \neq$ v' peaks shifts to smaller values of v^+ with increasing v'. In addition, the sum of the intensities of the $v^+ \neq$ v' peaks relative to the v^+ = v' peak decreases monotonically as v' is increased from 4 to 6. The C $^1\Pi_u$, v' = 7, 8 ← X $^1\Sigma_g^+$, v' = 0 bands are overlapped by the much more intense B' $^1\Sigma_u^+$, v' = 1,2 ← X $^1\Sigma_g^+$, v' = 0 bands (Namioka, 1964a,b), and were not studied. Although the C $^1\Pi_u$, v' = 9 Q(1) transition is blended with the D $^1\Pi_u$, v' = 1 Q(1) transition (Namioka, 1964a,b), some information can nevertheless be obtained regarding ionization of the C $^1\Pi_u$, v' = 9 level. The photoelectron spectrum following (3+1) ionization at this wavelength (Pratt et al., 1987a) is shown in the center frame of Figure 6. The D $^1\Pi_u$ state corresponds to the $1s\sigma_g 3p\pi_u$ Rydberg state, and the v' = 1 photoelectron spectrum is similar to that of the C $^1\Pi_u$, v' = 1 level. However, the small v' = 9 peak (at ~ 1.7 eV) almost certainly corresponds to ionization via the C $^1\Pi_u$, v' = 9 level. Although the v' = 9 peak is weak, it is interesting to note that no intensity is observed for v^+ = 5 - 12, which suggests that at this energy and inter-nuclear distance, the $2p\sigma_u 2p\pi_u$ state no longer plays an important role in the ionization process.

Photoelectron Spectra of D_2

Doubly excited states at the four photon energy are also expected to play a role in the (3+1) ionization via the C $^1\Pi_u$ states of the heavier isotopes of H_2. In particular, we have recently recorded the photo-electron spectra following (3+1) ionization via the C $^1\Pi_u$, v' = 0 - 4 levels of D_2 (Pratt et al., 1987b). The spectra for the C $^1\Pi_u$, v' = 0 - 3 ← X $^1\Sigma_g^+$, v" = 0, Q(3) transitions are shown in Figure 7. As in H_2, the v^+ = v' peak dominates each spectrum. The most striking difference between the H_2 and D_2 C $^1\Pi_u$ photoelectron spectra is that for v' = 3 and 4, the H_2 spectra extend to v^+ = v' + 6, while the D_2 spectra show significant intensity only for v^+ - v' ≤ 2. Although there are several possible explanations for this observation, they are all consistent with the model involving the $2p\sigma_u 2p\pi_u$ doubly excited state. The doubly excited potential curves for H_2 and D_2 will be nearly identical, and the mass effect on the electronic autoionization width should be small. However, because of the difference in vibrational spacings in the C $^1\Pi_u$ and X $^1\Sigma_g^+$ states, the four photon energies for v' = 3 and 4 are smaller (by ~ 0.26 and 0.34 eV, respectively) in D_2 than in H_2, which will make excitation to the doubly excited curve energetically less favorable. In

9

addition, for lower values of v', the smaller range of R sampled by the D_2 vibrational wavefunctions in the C $^1\Pi_u$ state will decrease the vibrational overlap with the $2p\sigma_u2p\pi_u$ state. Finally, if the D_2 is excited to the same position on the doubly excited potential curve as H_2, the D_2 wave-packet will propagate at only $1/\sqrt{2}$ of the speed of the H_2 wave-packet. Thus, with the same autoionization rate, a much narrower envelope is expected for D_2. Of course, detailed calculations of the C $^1\Pi_u$ photoelectron spectra are necessary to determine the validity of these arguments for D_2. However, at least qualitatively, it appears that the D_2 spectra can be explained in a manner consistent with the H_2 spectra.

Conclusions

The (3+1) ionization of H_2 via the C $^1\Pi_u$ state has been discussed in light of the existing experimental and theoretical data. It does not appear that the existing experimental data on the (3+1) ionization of H_2 via the C $^1\Pi_u$ state can be explained in terms of the simple direct photoionization of the Rydberg electron into the H_2^+ X $^2\Sigma_g^+$ continuum. As suggested by Chupka (1987) and Hickman (1987a,b), excitation and subsequent autoionization of the $2p\sigma_u2p\pi_u$ doubly excited state appear to strongly influence the vibrational branching ratios, particularly for C $^1\Pi_u$, v' = 3 - 6. Although a direct experimental study of the $2p\sigma_u2p\pi_u$ state remains to be performed, the present data serve to bracket the position of this state. When coupled with more detailed calculations of the vibrational branching ratios, these data should improve our understanding of the doubly excited states of H_2.

TWO COLOR (2+1) REMPI-PES OF H_2 VIA E,F $^1\Sigma_g^+$

The next level of sophistication in REMPI-PES studies of molecular photoionization utilizes an independently tunable laser to photoionize

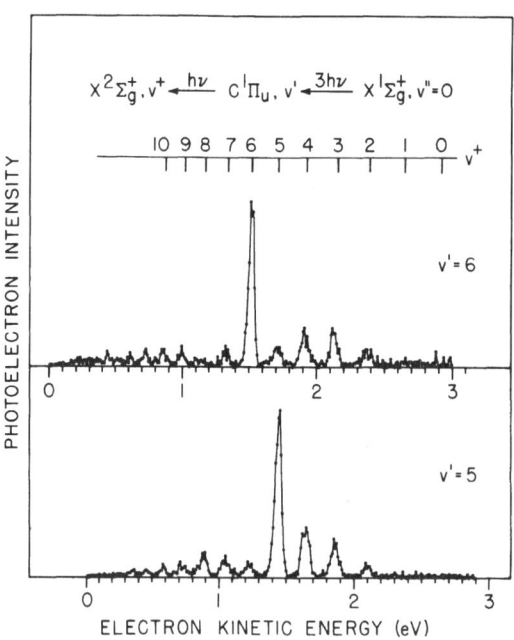

Figure 5. Photoelectron spectra of H_2 determined along the laser polarization axis ($\theta = 0°$) at the wavelengths of the resonant three-photon C $^1\Pi_u$, v' = 5, 6 ← X $^1\Sigma_g^+$, v" = 0, Q(1) transitions.

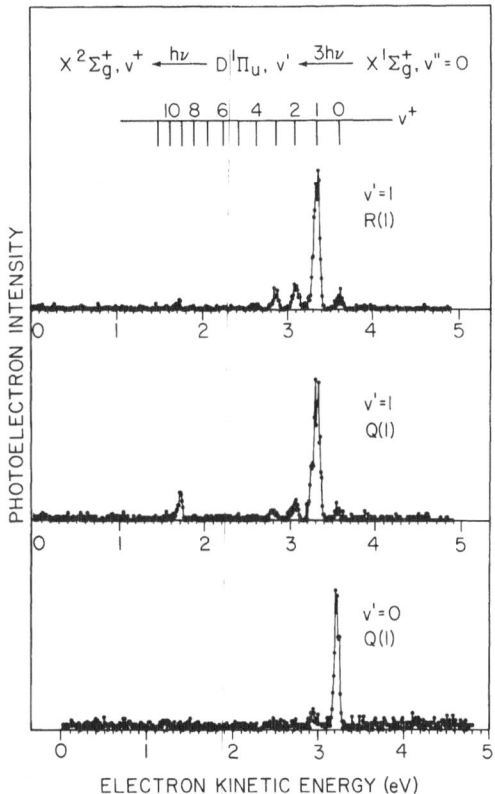

Figure 6.   REMPI-PES spectra of H_2 determined at the wavelengths of the resonant three photon $D\ ^1\Pi_u$, $v' = 0 \leftarrow X\ ^1\Sigma_g^+$, $v'' = 0$, $Q(1)$; $D\ ^1\Pi_u$, $v' = 1 \leftarrow X\ ^1\Sigma_g^+$, $v'' = 0$, $Q(1)$; and $D\ ^1\Pi_u$, $v' = 1 \leftarrow X\ ^1\Sigma_g^+$, $v'' = 0$ $R(1)$ transitions.

the excited state.   Such two-color REMPI processes permit one to examine the interaction and decay mechanisms of autoionizing states in great detail.   Furthermore, compared to the established VUV techniques used to study such processes, two-color REMPI-PES has several useful advantages: ability to select the quantum state of the excited target, in many cases; very high wavelength resolution; and access to nonoptical channels.

We have just completed initial measurements combining a two-step excitation process with photoelectron energy analysis to investigate rotational and vibrational autoioinization processes in molecular hydrogen.   The 4th anti-Stokes component of a Raman-shifted, doubled dye laser provides 5-20 μJ of light at ~193 nm, which populates the state H_2 E,F $^1\Sigma_g^+$, $v'=2$, $J'=1$ through a two-photon transition.   A third photon at ~ 400 nm (the 1st anti-Stokes of an excimer pumped dye laser) then excites from this level to a region near the thresholds for production of H_2^+ $^2\Sigma_g^+$, $v^+=2$, $N^+=1,3$.   The kinetic energies of the resulting photo-electrons are determined by time-of-flight analysis in a magnetic bottle electron spectrometer.   The high collection efficiency (50%) of the magnetic bottle permits us to follow individual ionic vibrational levels,

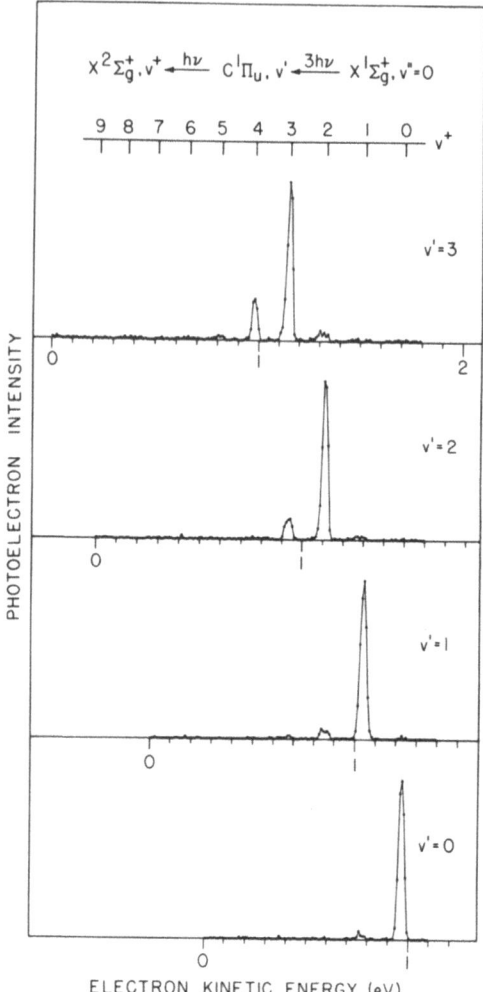

Figure 7. Photoelectron spectra of D_2 determined along the laser
polarization axis at the wavelengths of the resonant three
photon C $^1\Pi_u$, $v' = 0 - 3 \leftarrow X$ $^1\Sigma_g^+$, $v'' = 0$, $Q(3)$ transitions.

through gated detection of individual photoelectron peaks, as the
excitation wavelength of the H_2^+ $^2\Sigma_g^+$, v^+, N^+ $\leftarrow$ H_2 E,F $^1\Sigma_g^+$, $v'=2$, $J'=1$
transition is scanned.

This experimental approach makes it possible to directly determine
the final vibrational states produced by vibrational autoionization as a
function of position within the autoionization profile. Analogous
experiments on NO have also been performed (Achiba and Kimura, 1984 and
Kimman et al., 1986). Moreover, the vibrational branching ratios between
the thresholds for production of H_2^+ $^2\Sigma_g^+$, $v^+=2$, $N^+=1,3$ reflect the
competition between the rotational and vibrational autoionization
mechanisms. Thus it is now possible to measure quantities that
characterize autoionization dynamics at the level at which they are
calculated by the most sophisticated theories, see, e.g., Raoult and
Jungen (1981).

REFERENCES

Achiba, Y. and Kimura, K., 1984, Absorption and ionization of super-excited Rydberg states studied by two-color laser multiphoton ionization technique combined with photoelectron spectroscopy. In Book of Abstracts, International Conference on Multiphoton Processes III, Crete, P13.

Berry, R. S., 1966, Ionization of molecules at low energies. J. Chem. Phys. 45, 1228.

Bonnie, H.J.M., Eenschuistra, P. J., Los, J., and Hopman, H. J., 1986, Influence of the vibrational quantum number of the resonant state in multiphoton ionization/dissociation of hydrogen molecules. Chem. Phys. Lett. 125, 27.

Bonnie, H.J.M., Verschuur, J.W.J., Hopman, H. J., and van Linden van den Heuvell, H. B., 1986, Photoelectron spectroscopy on resonantly enhanced multiphoton dissociative ionization of hydrogen molecules. Chem. Phys. Lett. 130, 43.

Chupka, W. A., 1987, Photoionization of molecular Rydberg states: H_2, $C\ ^1\Pi_u$ and its doubly excited states. J. Chem. Phys. in press.

Cornaggia, C., Giusti-Suzor, A., and Jungen, Ch., 1987, Photoionization of the E,F excited state of H_2: Calculation of the vibrational and angular distributions of the photoelectrons. To be published.

Dill, D., 1972, Angular distributions of photoelectrons from H_2: Effects of rotational autoionization. Phys. Rev. A 6, 160.

Dill, D., 1976, A primer on photoelectron angular distributions. In Photoionization and Other Probes of Many Electron Interactions, ed. F. Wuilleumier, pp. 387. New York, New York: Plenum.

Dixit, S. N., Lynch, D. L., and McKoy, V., 1984, Three-photon resonant four-photon ionization of H_2 via the $C\ ^1\Pi_u$ state. Phys. Rev. A 30, 3332.

Dixit, S. N., and McKoy, V., 1986, Ionic rotational selection rules for (n+1) resonant enhanced multiphoton ionization. Chem. Phys. Lett. 128, 49.

Fano, U. and Dill, D., 1972, Angular momentum transfer theory of angular distributions. Phys. Rev. A 6, 185.

Guberman, S. L., 1983, The doubly excited autoionizing states of H_2. J. Chem. Phys. 78, 1404.

Hickman, A. P., 1987a, Photoionization and photodissociation of H_2 through autoionizing states. Bull. Am. Phys. Soc. 32, 1252.

Hickman, A. P., 1987b, Non-Franck-Condon distribution of final states in photoionization of H_2 ($C\ ^1\Pi_u$). Submitted for publication.

Huber, K. P., and Herzberg, G., 1979, Molecular Spectra and Molecular Structure IV. Constants of Diatomic Molecules. New York, New York: Van Nostrand Reinhold.

Kruit, P., and Read, F. H., 1983, Magnetic field paralleliser for 2π electron-spectrometer and electron-image intensifier. J. Phys. E 16, 313.

Kimman, J., Verschuur, J.W.J., Lavollee, M., van Linden van den Heuvel, H. B., and van der Wiel, M. J., 1986, Branching ratios for autoionization of v = 3 Rydberg states in nitric oxide. J. Phys. B 19, 3909.

Namioka, T., 1964a, Absorption spectra of H_2 in the vacuum-ultraviolet region. I. The Lyman and the Werner bands. J. Chem. Phys. 40, 3154.

Namioka, T., 1964b, Absorption spectra of H_2 in the vacuum ultraviolet region. II. The B' - X, B" - X, D - X, and D' - X bands. J. Chem. Phys. 41, 2141.

O'Halloran, M. A., Pratt, S. T., Dehmer, P. M., and Dehmer, J. L., 1987, Photoionization dynamics of H_2 $C\ ^1\Pi_u$: vibrational and rotational branching ratios. J. Chem. Phys. in press.

Pratt, S. T., Dehmer, P. M., and Dehmer, J. L., 1984, Photoionization of excited molecular states. H_2 C $^1\Pi_u$. Chem. Phys. Lett. 105, 28.

Pratt, S. T., Dehmer, P. M., and Dehmer, J. L., 1986, Photoionization dynamics of excited molecular states. Photoelectron angular distributions and rotational and vibrational branching ratios for H_2 C $^1\Pi_u$, v' = 0 - 4. J. Chem. Phys. 85, 3379.

Pratt, S. T., Dehmer, P. M., and Dehmer, J. L., 1987a, Photoelectron studies of resonantly enhanced multiphoton ionization of H_2 via the B' $^1\Sigma_u^+$ and D $^1\Pi_u$ states. J. Chem. Phys. 86, 1727.

Pratt, S. T., Dehmer, P. M., and Dehmer, J. L., 1987b, Photoionization dynamics of excited molecular states. D_2 C $^1\Pi_u$. Submitted to J. Chem. Phys.

Raoult, M. and Jungen, Ch., 1981, Calculation of vibrational preionization of H_2 by multichannel quantum defect theory: Total and partial cross sections and photoelectron angular distributions. J. Chem. Phys. 74, 3388.

Xu, E., Tsuboi, T., Kachru, R., and Helm, H., 1987, Four-photon ionization and dissociation of H_2. Post-deadline paper, 18th Annual Meeting of the APS Division of Atomic, Molecular, and Optical Physics, Cambridge, MA.

DOUBLE PHOTOIONIZATION NEAR THRESHOLD AND VIA CORE HOLE STATES

RECENT EXPERIMENTAL RESULTS

I. Nenner[1][2], J.H.D. Eland[3], P. Lablanquie[1],
J. Delwiche[4], M.-J. Hubin-Franskin[4], P. Roy[5],
P. Morin[1][2] and A. Hitchcock[6]

1) LURE, Laboratoire CNRS, CEA et MEN, Univ. Paris-Sud
   91405 Orsay Cedex, France
2) Département de Physico-Chimie, CEA, C.E.N. de Saclay
   91191 Gif Sur Yvette Cedex, France
3) Physical Chemistry Laboratory, Oxford OX1 3Q2
   United Kingdom
4) Maître de Recherche FNRS, Université de Liège, Bât. B6
   Sart Tilman, 4000 Liège 1, Belgium
5) Los Alamos National Laboratory, Los Alamos, New Mexico
   87545, USA
6) Department of Chemistry, McMaster University, Hamilton
   Canada

ABSTRACT

Double ionization of atoms and molecules by single photon excitation
of the neutral species from threshold up to the core ionization edges are
reviewed, using recent experimental results on argon, carbon disulphide,
methyl bromide and carbon monoxide, obtained with synchrotron radiation.

INTRODUCTION

Double Photoionization (DPI) of both atoms and molecules has a
significant probability near threshold (i.e. below core ionization edges)
because of electron correlation. Consequently, the measurement of double
ionization cross sections provides an important testing ground for
theoretical models which include many-electron effects.

In the immediate vicinity of DPI onset, more fundamental aspects of
the theory have to be tested because the two electrons are highly correlated in the Coulomb field of the ion. In the original Wannier theory
(Wannier, 1953) and subsequent developments (Greene and Rau, 1982, 1983;
Stauffer, 1982; Feagin, 1984; Read, 1985) the dynamics of two-electron
escape is subject to certain constraints whose effects can in principle
be verified experimentally. Those are that the excitation cross sections
should be proportional to E^n (with n > 1), E being the excess energy
above threshold; the energy distribution of electrons at a given excitation energy should be essentially flat and the outgoing two-electron wave
function should be restricted to special symmetries.

The choice of atoms whose double photoionization can be used to probe fundamental theory is essentially restricted to helium or iso-electronic ionic species, because only there is double photoionization a purely direct process. Unfortunately the double ionization cross section of helium is extremely small and lies at an inconveniently short wavelength, causing some experimental difficulty. For all other atoms the more complex electronic structure also allows indirect double photoionization via resonances, which can be doubly excited neutral species with an excited doubly charged ion core, or singly charged ion states embedded in a double ionization continuum. The study of these indirect processes is, of course, interesting in itself, but complicates the investigation of direct double ionization. The choice of argon as our first experimental subject, reported in this paper, was made mainly on experimental grounds.

At higher photon energies, when core excitation becomes accessible, the double ionization yield increases accordingly because of the Auger effect and related phenomena. In low Z elements, the most probable decay of a core vacancy is the filling of the hole by one electron and ejection of another rather than by x-ray emission and this Auger process dominates even when the core electron is initially promoted to an unoccupied level. When, as usual, this electron remains a spectator in the transition, the production of singly charged ions with excited configurations is favoured. Higher order processes, in which a second electron is ejected in the continuum, produce doubly (or triply) charged species with valence multi-holes, which are much less efficiently populated by simple valence electron correlation (direct DPI). The capability of producing such "hot" doubly or multiply charged species is of great interest in the study of many-body effects in atoms.

The extension of such studies to molecules is also of great interest; electron correlation both in the neutral species and in the final state is as important as in atoms, and is often most accessible experimentally, providing a severe test of molecular many-body theories. Furthermore, because (in the independent particle model) bonding molecular orbitals are more strongly and anti-bonding orbitals less strongly bound than the atomic orbitals from which they arise, an entirely new double ionization mechanism, akin to the Auger effect, can be envisaged when a single electron is ejected from a molecular orbital in the continuum and the hole relaxes, eventually ejecting a second electron from the same shell. This new indirect double ionization process, never experimentally established, may well account for large DPI cross sections in molecules, in the intermediate energy range between threshold and the core ionization region.

Core ionization/excitation of molecules also introduces new phenomena because the original core hole is localized on a particular atom. This localization affects both the population of the ion states after electron ejection, and also may introduce a degree of specificity into the process of multiply charged ion dissociation which is the normal final act in the drama of molecular core hole production.

The choice of suitable molecular subjects is very wide; we originally chose CS_2 because its large direct double ionization cross section was already known from earlier work (Lablanquie et al., 1985) and the first elements of DPI dynamics above threshold are available (Roy et al., 1987). We shall report results on CH_3Br (Nenner et al., 1987) and CO (Hitchcock et al., 1987) also, because the core excitation and ionization in this molecule exhibit resonance phenomena particularly clearly in the vicinity of core edges and some direct comparison of the population of doubly charged ionic states (in the CO case) can be made between the threshold region (Lablanquie et al., 1987) and core ionization edges (Hitchcock et al., 1987).

DOUBLE PHOTOIONIZATION NEAR THRESHOLD

Argon

The double photoionization of argon:

$$Ar + h\nu \rightarrow Ar^2 + e_1 + e_2$$

has a threshold near 43.5 eV (Moore, 1949), an energy at which we can obtain tunable synchrotron light freed from all higher order contribu-tions by use of an aluminium filter. The first threshold corresponds to formation of argon ions in the ^3P state, but at higher photon energies the Ar^{2+} states ^1D or ^1S, which also arise from the $2p^4$ configuration, could be populated too. To determine the final state populations at a fixed incident photon energy it is necessary to obtain a spectrum, analogous to a photoelectron spectrum, with the sum of the two ejected electron energies as abscissa. The two electrons must therefore be energy analyzed and detected in coincidence to ensure that they originate from single events, and this imposes severe experimental constraints. The coincidence count rate is proportional to the product of two overall detection efficiencies; the use of two conventional electrostatic analysers, each with typical efficiencies of 10^{-3}, would lead to an unusably low count rate. To circumvent this problem we have chosen to select one electron of near zero energy, for which an analyzer of high efficiency is available, while scanning the energy of the other. The zero electron energy selection was a steradiancy analyzer with electron timing from the light pulse signal to remove fast electron signals, while our scanning analyzser was a 127° condenser, set to a bandpass of 0.5 eV for high detection efficiency. Even so, the coincidence count rate did not exceed 1 count per minute at the chosen photon energy of 56 eV, selected to be well above all three Ar^{2+} thresholds.

The results of this first photoelectron-photoelectron coincidence (PEPECO) experiment are presented in Fig. 1 which also shows results of much less difficult experiments in which electrons of selected energies were detected in coincidence with Ar^+ ions and Ar^{2+} ions. The $e^- - e^-$ coincidence spectrum demonstrates directly that Ar^{2+} ions are formed by photoionization at this photon energy in the ^3P state exclusively. The lack of any step structure in the $Ar^{2+} - e^-$ coincidence curve is also consistent with this conclusion and supports it strongly. The observa-tion that only one state of Ar^{2+} is populated by photoionization is important both theoretically and for the interpretation of other experi-ments. On the theoretical side, it is the first direct experimental test of the wavefunction symmetry aspects of the Wannier-Rau theory. Accord-ing to the theory, double ionization is dominated by two-electron final states near the Wannier point, at which the two electrons depart with equal but opposite momenta and equal radius vectors at all times. Only a restricted class of two-electron wavefunctions are free from a node at this point, the symmetries of these favoured wavefunctions being such that all their quantum numbers LSΠ (all their total orbital angular momentum, spin and parity) are either even, or odd. When these favoured two-electron wavefunction symmetries are combined with the symmetries of the accessible states of Ar^{2+} it transpires that only one pairing, that of Ar^{2+} ($^3P^e$) with a $^3P^0$ outgoing wave leads to a combined symmetry whose production from neutral argon is dipole allowed. Our observation of the selective population of Ar^{2+} (^3P) thus confirms the applicability of both the Wannier-Rau wavefunction symmetry rule and also the dipole selection rule to argon double photoionization.

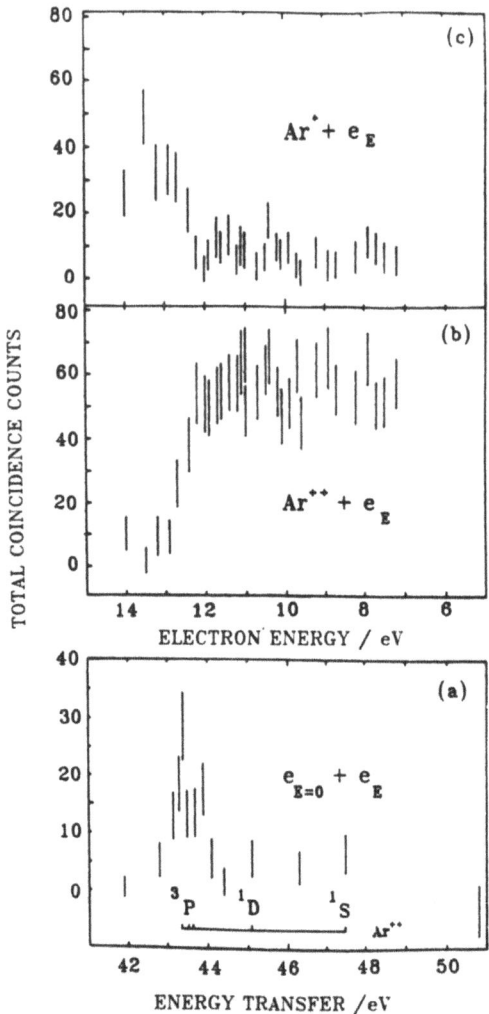

Fig. 1. Spectra of coincidences at 56 eV photon energy between electrons of different energies selected by the electrostatic analyzer and (a) low energy electrons (b) Ar^{2+} ions (c) Ar^+ ions (from Lablanquie et al., 1987).

If only one single state of Ar^{2+} is formed at all wavelengths, measurements of the yield of Ar^{2+} as a function of wavelength should enable us to test the threshold behaviour and also to search for indirect ionization processes. However, a more subtle test is provided by the yield of near-zero energy electrons in coincidence with Ar^{2+} ions as a function of wavelength, because, according to the theory, direct double ionization, which should produce electrons with a flat energy distribution, always gives some electrons of near-zero energy. Furthermore, (the theory predicts that) the total double ionization yield, which corresponds to the integral of the electron energy distribution, should be proportional to a power n (= 1.056) of the excess energy above the threshold. If electrons near zero energy are detected in a constant bandwidth, their yield should follow the (n-1)th power of the excess energy since the width of the total distribution is directly proportional to the excess energy. This makes such a measurement most sensitive to, and a better test for the value of the Wannier exponent n.

Our results for the yield of low energy electrons in coincidence with Ar^{2+} ions as a function of wavelength are shown in Fig. 2, together with a solid curve which portrays the expected 0.056th power behaviour. The highly structured nature of the data, and also equivalent structure of the Ar^{2+} yield curve (not shown), demonstrates at once that indirect as well as direct double ionization occurs. We estimate that at some wavelengths indirect processes contribute up to 20% of the total double ionization, but that the direct process, foreseen in the Wannier theory, is dominant. Because of the existence of the indirect process, we can only suggest that the form of the data is not inconsistent with the theoretically expected exponent (n = 1.056) for direct double ionization.

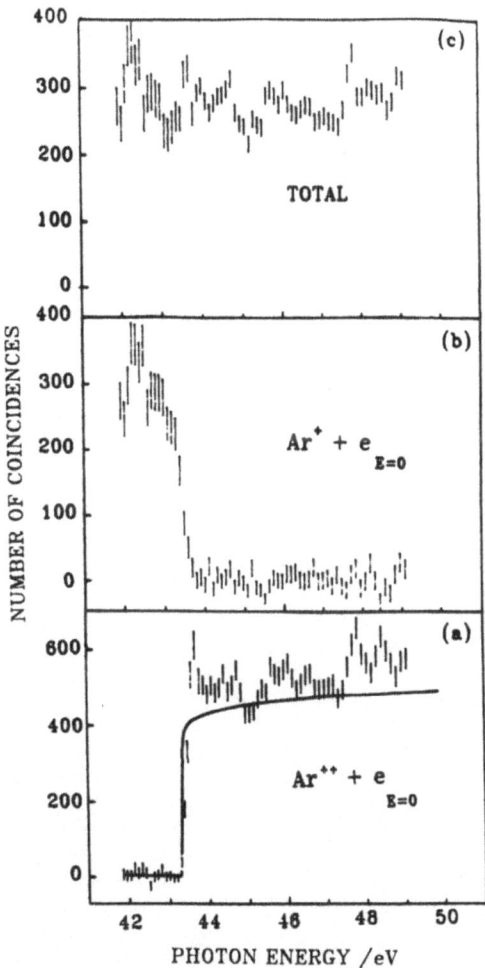

Fig. 2.   Spectra of coincidences between low-energy (E < 0.5 eV) electrons and (a) Ar^{2+} ions with a solid line showing the theoretical 0.056 power-law behaviour (b) Ar^+ ions, both as functions of photon energy (from Lablanquie et al., 1987).

19

Because only one single Ar^{2+} state is populated, the $Ar^{2+} - e^-$ coincidence data in Fig. 1 also constitute a test of the prediction that in the threshold region the electron energy spectrum in direct double ionization should be flat. At a photon energy much above threshold, by contrast, this spectrum is expected to have symmetrical peaks just above zero energy and just below the maximum available energy, showing the production of one fast and one slow electron as the most probable process. Our data (Fig. 1b) at 12.5 eV above threshold show a flat spectrum with no peak visible. From the lack of any peak near the maximum available energy, therefore, we conclude that the region of validity of the description of threshold behaviour extends at least 12.5 eV above threshold, which makes it much wider than in double photo-ionization of H^- or in single ionization of helium by electron impact. Feagin (1984) has given a theoretical model which suggests that the range of validity of the threshold description should increase both with the charge and the mass of the residual ion, but a firm quantitative predic-tion is lacking. This aspect of our results definitely requires further experimental work on other atoms, and also theoretical investigation.

The absence of peak structure at energies other than the maximum available in the electron spectrum of Fig. 1b is also significant. It demonstrates that although a resonance of neutral argon at 56.35 eV (Madden et al, 1969) is almost certainly populated at our nominal photon energy of 56 eV (0.5 eV bandwidth), it decays following the same two-electron dynamics as direct ionization. In particular, the absence of superexcitd states of Ar^+ do not play a significant role. If such states were involved in a two-step, Auger-like process, the signature of each would be two peaks, symmetrically disposed about the mean electron energy of 6.25 eV, in the electron spectrum.

To summarise the conclusions of this section, double photoionization of argon by single photons is well dscribed by the Wannier-Rau theory of direct ionization forming correlated electron pairs in the final state, at incident energies up to at least 12.5 eV above threshold, even though the double ionization cross-section contains substantial contributions from superexcited resonant states of neutral argon.

Carbon Disulphide

In double ionization of molecules near threshold we must expect several differences frm atomic behaviour. First, many more states of the product double charged ion are accessible, both electronic states and those arising from nuclear motion. Where doubly charged ion states are bound, the equilibrium bond lengths and angles are usually very different from those in the neutral molecules, so extended vibrational envelopes of the electronic bands are to be anticipated. In many cases the doubly charged ion states are repulsive in the Franck-Condon zone, and in such cases broad continuous bands will be seen. When we detect long-lived doubly charged ions, however, we are selectively observing just a few bound states. From detailed studies of CS_2, for instance, involving Auger spectroscopy, double charge transfer spectroscopy and large config-uration interaction calculations (Millié et al., 1986), we can reasonably think that only three electronic states of CS_2^{2+} are bound in the Franck-Condon zone, namely $^3\Sigma_g^-$, $^1\Delta_g$ and $^1\Sigma_g^+$, all arising with the valence electron configuration $\pi_u^4 \pi_g^2$. If the two-electron wavefunction symmetry rule of the Wannier-Rau theory, which we have shown to be applicable to argon, can be generalised to centro-symmetric molcules such as this, it should take the form that "favoured final states of doubly charged centro-symmetric ions will be ^3L_g or ^1L_u" where L is an arbitrary angular momentum. This would mean, in the case of CS_2, that we should expect to observe formation of CS_2^{2+} ($^2\Sigma_g^-$) alone, as a long-lived doubly charged

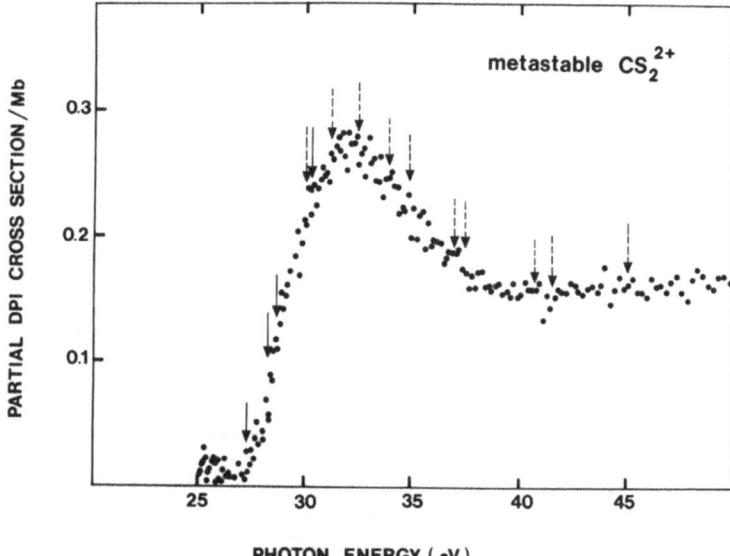

Fig. 3. Double photoionization cross section for metastable CS_2^{2+} ions (from Lablanquie et al., 1985). Arrows correspond to experimental locations of CS_2^{2+} states; solid and dashed arrows correspond to bound and dissociative states respectively.

ion state in direct photoionization near threshold. We can attempt to interpret the yield of CS_2^{2+} as a function of wavelength, Fig. 3, with these ideas in mind.

First, on a broad energy scale, the cross section for CS_2^{2+} formation rises to a maximum at about 33 eV in a peak with a half-width of about 5 eV, and then levels off at higher photon energies. Similar behaviour is seen in the yield curves for doubly charged rare gas ions, but the peaks are usually much broader there and less pronounced. Since t direct double ionization mechanism predicts only a monotonic rise in cross section with increasing energy, we have to interpret this peak as evidence of some sort of resonance. Its centre does not coincide with any state of CS_2^{2+} located in our detailed spectral study (Millié et al., 1986), nor with any known state of CS_2^+ (Roy et al., 1987). A similar feature, though less pronounced, does appear however in the yield curve of CS_2^+ and in the total absorption cross section. We therefore consider that it is probably a resonance of neutral CS_2, similar to the resonances seen in the Ar^{2+} yield and threshold electron yield. At this energy it must be at least doubly excited compared to the CS_2 ground state configuration, and if only doubly excited it probably needs one hole in the sulphur 3s shell.

When the CS_2^{2+} yield near threshold is examined the onset can be accurately determined as 27.3 eV, and is attributed to formation of the CS_2^{2+} $^3\Sigma_g^-$ ground state. The structures observed above the threshold do not coincide in any way with the locations of known states of CS_2^{2+} arrows on the diagram. They may be manifestations of molecular effects, or may result from the overlay of many individual resonances.

The understanding of the dynamics of the two-electron ejection near threshold from a neutral molecule is much more difficult than from an atom essentially because of the higher density of doubly charged ionic states and their intrinsic "width" due to large Franck-Condon envelopes. Nevertheless, a detailed study of the photoelectron spectrum in CS_2 (Roy et al., 1987) at selected photon energies revealed not only a number of satellite lines due to simultaneous ionization and excitation of valence electrons but also some evidence of DPI events. This is illustrated in the two PES of Fig. 4 obtained at two angles with respect to the main polarization axis of the light. Below 28 eV binding energies most satellite bands have been identified as configuration interaction states of $CS_2{}^+$ on the basis of ab initio SCF-CI calculations and measurements of electron angular distributions. The most intense ones are found in the 20-28 eV binding energy region and originate from inner valence ionization. Above 28 eV, the PES especially at $\theta = 0°$ shows additional structures superimposed on a continuum. However, the calculations (not shown here) do not predict any significant intensity for high lying satellite lines. In addition, those features correspond to the opening of DPI channels as determined by independent experiments (Millié et al., 1986) (see fig. 4). Considering that the photon energy of 65 eV is sufficiently high (30-40 eV) compared to any of these DPI states, the partitioning in energy of the two electrons is no longer governed by the Wannier-Rau conditions. It is also reasonable to expect that autoionization of a doubly excited state is unimportant. Compared to direct DPI, one should expect that one electron is ejected with essentially the excess energy whereas the other has almost zero kinetic energy. This bimodal distribution allows the rather easy observation of the fast electron in a conventional PES. This has been observed in neon by Carlson (1966). In the present CS_2 case, the features of interest (see Fig. 4) could originate from the opening of the direct DPI channel. This interpretation is also supported by the fact that the integrated PES intensity from 28 eV up, as a function of photon energy, follows faithfully (not shown, see Roy et al., 1987) the shape of the total DPI cross

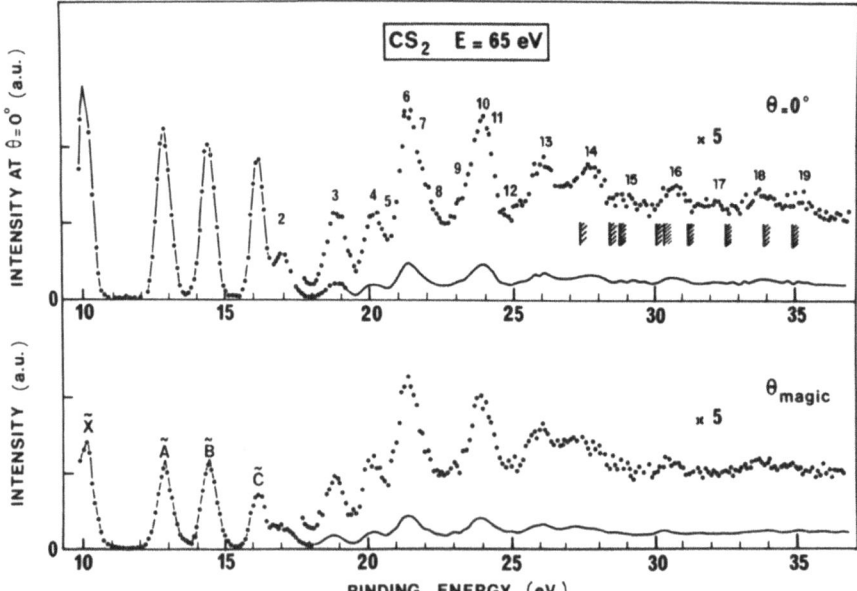

Fig. 4. Photoelectron spectra of CS_2 recorded at 65 eV photon energy at two angles $\theta = 0$ and θ_{magic} with respect to the polarization electric vector of the incident light (from Roy et al., 1987). Experimental DPI onsets are indicated on the top spectra.

section obtained by mass spectroscopy (Lablanquie, 1985). It would be of interest to find out if this distribution can be understood using a valence Auger picture, which would be typical of inner valence <u>molecular</u> orbitals.

DOUBLE PHOTOIONIZATION VIA CORE HOLE STATES

The ejection of a core electron in the continuum followed by the ejection of an Auger electron causes a sharp increase in the photon impact double ionization cross section above core ionization edges. Other features are also found in the cross section curves because of the presence of resonances both in the discrete spectrum below the edge and in the continuum above the edge, as established by Electron Energy Loss Spectroscopy (EELS) and the Photoabsorption measurements for atoms and also for molecules. The effects are observed particularly in dissociative double photoionization events (Morin et al., 1986).

These recent findings contrast with the early measurements of Brehm and Defrênes (1978) who studied dissociative double ionization by electron impact, and never found any effect when the incident electron energy was high enough to open core ionization channels. The important characteristic of electron impact in this connection is that the energy distribution deposited in the system, which may also include core ionization, covers a wide range of underlying continua, all energetically accessible. The proportion of double ionization events actually associated with the Auger process represents only a very small fraction of the total and is not easily detectable (Hitchcock et al., 1978).

In photon impact experiments, on the other hand, the energy deposited in the system is better defined and more can be learned about the relaxation channels of the core hole, especially when the inner electron is promoted into a bound orbital. At photon energies sufficiently high above the edge, to avoid post collision effects, the Auger spectrum, which is easily detected in a conventional photoelectron spectrometer, reflects the population of specific doubly charged ion states.

For discrete resonances, evidence of double ionization is more difficult to observe in photoelectron spectra because it originates from more complex processes. The Auger decay in the presence of a core excited electron (so-called resonant Auger) produces a singly charged ion (excited) and an electron with a well defined kinetic energy. Further autoionization into a doubly charged ion will produce an additional electron with a well defined but low kinetic energy which is difficult to observe (Morin, 1987). By contrast, the resonant Auger decay may alternatively occur by <u>simultaneous</u> ejection of two electrons. In this case, their energy distribution is essentially flat from 0 to the maximum excess energy. This process appears only as a continuum in a photoelectron spectrum and is more likely to be detected by a threshold electron analyser. Both double resonant Auger and two step autoionization produce doubly charged ions which can be detected in a conventional mass spectrometer. Such measurements are largely insufficient for molecules because doubly charged molecular ions are generally unstable and usually dissociate into singly charged fragments. Therefore photoion-photoion coincidence (PIPICO) measurements are more appropriate to determine the number of positive charges of the initially formed parent ion.

We present in Fig. 5 the EELS of the methyl bromide molecule CH_3Br, in the region of the bromine 3d edge as obtained by Hitchcock and Brion

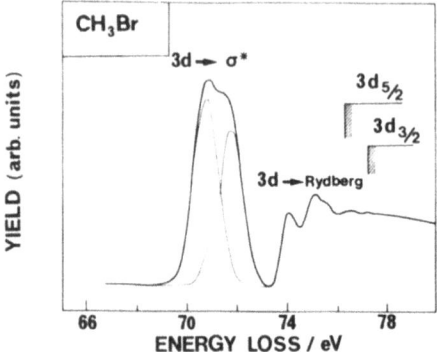

Fig. 5. Electron Energy Loss spectrum of CH_3Br near the bromine 3d edge
         (from Hitchcock and Brion, 1978).

(1978). Discrete resonances result from the promotion of a 3d electron
into the antibonding σ* orbital and into Rydberg orbitals. The continuum
resonance (3d → εf) is a shape resonance with a strong atomic character.

The threshold electron spectrum (Fig. 6) measured recently (Morin
and Nenner 1987, Nenner et al., 1987) in the region of discrete resonances shows features very similar to those seen in the absorption spectrum. This readily shows that doubly charged ions are produced by a
double resonant Auger process. However this spectrum does not contain
any information on the final state energies. Metastable CH_3Br^{2+} ions
have been found in small quantities only in conventional mass spectra
because of rapid fragmentation into singly charged fragments. The
photoion-photoion coincidence spectra (Morin and Nenner, 1987; Nenner
et al., 1987) confirms this instability but more importantly, the photon
energy dependence of each ion pair intensity brings additional information. This is illustrated in Fig. 7. The CH_i^+ + Br^+ pair resonates

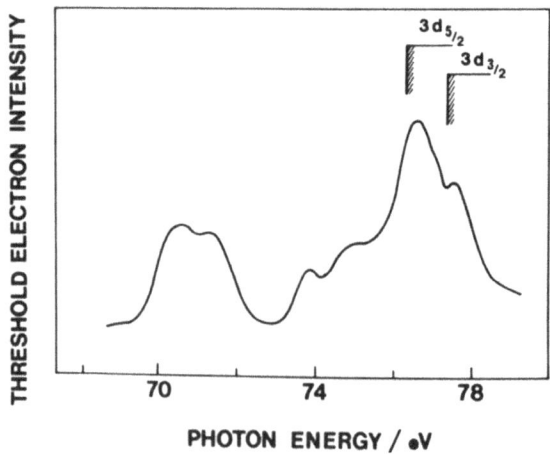

Fig. 6. Threshold electron spectrum of CH_3Br near the bromine 3d edge
         (from Morin and Nenner, 1987; Nenner et al., 1986).

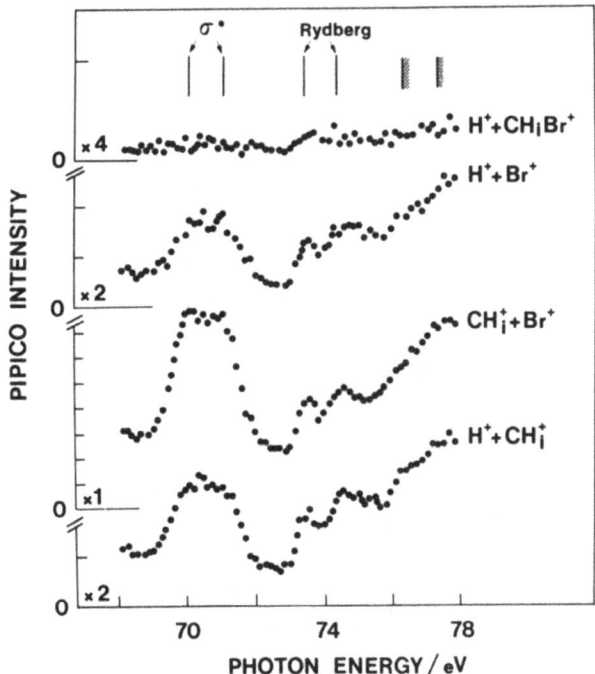

Fig. 7. Ion pair intensity by photoion-photoion coincidence in CH_3Br as a function of photon energy (from Morin and Nenner, 1987; Nenner et al., 1987).

strongly in the discrete resonance region. So, to a lesser extent, do the $H^+ + Br^+$ and the $H^+ + CH_i^+$ pairs. In contrast, the $H^+ + CH_iBr^+$ pair cross section remains essentially constant. The Br 3d core hole excitation favours only the production of certain doubly charged ions. In such a polyatomic molecule the localization of the excitation along the C-Br bond through discrete resonances eliminates any relaxation process leading to an ion with the two holes localized on the C-H bond. Nevertheless, the C-H bond breaking, represented by the $H^+ + Br^+$ and $CH_i^+ + H^+$ pairs, is rather interpreted as the result of secondary events due to the instability of the CH_3^+ ion produced after the initial $CH_3^+ - Br^+$ cleavage.

In simpler systems such as the CO molecule, there is no question of localization in different bonds, but information on the energies of the final doubly charged ions populated through core excitation can be obtained. According to Lablanquie et al. (1987) and Hitchcock et al. (1987), the dominant dissociation channel of CO^{2+} is $C^+ + O^+$ and from PIPICO measurements in optimized conditions for extracting kinetic energy release distributions (KERD's), it is possible to obtain insight into the CO^{2+} states populated. As shown in Fig. 8, the $C^+ + O^+$ PIPICO signal shows a quite different shape depending on the photon energy. Those energies have been selected to coincide with the C 1s → π* and O 1s → π* discrete resonances located at 287 eV and 534 eV respectively, with the C 1s → σ* and O 1s → σ* continuum shape resonances found at 350 and 550 eV respectively, and in the DPI threshold region at 42 and 70 eV. From a fitting procedure of these KERD's using a Monte-Carlo method taking into account the experimental geometry, applied fields and assumed multi-component KERD's, the average values were found to increase drastically from the threshold region into the core region. This means that

25

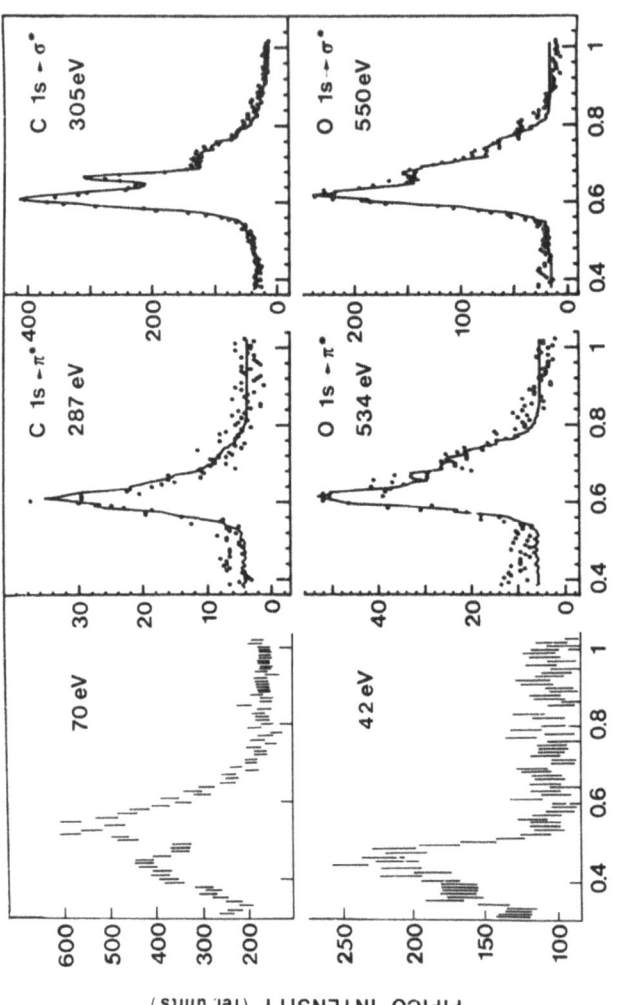

Fig. 8. (C^+ + O^+) photoion–photoion coincidence spectra at selected photon energies nesr threshold (42 and 70 eV) and near the C 1s edge on the π^* (287 eV) and σ^* (305 eV) resonances near the O 1s edge, on the π^* (534 eV) and σ^* (550 eV) resonance (from Lablanquie et al., 1987; Hitchcock et al., 1987).

core excitation produces almost exclusively dissociative doubly charged ions with very excited configurations. The excess energy is found primarily as kinetic energy of the fragments. More subtle differences are found in the KERD's between the C 1s and O 1s core excitation regions but a precise knowledge of the CO^{2+} spectroscopy, not yet available, is required to analyse them.

ACKNOWLEDGEMENTS

We are very pleased to thank the staff of LURE for their expert operation of the machines. Two of us (M.-J. H.-F. and J. D.) express their gratitude to the "Fonds National de la Recherche Scientifique", the "Fonds de la Recherche Fondamentale Collective" and the "Service de la Politique Scientifique" of Belgium for their financial support. We also gratefully acknowledge the support from NATO (Grant No. 484/87).

REFERENCES

Brehm, B., and DeFrenes, G., 1978, Int. J. Mass Spectrom. Ion Phys., 26:251.

Carlson, T.A., 1966, Phys. Rev., 156:42.

Feagin, J.M., 1984, J. Phys. B17:2433.

Greene, C.H., and Rau, A.R.P., 1982, Phys. Rev. Lett., 48:533.

Greene, C.H., and Rau, A.R.P., 1983, J. Phys. B., 16:99.

Hitchcock, A.P., and Brion, C.E., 1978, J. Electron. Spectrosc. Relat. Phenom., 13:193.

Hitchcock, A.P., Brion, C.E., and Van der Wiel, M.J., 1978, J. Phys. B 11:3245.

Hitchcock, A.P., Lablanquie, P., Morin, P., Lizon, A., Lugrin, E., Simon, M., Thiry, P., and Nenner, I., 1987, Phys. Rev. A, submitted for publication.

Lablanquie, P., Nenner, I., Millie, P., Morin, P., Eland, J.H.D., Hubin-Franskin, M.-J., and Delwiche, J., 1985, J. Chem. Phys., 82:2951.

Lablanquie, P., Eland, J.H.D., Nenner, I., Morin, P., Delwiche, J., and Hubin-Franskin, M.-J., 1987, Phys. Rev. Lett., 58:992.

Lablanquie, P., Delwiche, J., Hubin-Franskin, M.-J., Nenner, I., Eland, J.H.D., and Ito, K., 1987, J. Mol. Struct., in press.

Madden, R.P., Ederer, D.L., and Codling, K., 1969, Phys. Rev., 177:136.

Millie, P., Nenner, I., Archirel, P., Lablanquie, P., Fournier, P., and Eland, J.H.D., 1986, J. Chem. Phys., 84:1259.

Moore, C.E., 1949, "Atomic Energy Levels as Derived from Analyses of Optical Spectra, National Bureau of Standards", Circular No. 465, U.S. G.P.O., Washington, D.C.

Morin, P., De Souza, G.G.B., Nenner, I., and Lablanquie, P., 1986, Phys. Rev. Lett., 56:131.

Morin, P., and Nenner, I., 1987, Phys. Script., T17:171.

Morin, P., 1987, Giant resonances in atoms, molecules and solids, in: "Procedings of the NATO Advanced Study Institute, Les Houches", J.P. Connerade, J.M. Esteva and R. Karnartak, eds., Plenum, New York, p.291; LURE Activity Report (1987).

Nenner, I., Morin, P., Siumon, M., Lablanquie, P., De Souza, G.G.B., 1987, in: "Proceedings of D.I.E.T. III", M. Knokek and R. Stulen, eds., Springer, Berlin, to be published.

Read, F.H., 1985, in: "Electron Impact Ionization", T.D. Mark and G.H. Dunn, eds., Springer, Berlin, p.42-88.

Roy, P., Nenner, I., Millie, P., Morin, P., Roy, D., 1987, J. Chem. Phys. (in press).

Stauffer, A.D., 1982, Phys. Lett., 91A:114.

Wannier, G.H., 1953, Phys. Rev., 90:817.

MULTIPLE PHOTOIONIZATION STUDIED BY PHOTO-ION AND PHOTOELECTRON SPECTRA

Akira Yagishita*

Institut für Experimentalphysik, Universität Hamburg
Luruper Chaussee 149, D-2000 Hamburg 50, West Germany

INTRODUCTION

The application of synchrotron radiation to atomic and molecular physics in Japan started with the 1.3 GeV electron synchrotron at the University of Tokyo[1]. Interesting studies, though limited to photographic measurements, were done with the synchrotron. With the advent of the dedicated 400 MeV storage ring (SOR-RING) at the University of Tokyo[2], more demanding experiments, e.g. photoelectron spectroscopy[3,4], lifetime measurements by fluorescence[5,6], and mass spectrometry[7], could be started. Upon completion of the dedicated synchrotron radiation source (Photon Factory) at Tsukuba[8], possibilities for gas-phase research have dramatically widened.

At the Photon Factory, eight beam lines were completed for VUV and soft x-ray work by the end of 1982. Three beam lines among them were designed mainly for gas-phase experiments. A Grasshopper and Seya-Namioka monochromator are installed in the beam lines 11A and 12A respectively[9,10] and used for general purposes. The beam line 12B is equipped with a 6.65m off-plane Eagle spectrograph/monochromator for high resolution spectroscopy[11,12]. The undulator beam line[13] equipped with a 10m grazing incidence[14] and double-crystal monochromator has been opened to gas-phase experiments in May of this year. Detailed descriptions of the facilities of the Photon Factory and users' reports are given in ref.15.

From among more than ten areas of research on atomic and molecular physics, a review of new data from two areas is presented: multiple photoionization of rare-gas and alkaline-earth atoms; and dissociative ionization of the oxygen molecule via discrete and shape resonances. We are interested in studying decay channels of inner-shell excited states through resonant photo-ion spectroscopy. The techniques employed are time-of-flight (TOF) mass spectrometry, coincidence measurements of threshold photo-electrons with photo-ions, and photo-ion spectroscopy. For the atoms, a preliminary interpretation of the photo-ion data will be presented. This interpretation is based in part on the absorption and photo- and Auger-electron spectra obtained at other SR Laboratories. For

*On leave from Photon Factory, National Laboratory for High Energy Physics, Oho-machi, Tsukuba-gun, Ibaraki 305, Japan.

29

the molecule, experimental aspects of the measurements will be emphasized.

DECAY OF Sr 4p RESONANCES

The total ion-yield spectrum (a sum of Sr^+- and Sr^{2+}-yield) of Sr (Fig. 1) displays the following main features[16]: an extremely strong $4p^6 5s^2 \rightarrow 4p^5 5s^2 (^2P_{1/2}) 4d$ transition; three main series converging to the $4p^5 5s^2 \ ^2P_{1/2,3/2}$, $4p^5 4d 5s \ ^2P_{3/2}$ ionization limits; broad structures between 28 and 29.5 eV. The assignment is based on the classification given by Mansfield and Newsom[17].

Figure 2 shows a typical photoelectron spectrum obtained at the $4p^5 5s^2 (^2P_{1/2}) 6d$ resonance energy[18]. A schematic representation of decay channels of the state is included in the figure. As illustrated in Fig. 2, electrons are ejected through both direct (Ⓓ) and resonance (Ⓐ) processes.

The direct processes are classified into two groups: (i) photo-emission from the valence shell, leading to photolines of Sr^+; (ii) photoionization of the subvalence 4p shell, followed by Auger decays resulting in Sr^{2+}. The two channels are written as follows:

$$h\nu + Sr \quad \xrightarrow{\ \ Ⓓ\ \ } \quad Sr^+ \ 4p^6 nl \qquad + \ e_{photo} \qquad (i)$$

$$\xrightarrow{\ \ Ⓓ\ \ } \quad Sr^{+*} \ 4p^5 nln'l' + e_{photo}$$
$$\xrightarrow{\ \ Ⓒ\ \ } \quad Sr^{2+} + e_{Auger} + e_{photo} \qquad\qquad (ii)$$

The decay processes via the resonance state are classified into two groups: (iii) autoionization to the states of Sr^+; (iv) autoionization to the intermediate states of Sr^+ followed by Auger decays producing Sr^{2+}

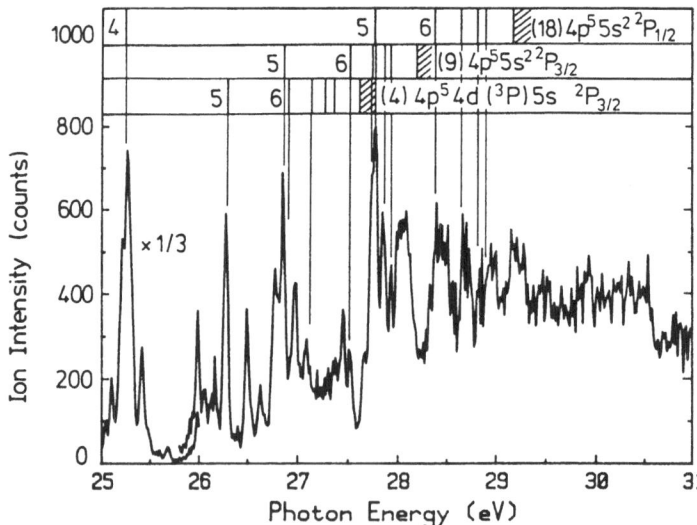

Fig. 1.   Total ion-yield spectrum of Sr, from Nagata et al[16]. Main nd Rydberg series converging to the limits (4), (9), and (18) are indicated at the top. Following Mansfield and Newsom[17], the numbers are used to represent the series limits.

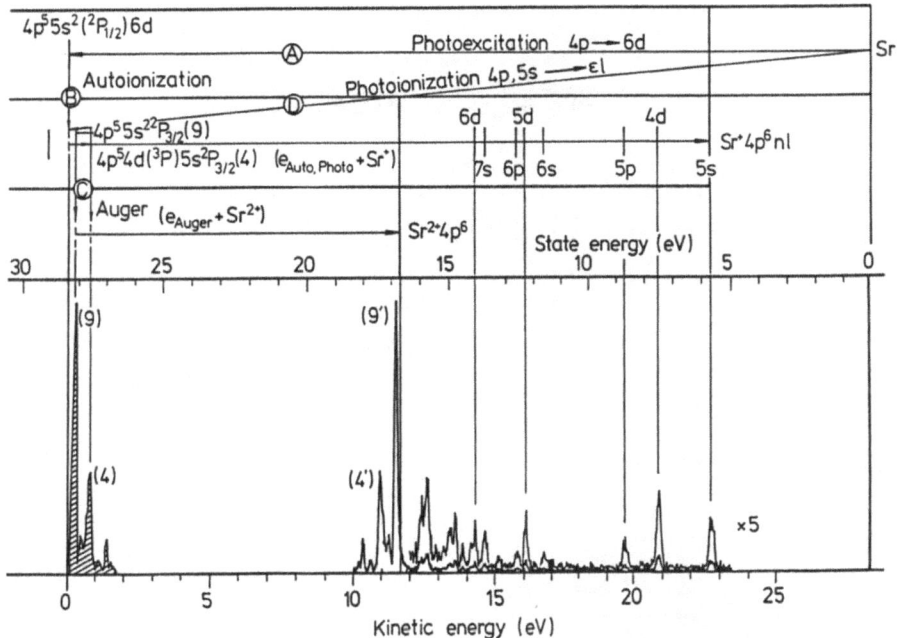

Fig. 2. Sr photoelectron spectrum measured at the $4p^5 5s^2 (^2P_{1/2})6d$ resonance energy, from Yagishita et al[18] and schematic of decay channels of the state. Peaks (4') and (9') are Auger peaks from the ionization continua (4) and (9), respectively. Shaded peaks (4) and (9) are not measured ones, but are drawn for understanding the first step of the two-step processes.

(two-step autoionization). Those resonant processes are described as follows:

$$h\nu + Sr \longrightarrow \text{(A)} \longrightarrow Sr^* \ 4p^5 5s^2 nd$$

$$\longrightarrow \text{(B)} \longrightarrow Sr^+ \ 4s^6 n'l' + e_{Auto} \qquad \text{(iii)}$$

$$\longrightarrow \text{(A)} \longrightarrow Sr^* \ 4p^5 5s^2 nd$$

$$\text{step 1} \longrightarrow \text{(B)} \longrightarrow Sr^{+*} \ 4p^5 n'l'n''l'' + e_{Auto} \qquad \text{(iv)}$$

$$\text{step 2} \longrightarrow \text{(C)} \longrightarrow Sr^{2+} \ 4s^6 + e_{Auger} + e_{Auto}$$

Production of Singly Charged Ion

Figure 3(a) shows an Sr^+-yield spectrum by Nagata et al[16] and Fig. 3(b) the sum of constant ionic state (CIS) spectra for three exit channels by Yagishita et al[18]. From Figs. 3(c), (d) and (e), the enhanced intensities of the Sr^+ 5s, 4d and 5p photoelectron lines at the resonance energies are clearly to be seen. The intensities of these peaks at off-resonance energies (channel (i)) are negligible compared with those at resonance (channel (iii)).

Below the ionization limit (4), we find a good correspondence between the Sr^+-yield spectrum and the sum of the CIS spectra, although not all the exit channels are included in the sum. Above the limit (4),

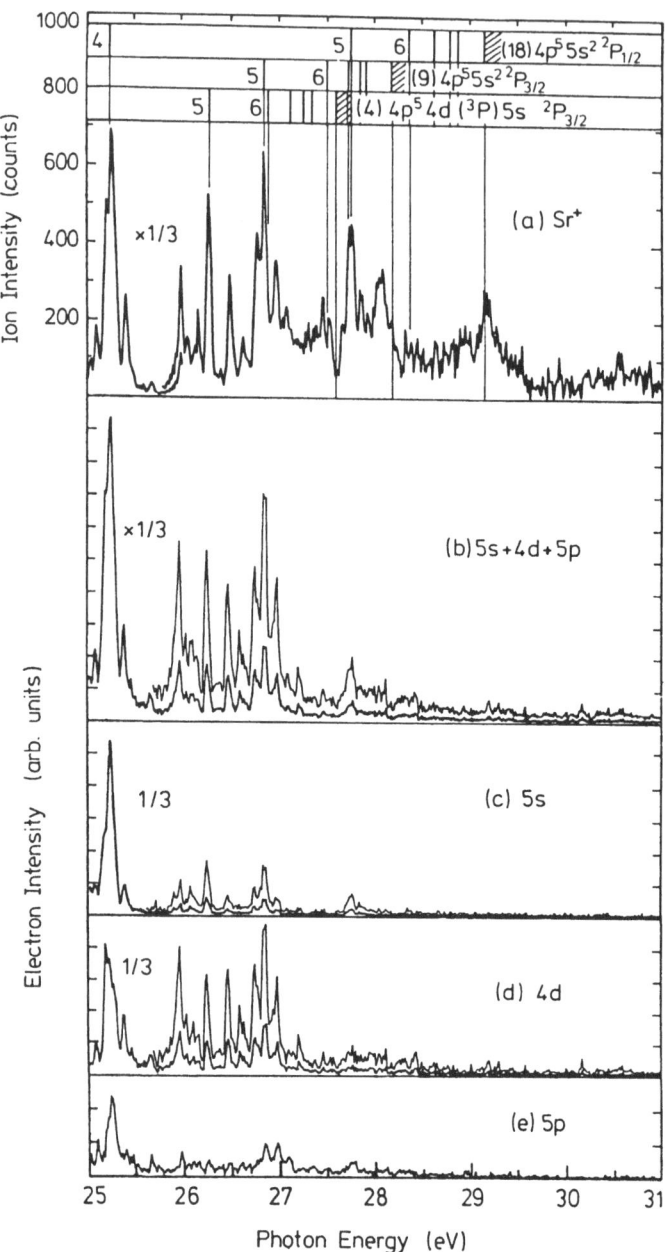

Fig. 3. (a) Sr[+]-yield spectrum, from Nagata et al[16]. (b) Sum of
relative cross sections of the main (5s) and of two satellite
(4d, 5p) exit channels, from Yagishita et al[18]. (c), (d) and
(e) are cross sections for Sr[+] 5s, 4d and 5p ionic states,
respectively.

we can not find any correspondence between them, because the main decay
channel changes from channel (iii) to channel (iv).

Decay channels of the $4p^5 5s^2 (^2P_{1/2})4d$ state. Detailed information
on channel (iii) is obtainable from the photoelectron and photo-ion data.
The strongest $4p^5 5s^2 (^2P_{1/2})4d$ resonance serves as a good example:

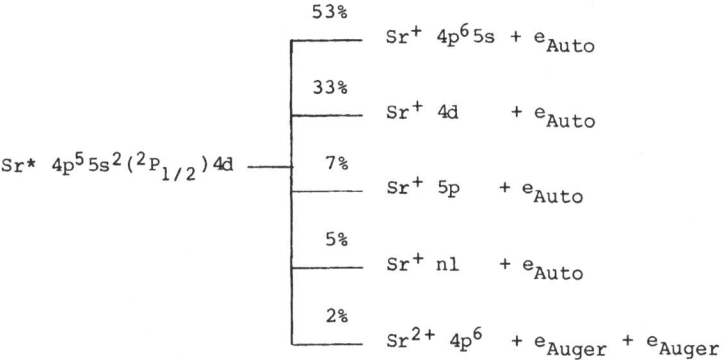

The errors of these values are less than 10%. The intensity for the 4d satellite, produced via so-called resonance Auger (the excited 4d electron remains as a spectator), is about 62% of the main line, generated via direct recombination (the 4d electron participates). In Ca, the 3d satellite is ten times weaker than the main line at the strongest $3p \rightarrow 3d$ resonance[19]. The difference in the satellite intensities suggests that the 4d electron and 4p hole wavefunctions of Sr overlap less than the 3d electron and 3p hole of Ca.

4d satellite enhancement between the strongest resonance and the ionization limit (4). In general a resonance state does not enhance a particular excited level of the final ionic state[19]. But even in more complicated cases, we can predict an intensity distribution of satellites to some extent. These predictions are based on the character of the resonance state. Mansfield and Newsom have identified the absorption line at 26.29 eV as a transition to the $4p^5 4d(^3P)5s$ $^2P_{3/2}$ 5d state belonging to the nd series converging toward the ionization limit (4)[17]. As according to their calculation the ionic state mainly contains two components: $4p^5 4d(^3P)5s^2P$ (47%) $+ 4p^5 5s^2$ ^2P (36%), we can predict the relative satellite intensities; the 4d line should be the strongest, the 5p the weakest, among the three photolines (Fig. 3(c), (d) and (e)). This simple interpretation, based on the core overlap, for the decay channels is applicable to the other resonance states.

Production of Doubly Charged Ion

Figure 4(a) shows an Sr^{2+}-yield spectrum by Nagata et al[16] and Fig. 4(b) the sum of Auger-yield spectra for three Auger transitions by Yagishita et al[18]. The intensities of the Auger peaks (4') and (9') are considerably enhanced at the 4p resonance energies (Fig. 4(c) and (d)). The Auger peak (18') appears above the ionization limit (18), therefore it is not to be seen in Fig. 2. Because the intensities for the other Auger transitions contribute very little to the Sr^{2+}-yield, the sum of the Auger yields reproduces the Sr^{2+}-yield spectrum except for a few discrepancies described below.

Below the ionization limit (9). The peak of the Auger-yield originating from the $5p^5 5s^2(^2P_{1/2})5d$ state in Fig. 4(b) is by far the strongest; however, in the Sr^{2+}-yield spectrum (Fig. 4(a)) the peak is not stronger than the others. This discrepancy is caused by the post-collision interaction (PCI) effect; the energy of e_{Auto} in channel (iv) is only 0.15 eV in this case, therefore the Auger electron in (iv) is affected by the slowly moving electron. A good demonstration of the PCI effect is shown in Fig. 5; the Auger peak (4') is broadened and shifted toward higher kinetic energies. A fraction of the peak overlaps the

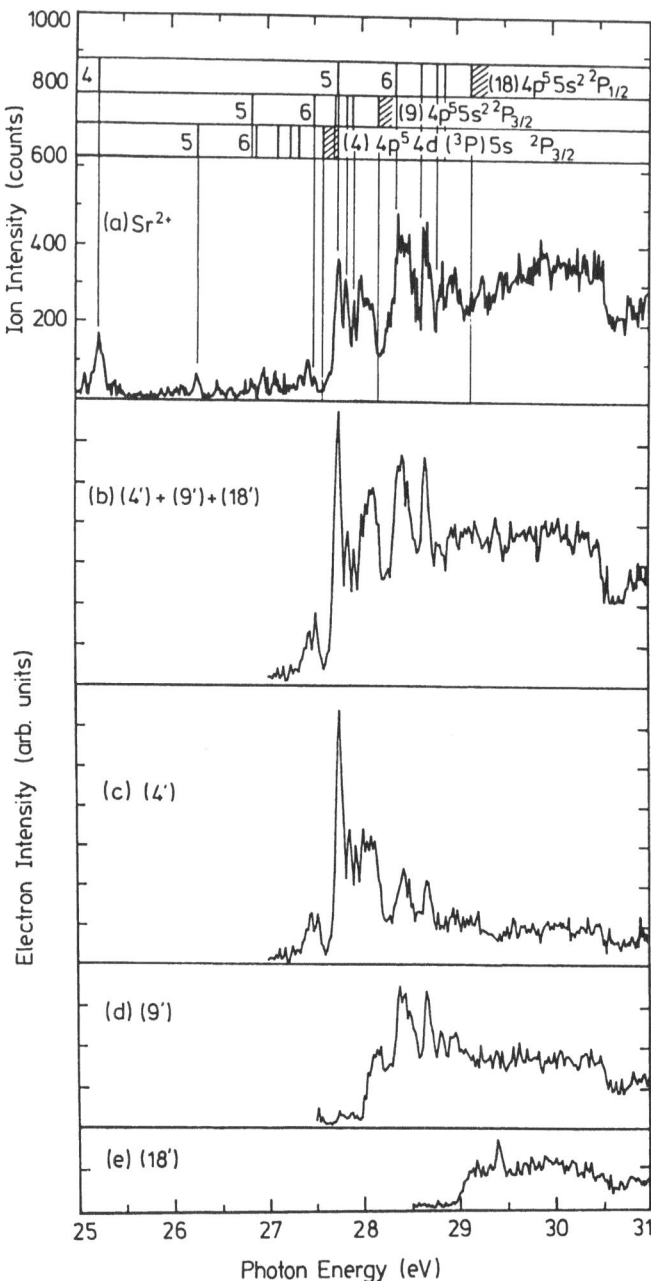

Fig. 4. (a) Sr^{2+}-yield spectrum, from Nagata et al[16]. (b) Sum of Auger-yield spectra for the (4'), (9') and (18') Auger transitions, from Yagishita et al[18]. (c), (d) and (e) are the Auger-yield spectra for the (4'), (9') and (18') Auger, respectively.

region of the singly charged ion (shake–down). Because of the limited energy window the Auger-yield spectrum (Fig. 4(c)) also encompasses that portion of the Auger profile corresponding to the shake–down processes. Therefore the shake–down part must be subtracted from the Auger-yield spectrum in order to compare it with the Sr^{2+}-yield. The shake–down part contributes to the Sr^{+}-yield (see Fig. 3(a)).

From the Auger- and photo–electron spectrum of Fig. 5 and the ion-yield spectra of Fig. 3(a) and Fig. 4(a), the decay channels of the $4p^5 5s^2 (^2P_{1/2}) 5d$ state have been determined, taking shake–down into account:

$$Sr* \ 4p^5 5s^2 (^2P_{1/2}) 5d$$

- 30% $\longrightarrow$ $Sr^+ \ 4p^6 nl + e_{Auto}$
- 70% $\longrightarrow$ $Sr^{+*} \ 4p^5 4d (^3P) 5s + e_{Auto}$ (0.15 eV)

$$\downarrow$$

$$Sr^{2+} \ 4p^6 + e_{Auger} (4')$$

The errors of these values are 20%. Two-step autoionization decay is the predominant channel for the resonance state; this means auto–ionization decay induced by core rearrangement is faster than the Auger-type autoionization. Because core rearrangement is likely to occur between similar core configurations, the result is not too surprising if we consider that the ionic state (4) contains a strong $4p^5 5s^2$ component[17].

Above the ionization limit (9). The ionization continua (4) and (9) are intermediate states of two-step autoionization. We have concluded from the ion-yield data that the $4p^5 5s^2 (^2P_{1/2}) 6d$ state mainly autoionizes via the continuum (9)[16]. However, there is also considerable decay via the continuum (4) (see Fig. 2 and Fig. 4(c)). The decay channels of the $4p^5 5s^2 (^2P_{1/2}) 6d$ state can be summarized as follows:

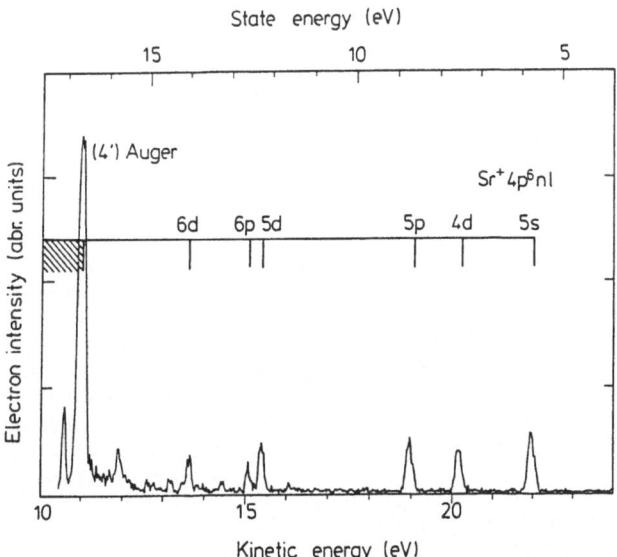

Fig. 5. Sr photoelectron spectrum measured at the $4p^5 5s^2 (^2P_{1/2}) 5d$ resonance energy, from Yagishita et al[18].

35

$$\text{Sr*} \; 4p^5 5s^2 (^2P_{1/2}) 6d \; \underset{\underset{47\%}{\overset{33\%}{\rule{0pt}{3.5em}}}}{\overset{20\%}{\rule{0pt}{0pt}}}$$

$$\begin{array}{l}
\xrightarrow{} \text{Sr}^+ \; 4p^6 nl + e_{\text{Auto}} \\[1em]
\xrightarrow{} \text{Sr}^{+*} \; 4p^5 4d(^3P)5s + e_{\text{Auto}} \\[1em]
\xrightarrow{} \text{Sr}^{+*} \; 4p^5 5s^2 \; ^2P_{3/2} + e_{\text{Auto}}
\end{array} \; \to \text{Sr}^{2+} + e_{\text{Auger}} \; (4',9')$$

The errors of these values are 20%. Two-step autoionization for the 6d resonance state is more predominant than for the $4p^5 5s^2 (^2P_{1/2})5d$ state. This tendency is understandable, because for the former state two intermediate channels (4) and (9) are open, but for the latter only one (4). The autoionization induced by the core rearrangement from the core (18) for the 6d resonance to the ionic state (9) may occur preferentially, because the states (18) and (9) strongly resemble each other.

Near the ionization limits. Close to the ionization limit (18), we can see the shake-down effect in the Sr^+- and Sr^{2+}-yield spectra: there is a dip at the limit in the Sr^{2+}-yield spectrum (Fig. 4(a)); complementary to the dip is a peak in the Sr^+-yield spectrum (Fig. 3(a)). The effect is also expected to occur at the limits (4) and (9), but the dip obscured by the resonance structures is very hard to recognise. In the Auger-yield spectrum, we cannot see the effect, for the same reason as mentioned in the previous section.

The Auger-yield in Fig. 4(c) does not vanish below the ionization limit (4). If a 4p excited high Rydberg state autoionizes to a Rydberg state of a singly charged ion, the energy of e_{Auto} (in channel (iii)) is very close to the energy of e_{Auger} (in channel (iv)). In this case the energy window cannot distinguish e_{Auto} from e_{Auger}. This explains the onset of the Auger-yield below the ionization limit (see Fig. 4(c) and also Fig. 4(d) and (e)). The low energy onset provides evidence for the population of Sr^+ $4p^6 nl$ states with large n via the autoionisation of the highly excited states of the 4p electron.

The two processes mentioned above do not contribute to the Sr^{2+}-yield spectrum, but contribute to the Auger-yield spectrum. This explains the relationship between the ion-yield spectrum and Auger-yield spectrum. Above the limit (4), the sum of the Auger-yield spectra (Fig. 4(b)) is very similar to the sum of the ion-yield (Fig. 1), but not to the Sr^{2+}-yield (Fig. 4(a)).

DECAY PROCESSES OF THE 2p EXCITED STATES IN Ar and Ca

One of the merits of the photo-ion measurements by the time-of-flight (TOF) method is the possibility of determining relative photoionization cross sections more easily than by absorption spectroscopy. The cross sections determined by TOF are very reliable for zero levels and linearity, bcause crossed-beam and ion-counting techniques are used. For orbital angular momenta larger than two, the effective potential for certain atoms has two wells. As is well known, the delayed maximum observed in the Xe 4d photoionization cross section is understood qualitatively in terms of a collapse of the f-symmetric final state wavefunction (e.g. see ref. 20). As an example of the trapping of the 3d wavefunction in the inner potential well (collapse), we show the photoionization cross sections of Ar and Ca together with the results of the Hartree-Fock-Slater calculation performed by Yeh and Lindau[23] in Figs. 6 and 7. Changing the target from Ar (z = 18) to Ca (Z = 20), the weak and narrow lines assigned to the 2p → 3d transitions change to extremely strong and broad bands. This highlights the strong increase in

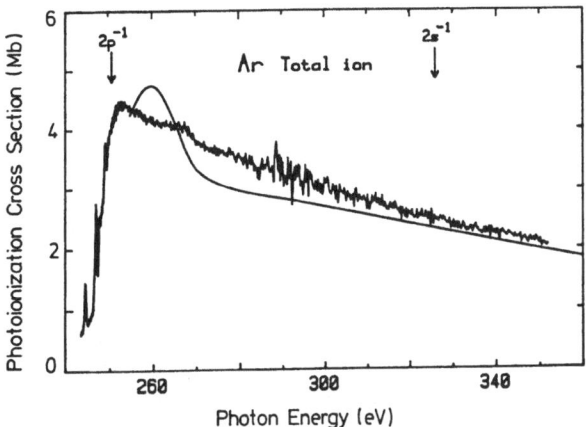

Fig. 6. Ar photoionization cross section measured with a 1.5 eV bandwidth, from Hayaishi et al[2]. Relative scale has been normalized
to the absolute values of Marr and West[22]. Solid curve: HFS
calculation by Yeh and Lindau[23].

over-lap between the 2p and 3d wavefunctions caused by the collapse of
3d wavefunction in 2p excited Ca[25]. Unfortunately we cannot compare the
experimental data with theory below the 2p ionization energy of Ca. But
as in the case of Xe 4d photoionization[20], we cannot expect good agreement between the experiment and calculations using a one-electron model.

Charge-state Distribution of Photo-ions Produced via the 2p Shell Excitation

Figures 8 and 9 show the ion-yield curves of Ar and Ca respectively,
for each charge state up to 3^+ near the 2p ionization edges. The charge-
state distribution of photo-ions strongly changes from Ar to Ca: in Ar,
below the 2p ionization edges, the yield of the singly charged ion Ar^+ is

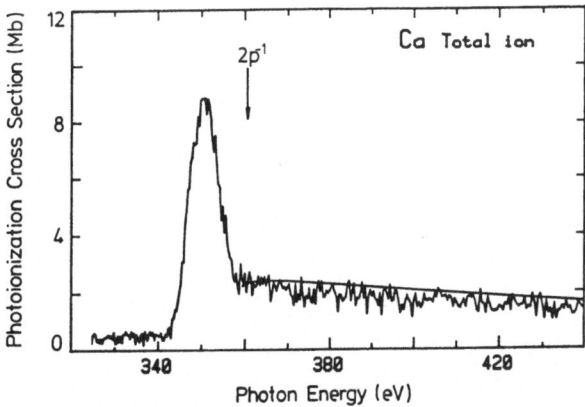

Fig. 7. Ca photoionization cross sections measured with a 4 eV bandwidth, from Nagata et al[24]. Experimental data have been normalized to the HFS calculation of Yeh and Lindau[23] represented by a
solid curve at the maximum point.

37

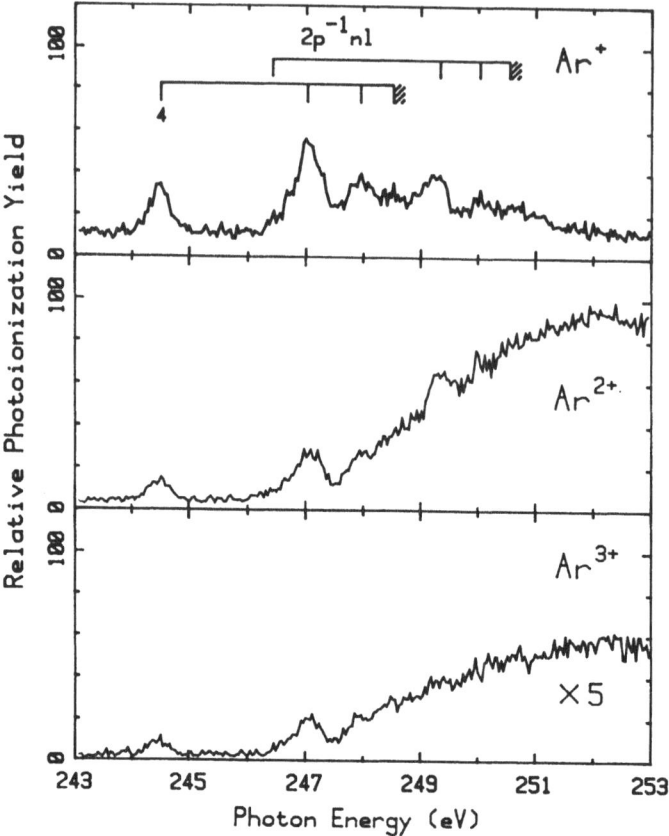

Fig. 8. Ar^+-, Ar^{2+} and Ar^{3+}-yield spectra measured with 0.5 eV bandwidth, from Hayaishi et al[21].

the strongest; below the ionization edges of Ca, the Ca^{2+} yield prevails, and the Ca^+ yield does not show any structure. The distributions are summarized as follows:

$$Ar^* \ 2p^{-1}(^2P_{3/2})3d \ \begin{cases} \xrightarrow{62.5\%} Ar^+ \\ \xrightarrow{32.5\%} Ar^{2+} \\ \xrightarrow{5\%} Ar^{3+} \end{cases}$$

and

$$Ca^* \ 2p^{-1}(^2P_{3/2})3d \ \begin{cases} \xrightarrow{\approx 0\%} Ca^+ \\ \xrightarrow{75\%} Ca^{2+} \\ \xrightarrow{25\%} Ca^{3+} \end{cases}$$

The errors of these values are less than 10%.

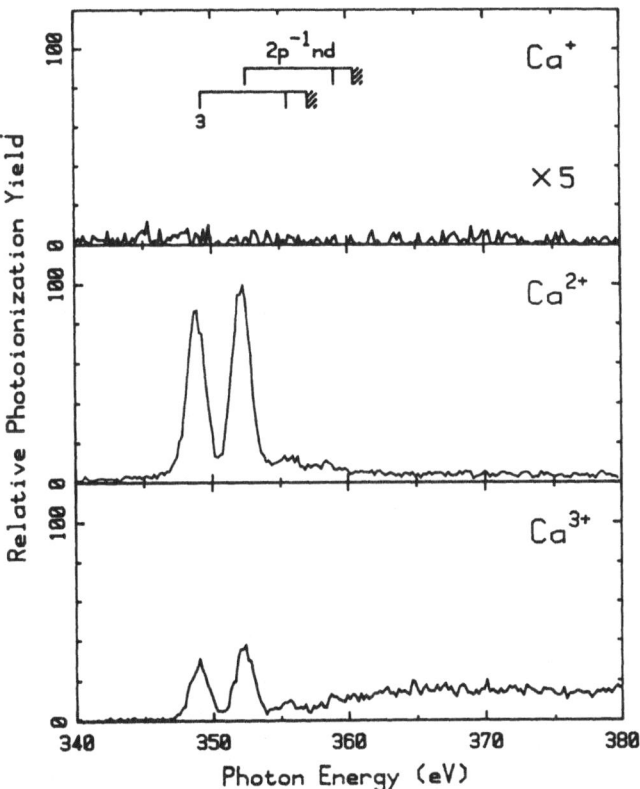

Fig. 9. Ca^+-, Ca^{2+}- and Ca^{3+}-yield spectra measured with 2.0 eV band-width, from Nagata et al[24].

Interpretation of the Charge-state Distributions

In order to give a simple interpretation of the distributions, we show the energy diagrams of Ar and Ca in Fig. 10.

Below the 2p ionization limits. In Ar, the final charge states are mainly determined by the ratio of the autoionization channels (1) and the two-step autoionization channels (1) → (2) in Fig. 10. Although the ratios for the resonance Auger final states are not available, we can estimate the distribution using the ratios for normal Auger transitions. From Werme et al[26], the intensities for the $L_3,M_{23}M_{23}$ Auger comprise 75% of all the L_3MM Auger decays. If we do not take Auger shake-off processes into account, this ratio gives the distribution of Ar^+ (75%) and Ar^{2+} (25%). These values are in reasonable agreement with the experimental data, Ar^+ (62.5%) and Ar^{2+} (32.5%). The Auger shake-off processes add to the Ar^{2+} yield and give rise to the Ar^{3+}.

In Ca, almost all final states of the resonance Auger are the auto-ionizing states, which decay to Ca^{2+}; (1) → (2) in Fig. 10. The two 4s electrons which characterize the chemical properties of Ca are respon-sible for these decays. Therefore, we cannot see any structure in the Ca^+-yield spectrum. Some final states of the resonance Auger (for example, $3s^{-1}3p^{-1}3d$ and $3s^{-2}3d$ states) can decay to Ca^{3+} (see Fig. 10). The final states of Auger shake-off processes also decay to Ca^{3+} (for example, $3p^{-2}4s^2 \to 3p^{-1}4s^0+e$). The two decay-channels mentioned above contribute to the Ca^{3+}-yield, which comprises 25% of the total ion-yield.

39

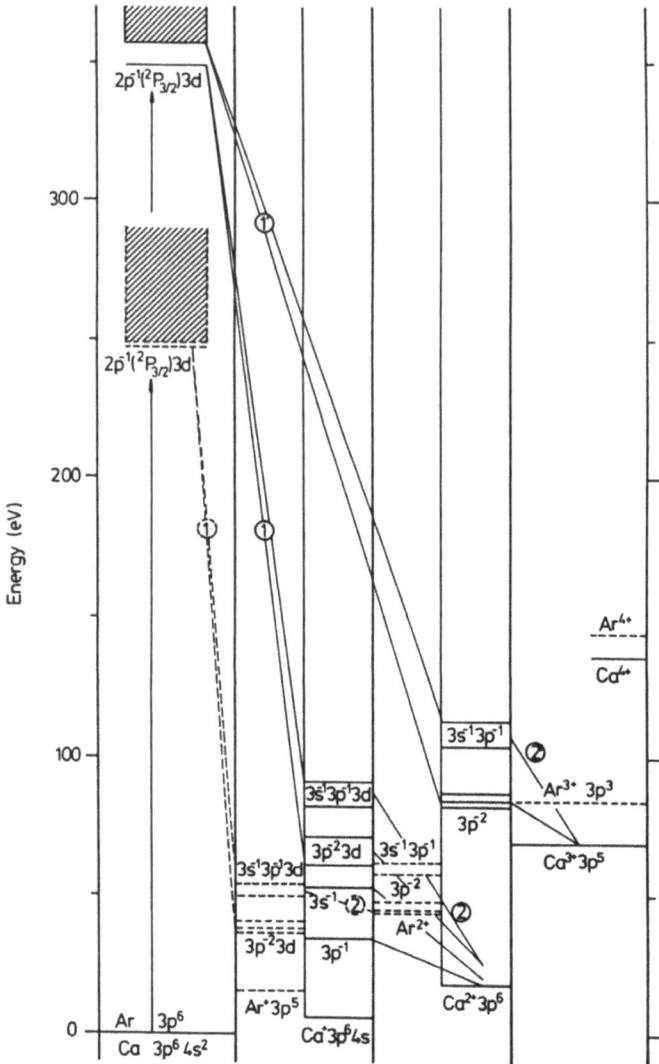

Fig. 10. Energy diagram of Ar and Ca. Energy levels of Ar^{2+} and Ar^+ were drawn based on the data of refs. 26 and 27, respectively. Energy levels of Ca^{2+} and Ca^+ were drawn based on the data of refs. 28 and 29, respectively.

Above the 2p ionization limits. In simple cases, normal Auger decay leads to doubly charged ions and Auger shake-off decay to triply charged ions. This holds for Ar, but not for Ca (see Figs. 8 and 9). Almost all final states of the normal Auger decay of Ca autoionize to Ca^{3+} (① → ② in Fig. 10). Therefore the distribution of the charge states is dominated by Ca^{3+}. The Auger shake-off accompanying three-electron ejection is energetically possible, but too weak to be detected by the TOF. Below the 2s ionization limit the intensities for Ca^{4+} were comparable to the noise level.

DECAY OF THE Kr AND Sr 3d EXCITED STATES

Now let us turn to 3d-core excitation. Figures 11 and 12 show the Kr and Sr ion-yield curves for each charge stage up to 4+ near the 3d-ionization edges, respectively. The charge-state distribution of photo-ions differs considerably for Kr and Sr: in Kr, the Kr^{2+} yield is abundant; in Sr, the Sr^{3+} yield is the strongest. The Sr^+ yield spectrum does not show any structure. These distributions can be explained in the same manner as those in the previous section.

Comparison Between the Photoelectron and Photo-ion Data of Kr

In Kr, the electrons ejected from the $3d^{-1}(^2D_{5/2})5p$ state have been measured, over the whole energy range, by two groups[32,33]. The data offer very detailed information on the decay channels of the excited states (1) and (2) in Fig. 13. Table 1 shows the comparison between the photo-electron and photo-ion data.

The agreement between the data of Morin et al[33] and Hayaishi et al[7] is excellent; however, the data of Heimann et al[32] strongly differ from both of them. One reason for the big difference is considered to be due to the poor energy resolution of the electron spectra reported by Heimann et al[32]. In their spectra many lines due to the second step of the two-step autoionization are not resolved. Therefore the shake-off contributions to the continuum part of the spectrum are overestimated. Here we want to emphasize the importance of high-resolution photoelectron

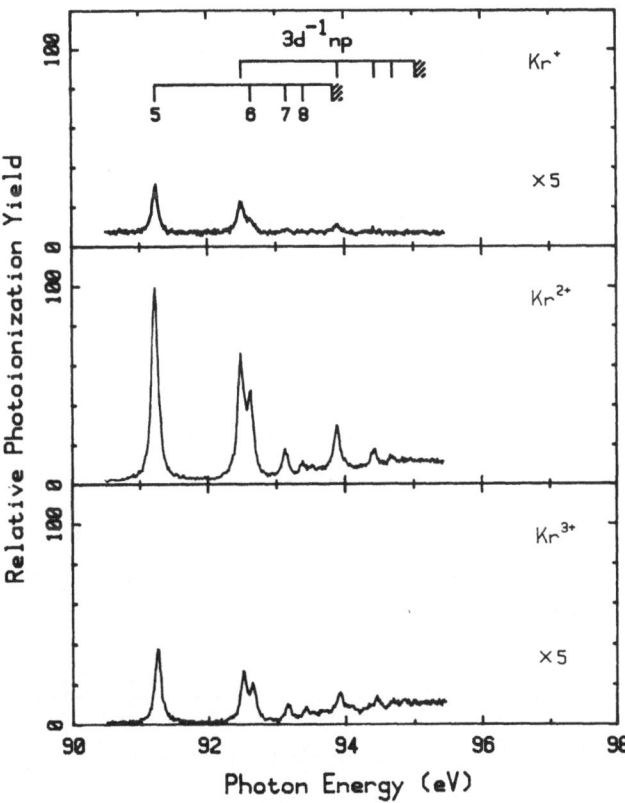

Fig. 11. Kr^+-, Kr^{2+} and Kr^{3+}-yield spectra, from Hayaishi et al[7].

41

Table 1. Decay channels of the $3d^{-1}(^2D_{5/2})5p$ state

Product ions	Investigator		
	Heimann et al[32]	Morin et al[33] a)	Hayaishi et al[7] (10% errors)
Kr^+	Autoionization 53% + nonresonant (5s+5p)	Autoionization 8%	5%
Kr^{2+}	Two-step 16%	Two-step 77%	
	Auger shake-off 37%	Auger shake-off 10%	88%
Kr^{3+}		Auger shake-off 5%	7%

a) Morin et al have determined the ratio of Kr^{2+} (shake-off) to Kr^{3+} (shake-off) by coincidence measurements between threshold electrons and photo-ions.

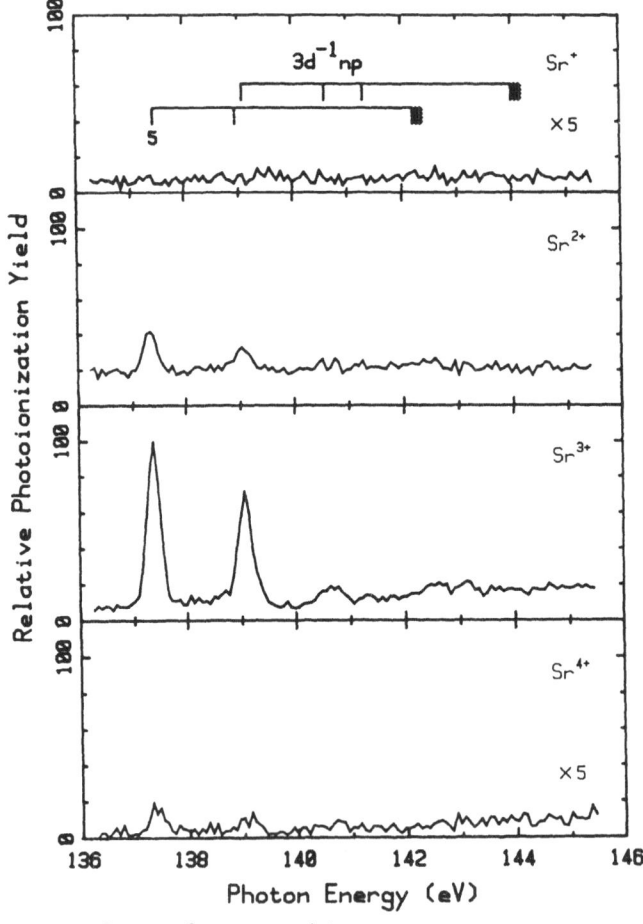

Fig. 12. Sr^+-, Sr^{2+}-, Sr^{3+}- and Sr^{4+}-yield spectra, from Koizumi et al[30]. By the assignments of Mansfield and Connerade[31], the main component of the first peak at 137.5 eV has a $3d^9_{5/2}5s^25p$ configuration.

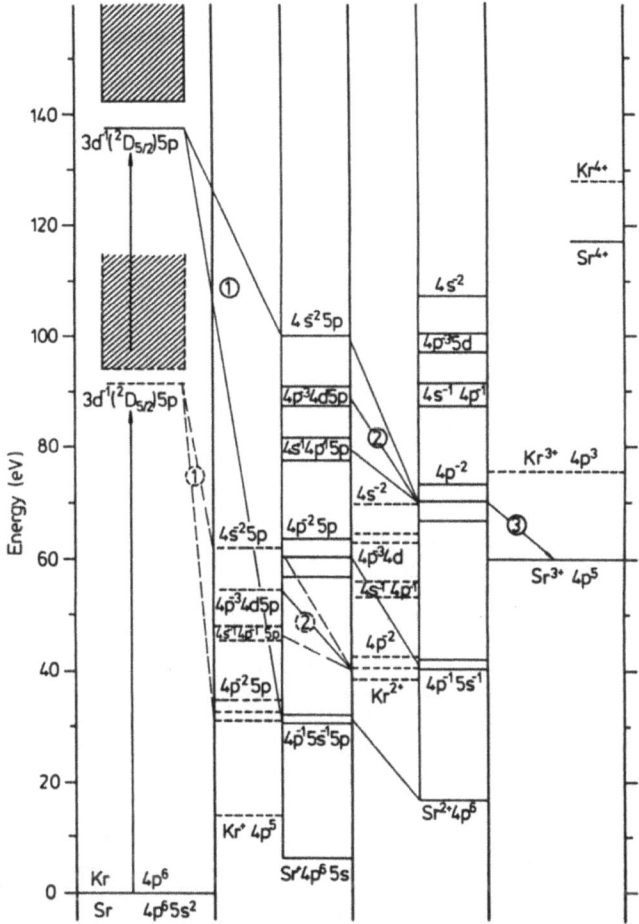

Fig. 13. Energy diagram of Kr and Sr. Energy levels of Kr^{2+} and Kr^+
were drawn based on the data of ref. 34. Energy levels of Sr^{2+}
were drawn based on the data of ref. 35. To determine the
energy levels of Sr^+, energy shifts of resonance Auger lines
compared to normal Auger lines were assumed to be the same
values as for Kr.

data for the determination of the ratio of two-step versus Auger shake-
off for the creation of Kr^{2+}.

Three-step Autoionization of the $3d^{-1}5p$ Excited States

Since Sr has two 5s valence electrons in addition to the Kr core,
many final states of the two-step autoionization process can autoionize
to Sr^{3+}, ① → ② → ③ in Fig. 13. To estimate the charge-state
distribution, we use the data for the normal Auger M_3NN from Mehlhorn
et al[35]. The $3d^{-1}$ $4p^{-2}$, $4p^{-1}5s^{-1}$ and $5s^{-2}$ Auger transitions, which
correspond to the $3d^{-1}5p$ $4p^{-2}5p$, $4p^{-1}5s^{-1}5p$ and $5s^{-2}5p$ in the resonance
Auger, comprise about 25% of all the M NN Auger decays. In the resonance
case, thus we expect 75% to contribute to the Sr^{+3} yield through three-
step autoionization. Comparing with the experimental result of Sr^{2+}
(20%) and Sr^{3+} (80%), the estimate of Sr^{2+} (25%) and Sr^{3+} (75%) is

43

satisfactory. Note we are neglecting the Auger shake-off processes, the final states of which mainly decay to Sr^{3+} (see Fig. 13). The Auger shake-off probability of Sr is not expected to differ strongly from that of Kr, therefore we conclude that three-step autoionization is the main decay-channel of the 3d excited states, leading to Sr^{3+}.

DECAY PROCESSES FOLLOWING $3\sigma_g$ AND 1s SHELL EXCITATION OF O_2

The decay channels for molecular excited states are generally very complicated in comparison with those for atomic excited states, because a lot of dissociation paths open for the excited state and Auger final states. Here we mention the dissociation paths of O_2 following the sub-valence $3\sigma_g$ and core 1s shell excitation.

Dissociation of the O_2^+ $B^2\Sigma_g^-$ State

Figure 14 shows the TOF spectra of O^+ ions measured by Akahori et al[36]; a time-to-amplitude converter was started by threshold electrons which have energies less than 30 meV. Photon energy and the threshold electrons define the state of O_2^+. The time-of-flight spread, depending on kinetic energies and angular distributions, of the O^+ ions defines the dissociation paths. As can be seen from the potential curves of O_2^+ in Fig. 15, two dissociation paths to the $O^+(^4S^0)+O(^3P)$ and $O^+(^4S^0)+O(^3P)$ limits respectively are open for the vibrational levels $v' = 3$ and 6 of the $B^2\Sigma_g^-$ state. The pair of peaks well separated in Fig. 14, corresponding to O^+ generated with forward and backward initial velocities with respect to the axis of the TOF spectrometer, is assigned to the $O^+(^4S^0)+O(^1D)$ limit, and the pair of barely resolved peaks is assigned to the $O^+(^4S^0)+O(^1D)$ limit. The results of a thorough analysis are presented in ref. 36.

O_2 $X^3\Sigma_g^- \rightarrow O_2^+$ $B^2\Sigma_g^-$ Photoionization Cross Section

To measure the kinetic energy of photoelectrons just above the ionization threshold by an electrostatic energy analyzer is difficult. It is much easier to measure the atomic ions generated via

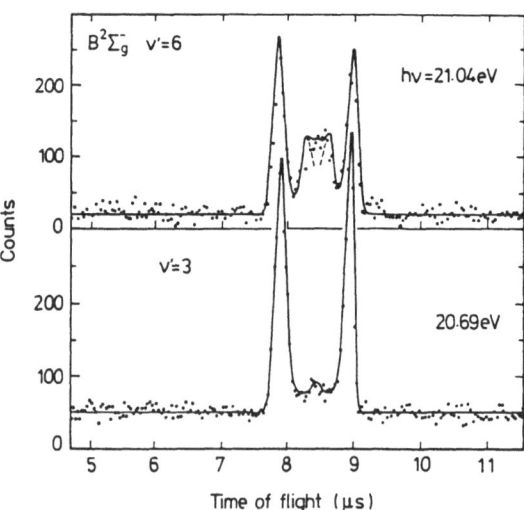

Fig. 14. Time-of-flight spectra of O^+ corresponding to vibrational levels $v' = 6$ (upper) and $v' = 3$ (lower) of the O_2^+ $B^2\Sigma_g^-$ state, from Akahori et al[36].

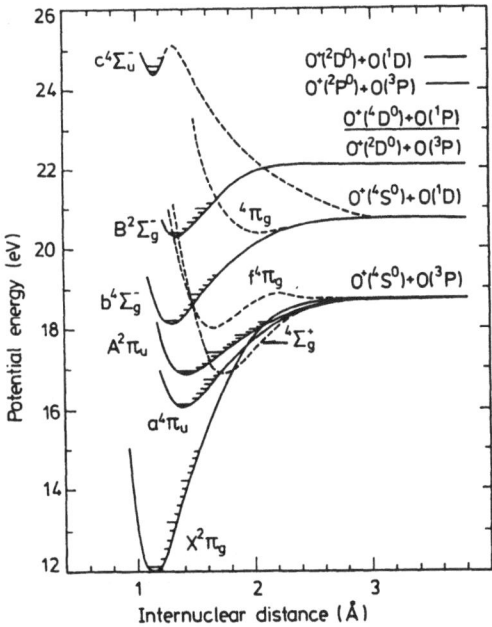

Fig. 15. Potential curves for O_2^+. Full curves were taken from Gilmore[37] and broken curves from Beebe et al[38].

dissociation. The kinetic energy of O^+ is half of the kinetic energy released in the O_2^+ $B^2\Sigma_g^- \to O^+(^4S^0)+O(^3P)$ dissociation. This energy of ≈ 1 eV is independent of the photon energy.

In order to determine the photoionization cross section for the O_2^+ $B^2\Sigma_g^-$ state close to the threshold, we have recently measured the O_2^+ ions using a position-sensitive parallel-plate analyzer[39]. Figure 16 shows the photo-ion spectra for photon energies close to the threshold 20.30 eV of the O_2 $X^3\Sigma_g^- \to O_2^+$ $B^2\Sigma_g^-$ ionization. The peaks around the kinetic energy of 1 eV correspond to the dissociation to the $O^+(^4S^0)+O(^3P)$ limit. Since the fluorescence decay of the state is negligible in this energy region[36], the origin of the ≈ 1 eV peaks is only the dissociation. The vibrational levels of the O_2^+ $B^2\Sigma_g^-$ state cannot be resolved because of limited analyzer resolution 0.25 eV (FWHM). Figure 17 shows the cross section obtained by photo-ion spectroscopy together with the theoretical cross section given by Raseev et al[40]. The vibrational levels of the $B^2\Sigma_g^-$ state located above the second limit $O^+(^4S^0)+O(^1D)$ can also dissociate to this limit; for example, 15% of the $B^2\Sigma_g^-$ ($v' = 6$) state dissociates to this limit[36]. Therefore, our cross section for the $B^2\Sigma_g^-$ state is somewhat too small except for the region close to the threshold. The gross feature of the cross section with a maximum around 21.7 eV is considered to be due to the σ_u shape resonance predicted by Raseev et al[40]. Morin et al have measured the cross section for photon energies above 21 eV by photoelectron spectroscopy, and pointed out the σ_u shape resonance with a maximum around 21.5 eV[41]. However, there is a difference between our cross section and theirs around 21 eV; theirs increases steeply starting from 21 eV. The difference may be resolved considering the transmission of their analyzer for the low-energy region (< 2 eV). In contrast to this, the O^+-ion yields are independent of the transmission, because the energy is constant over the whole photon-energy range.

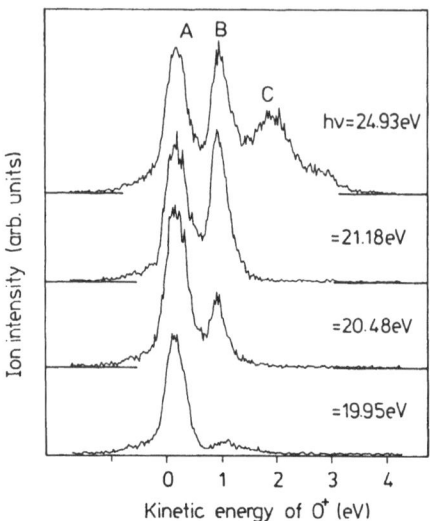

Fig. 16.   Kinetic energy spectra of O^+, from Ukai et al[39].   Peak A is composed of O_2^+ with thermal kinetic energies and O^+ with very small kinetic energies released in dissociation.   Peaks B and C correspond to the O_2^+ $B^2\Sigma_g^-$ → $O^+(^4S^0)+O(^3P)$ and O_2^+ $c^4\Sigma_g^-$ → $O^+(^4S^0)+O(^3P)$ dissociation, respectively.   The energy resolution 0.25 eV (FWHM) of the analyzer was mainly limited by the thermal motion of O_2, as a gas-cell was used.

Dissociation of O_2 1s Core Hole States

        Figures 18(a) and (b) show the TOF spectra of fragment ions for O_2 at the $1\sigma_g$ → $1\pi_g$ discrete and $1\sigma_g$ → $\varepsilon\sigma_u$ shape resonance, respec- tively[42].   The profile of the peak assigned to O^+ ions depends strongly on the photon energy, i.e., the excited states.   There are two reasons for the change of the profile; (1) the variation of kinetic energy dis- tribution for the fragment ion O^+, and (2) the variation of angular dis- tribution for O^+

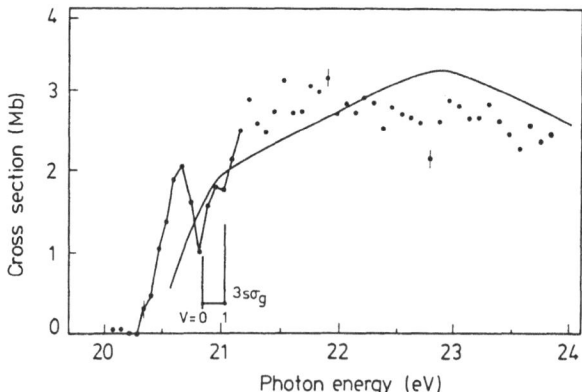

Fig. 17. O_2 $X^3\Sigma$ → O_2^+ $B^2\Sigma_g^-$ photoionization cross section, from Ukai et al[39].   Solid curve is the theoretical result of Raseev et al[40].   The experimental data are normalized to the theoretical at a photon energy of 22.0 eV.

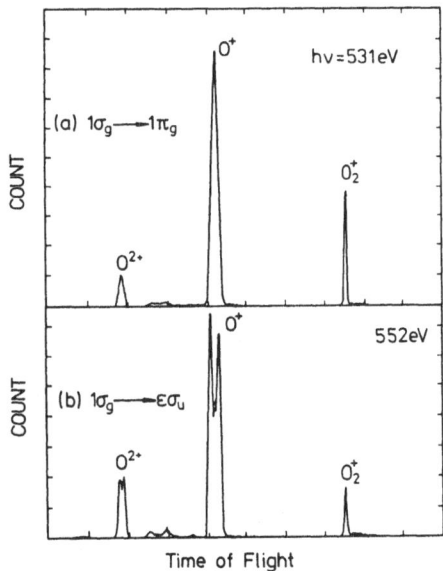

Fig. 18. Time-of-flight spectra of fragment ions for O_2 at the
$1\sigma_g \rightarrow 1\pi_g$ (upper) and the $1\sigma_g \rightarrow \varepsilon\sigma_u$ (lower) excitation.

The kinetic energies of O^+, i.e. the dissociation paths, are determined by the final states of the resonance and normal Auger decays.
Because there are two kinds of Auger decay to the different final states,
e.g. the $O_2^+ \; 1\pi_u^{-2}1\pi_g$ and $O_2^{2+} \; 1\pi_u^{-2}$ states as the final states of
resonance and normal Auger decays, respectively, different dissociation
paths are expected at the two resonances. As a result, the profile for
the fragment ions changes as mentioned in the previous sections.

When the molecular axis of O_2 is perpendicular to the direction of
polarization of synchrotron radiation, the $1\sigma_g \rightarrow 1\pi_g$ transition
preferentially ocurs. The resonance Auger decay and dissociation after
the decay occur much faster than the thermal motion of O_2. Therefore the
fragment ions are ejected in the direction perpendicular to the axis of
the TOF spectrometer which is parallel with the direction of polarization. As a result, there is no spread of the flight time of O^+.
However, the molecular axis is aligned with the axis of the TOF for the
$1\sigma_g \rightarrow \varepsilon\sigma_u$ transition, and the O^+ ions are ejected in the forward and
backward direction with respect to the axis of the TOF. Consequently,
the profile for these ions becomes broad and splits into two peaks (see
Fig. 18(b)).

In principle one can determine the energies and angular distributions of the fragment ions by analyzing the profile of the TOF spectrum.
For an example see ref. 43. However, there are many ambiguities in the
deconvolution procedure. To deconvolute the profiles, one must use a lot
of parameters for kinetic energies corresponding to dissociation paths,
relative intensities of each kinetic-energy component, and represent
angular distributions. The angle-resolved photo-ion spectroscopy
described in ref. 39 directly determines the energy and angular distribution from the data. In the future we plan to investigate the dissociation processes following inner-shell excitation using a photo-ion energy
analyzer modified to separate the mass of ions.

CONCLUSION

For the examples discussed here multi-step autoionization predomin-
ates for the production of multiply charged photo-ions, among the decay
modes of the inner-shell excited states. From the data on intensities of
normal Auger lines, the charge-state distribution of the photo-ions,
produced through multi-step autoionization, can be estimated. However,
electron spectra with high resolution over the whole energy range are
indispensable, as we have demonstrated in the case of Sr 4p and Kr 3d
excitation. Together such data allow for the full interpretation of the
decay channels of the inner-shell excited states.

To study the dissociation channels following inner-shell excitation
of molcules, the conventional photo-ion TOF technique is not sufficient.
Coincidence measurements between Auger electrons and energy- and angle-
resolved photo-ions are desirable for a full understanding of the dissociation channels.

ACKNOWLEDGEMENTS

This manuscript was prepared during the author's stay at the
Institut für Experimentalphysik, Universität Hamburg, as a fellow of the
Alexander von Humboldt Stiftung. It is a great pleasure for the author
to acknowledge the hospitality of Prof. B. Sonntag and the members of his
research group and his critical reading of the manuscript. The author
would also like to thank his co-workers at the Photon Factory for their
essential contributions to the results presented here. The author is
especially indebted to Dr. T. Hayaishi for communication of his encouraging results in advance of publication.

REFERENCES

1. M. Nakamura, M. Sasanuma, S. Sato, M. Watanabe, H. Yamashita,
   Y. Iguchi, A. Ejiri, S. Nakai, S. Yamaguchi, T. Sagawa, Y. Nakai,
   and T. Oshio, Phys. Rev. Lett. 21:1303 (1968).
2. T. Miyahara, H. Kitamura, S. Sato, M. Watanabe, S. Mitani,
   E. Ishiguro, T. Fukutani, T. Ishii, Shi. Yamaguchi, M. Endo,
   Y. Iguchi, H. Tsujikawa, T. Sugiura, T. Katayama, T. Yamakawa,
   Se. Yamaguchi, and T. Sasaki, Particle Accelerators 7:163 (1976).
3. H. Hanashiro, Y. Sasaki, T. Sasaki, A. Mikuni, T. Takayanagi,
   K. Wakiya, H. Suzuki, A. Danjo, T. Hino, and S. Ohtani, J. Phys.B
   12:L775 (1979).
4. H. Kanamori, S. Iwata, A. Mikuni, and T. Sasaki, J. Phys. B 17:3887
   (1984).
5. Y. Hatano, M. Ohno, N. Kouchi, H. Koizumi, A. Yokoyama, G. Isoyama,
   H. Kitamura, and T. Sasaki, Chem. Phys. Lett. 84:445 (1981).
6. K. Shinsaka, H. Koizumi, T. Yoshimi, N. Kouchi, Y. Nakamura,
   M. Toriumi, M. Morita, Y. Hatano, S. Asaoka, and H. Nishimura,
   J. Chem. Phys. 83:4405 (1985).
7. T. Hayaishi, Y. Morioka, Y. Kageyama, M. Watanabe, I.H. Suzuki,
   A. Mikuni, G. Isoyaa, S. Asaoka, and M. Nakamura, J. Phys. B
   17:3511 (1984).
8. K. Kohra and T. Sasaki, Nucl. Instrum. Meth. 208:23 (1983).
9. M. Yanagihara, H. Maezawa, T. Sasaki, Y. Suzuki, and Y. Iguchi,
   KEK Report 84-17.
10. Y. Yagishita, T. Hayaishi, Y. Itoh, T. Koizumi, T. Matsuo,
    J. Murakami, T. Nagata, Y. Sato, H. Shibata, M. Yoshino, and
    Y. Itikawa, KEK Report 86-6.
11. K. Ito, T. Sasaki, T. Namioka, K. Ueda, and Y. Morioka, Nucl.
    Instrum. Meth. A 246:290 (1986).

12. K. Ito, T. Namioka, Y. Morioka, T. Sasaki, H. Noda, K. Goto, T. Katayama, and M. Koike, Applied Optics 25:837 (1987).

13. H. Maezawa, Y. Suzuki, H. Kitamura, and T. Sasaki, Nucl. Instrum. Meth. A 246:82 (1986).

14. H. Maezawa, S. Nakai, S. Mitani, H. Noda, T. Namioka, and T. Sasaki, Nucl. Instrum. Meth. A 246:310 (1986).

15. Photon Factory Activity Report 1982/83, 1983/84, 1984/85, and 1985/86, National Laboratory for High Energy Physics.

16. T. Nagata, J.B. West, T. Hayaishi, Y. Itikawa, Y. Itoh, T. Koizumi, J. Murakami, Y. Sato, H. Shibata, A. Yagishita, and M. Yoshino, J. Phys. B 19:1281 (1986).

17. M.W.D. Mansfield and G.H. Newsom, Proc. R. Soc. Lond. A 377:431 (1981).

18. A. Yagishita, S. Aksela, Th. Prescher, M. Meyer, E.V. Raven, and B. Sonntag, to be submitted to J. Phys. B.

19. J.M. Bizau, P. Gérard, F.J. Wuilleumier, and G. Wendin, Phys. Rev. Lett. 53:2083 (1984).

20. A.F. Starace, in: "Handbuch der Physik, Vol. 31, W. Mehlhorn, ed., Springer, Berlin, p.1.

21. T. Hayaishi, A. Yagishita, E. Murakami, and Y. Morioka, Photon Factory Activity Report 1986, p.234, National Laboratory for High Energy Physics.

22. G.V. Marr and J.B. West, Atomic Data and Nuclear Data Tables 18:497 (1976).

23. J.J. Yeh and I. Lindau, Atomic Data and Nuclear Data Tables 32:1, (1985).

24. T. Nagata, Y. Itikawa, T. Hayaishi, Y. Itoh, T. Koizumi, Y. Sato, E. Shigemasa, A. Yagishita, and M. Yoshino, XV ICPEAC, Book of Abstracts, Brighton, U.K. (1987).

25. M.W. Mansfield, Proc. R. Soc. Lond. A 348:143 (1976).

26. L.O. Werme, T. Bergmark, and K. Siegbahn, Phys. Scripta 8:149 (1973).

27. S. Aksela, private communication.

28. W. Weber, Diploma Thesis, Universität Freiburg (1984).

29. B. Sonntag, private communication.

30. T. Koizumi, T. Hayaishi, Y. Itikawa, Y. Itoh, T. Matsuo, T. Nagata, Y. Sato, E. Shigemasa, Y. Yagishita, and M. Yoshino, XV ICPEAC, Book of Abstracts, p.6, Brighton, U.K. (1987).

31. M.W.D. Mansfield and J.P. Connerade, J. Phys. B 15:503 (1982).

32. P.A. Heimann, D.W. Lindle, T.A. Ferrett, S.H. Liu, L.J. Medhurst, M.N. Piancastelli, D.A. Shirley, U. Becker, H.G. Kerkhoff, B. Langer, D. Szostak, and R. Wehlitz, submitted to J. Phys. B.

33. P. Morin et al., private communication.

34. H. Aksela, S. Aksela, H. Pulkkinen, G.M. Bancroft, and K.H. Tan, Phys. Rev. A 33:3876 (1986).

35. W. Mehlhorn, B. Breuckmann, and D. Hausamann, Phys. Scripta 16:177 (1977).

36. T. Akahori, Y. Morioka, M. Watanabe, T. Hayaishi, K. Ito, and M. Nakamura, J. Phys. B 18:2219 (1985).

37. F.R. Gilmore, J. Quant. Spectrosc. Radiat. Transfer 5:369 (1965).

38. N.H.F. Beebe, E.W. Thulstrup, and A. Anderson, J. Chem. Phys. 64:2080 (1976).

39. M. Ukai, A. Kimura, S. Arai, P. Lablanquie, K. Ito, and Y. Yagishita, Chem. Phys. Lett. (to be published).

40. G. Raseev, H. Lefebvre-Brion, H. Le Rouzo, and A.L. Roche, J. Chem. Phys. 74:6686 (1981).

41. P. Morin, I. Nenner, M.Y. Adam, M.J. Hubin-Franskin, J. Delwiche, H. Lefebvre-Brion, and A. Giusti-Suzor, Chem. Phys. Lett. 92:609 (1982).

42. Y. Sato, A. Yagishita, T. Nagata, T. Hayaishi, M. Yoshino, T. Koizumi, Y. Itoh, T. Matsuo, H. Shibata, and T. Sasaki, Photon Factory Activity Report 1986, p.229, National Laboratory for High Energy Physics.

43. N. Saito and I.H. Suzuki, Chem. Phys. Lett. 129:419 (1986).

ELECTRON MOMENTUM SPECTROSCOPY OF MOLECULES:

A REVIEW OF RECENT DEVELOPMENTS

Erich Weigold

School of Physical Sciences
Flinders University
Adelaide, Australia 5042

INTRODUCTION

Conventional spectroscopic studies of atoms and molecules have for the most part concentrated on the determination of the energy level structure. Electron momentum spectroscopy, also formerly known as binary (e,2e) spectroscopy, has added a crucial further dimension of information, namely the determination of the momentum distributions for the electrons associated with the separate valence orbitals. Since its inception over a decade ago[1-3], it has developed into a technique which is providing increasingly sensitive tests of state-of-the art theoretical quantum chemical calculations. Measured orbital momentum distributions (more exactly the squares of the momentum space ion-neutral overlaps) have provided very sensitive tests of calculated orbital wavefunctions or ion-neutral overlaps.

In an EMS experiment the kinematics of the (e,2e) reaction is completely determined (see fig. 1). Of primary interest are the measured recoil momenta

$$\mathbf{p} = \mathbf{p}_0 - \mathbf{p}_A - \mathbf{p}_B \tag{1}$$

and the separation energy of the state $|\Psi_f >$ of the residual ion

$$\epsilon_f = E_0 - E_A - E_B. \tag{2}$$

If an electron is suddenly removed from a target system the recoil momentum of the residual system is equal and opposite to the momentum of the electron in the target system. The measurement of the distribution of ion recoil momenta for the same separation energy ϵ_i then gives a direct measurement of the momentum profile for that separation energy. The simplest illustration is provided by atomic hydrogen, where there is one orbital ψ_0 with separation energy $\epsilon_0 = 13.6eV$. The measurement of the hydrogen momentum profile was made by Lohmann and Weigold[4] using noncoplanar symmetric kinematics, i.e. fixing $E_A = E_B$ and $\theta_A = \theta_B = 45°$ and varying the recoil momenta by varying the out of plane azimuthal angle $\phi(= \phi_A - \phi_B - \pi)$. With this kinematics the differential cross section is proportional to the momentum profile, which in this case is simply the square of the momentum space ground state wavefunction, the solution of the Schrödinger equation for the hydrogen atom. In atomic units this is $| < \mathbf{p}|\psi_0 > |^2 = 8\pi^{-2}(1 + p^2)^{-4}$. As fig. 2 shows, not only do the experimental and theoretical profiles agree in great detail, but the fact that the momentum information is characteristic of the target

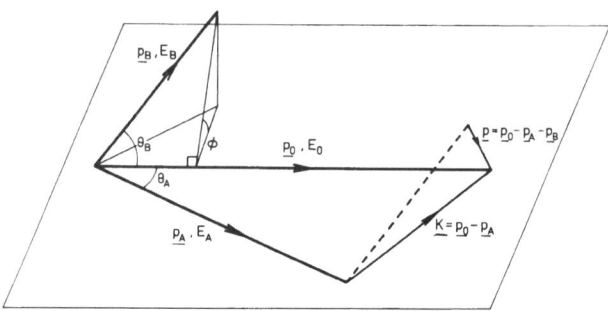

Fig. 1
Diagram showing the kinematic variables of a kinematically-complete (e,2e) knock-out reaction. The recoil momentum $\mathbf{p} = \mathbf{p}_0 - \mathbf{p}_A - \mathbf{p}_B$ and momentum transfer $\mathbf{K} = \mathbf{p}_0 - \mathbf{p}_A$ (where $E_A \geq E_B$) are indicated.

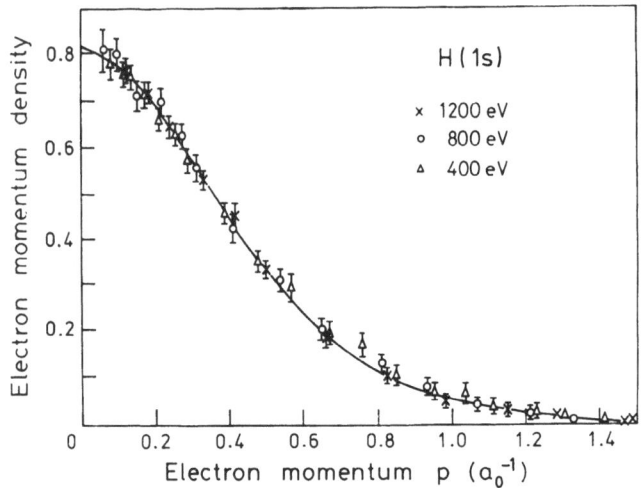

Fig. 2
The distribution of recoil momenta in the (e,2e) experiment on atomic hydrogen at different total energies (Ref. 4) compared with the square of the momentum space wave function for the ground state of hydrogen (solid curve).

itself, and does not depend on other aspects of the reaction, is confirmed by the independence of the measured profile from the incident energy.

The noncoplanar symmetric (e,2e) reaction has formed the basis of EMS[1,3]. In this geometry the momentum transfer K (see fig. 1) at any total energy $E = E_A + E_B$ is maximized and is independent of the azimuthal angle, which is varied to obtain the momentum profile. The separation energy is scanned by varying the incident energy E_0.

At high enough total energy the momentum profile is given to a good approximation by the spherically averaged square modulus of the characteristic momentum-space orbital of the knocked-out electron. This is the orbital for which the leading configuration in a configuration-interaction expansion of the ion is a hole coupled to the target Hartree-Fock ground state. This approximation, the plane-wave impulse approximation (PWIA), is valid up to some maximum momentum, which at the energies normally used (≥ 1000eV) is at least 1 a.u. The criterion for sufficiently high energy is purely experimental. The structure information derived from the experiment must be independent of the total energy.

Inner-valence states of atoms and molecules usually consist of fragments of an orbital split by final-state configuration interaction. The characteristic orbital of each fragment is identified by its momentum profile. If all the fragments of a particular orbital are found in an experiment, the summed crossed section for them is equal to the cross section for the orbital, in the approximation that the Hartree-Fock configuration dominates the target ground state. The relative cross sections for the fragments, normalized to a sum of unity, are the spectroscopic factors or pole strengths. This interpretation is checked by the total-energy independence of the spectroscopic factors and the relative momentum profiles for different summed orbitals.

In the case of small molecues[3,5] these checks have always been obeyed when the momentum profiles are calculating using the PWIA. For inert-gas atoms spectroscopic factors for inner valence[2]S states have always been determined consistently at $p \sim 0$, independent of total energy, by relative cross sections within the corresponding symmetry manifold. However, the ultimate confirmation, the comparison of momentum profiles between orbitals, has required the distorted-wave impulse approximation (DWIA)[3,5]. Relative momentum profiles have been correctly calculated for the valence s and p manifolds of neon[6], argon[7] and xenon[8]. Momentum-profile shapes are correctly described for all measured momenta by the DWIA for these targets and for helium[9].

In the following section a brief outline is given of the general theory of the (e,2e) reaction. This is followed by a brief discussion of some relevant atomic examples before discussing some recent work on molecules.

GENERAL THEORY OF THE (e,2e) REACTION

The differential cross section in terms of the electron momenta p_0, p_A, p_B (in obvious notation) is

$$\frac{d^5\sigma}{d\hat{p}_A d\hat{p}_B dE_B} = (2\pi)^4 \frac{p_A p_B}{p_0} f_{ee} G_f(p), \qquad (3)$$

where the electron-electron collision factor f_{ee} is the square modulus of the appropriate half-shell Mott-scattering t-matrix element averaged over electron-spin degeneracies. In the noncoplanar symmetric geometry it is essentially independent of the angle ϕ. The cross section for molecules is spherically averaged (over all directions of **p** for convenience) because of the unobserved orientation of the target. In addition there is a vibrational average which is approximated by calculating electronic wave functions for the equilibrium target configuration of the nuclei. The sum over final-state electronic degeneracies and an average over initial-state electronic degeneracies is taken.

The structure factor $G_f(p)$, written in terms of magnetically degenerate initial and final states, is

$$G_f(p) = \sum_{M_0,M_f} |<\mathbf{p}\Psi_f : J_f M_f |\Psi_0 : J_0 M_0>|^2/(2J_0 + 1). \tag{4}$$

where the target electronic states $|\Psi_0 : J_0 M_0>$ constitute a basis for a representation of the symmetry group of the target which has rank J_0 and dimension $2J_0 + 1$. For an atomic target the symmetry group is SU(2) and J_0 is the total angular momentum quantum number. The vector index of the representation space is M_0 (magnetic quantum number). A corresponding notation using subscript f describes the electronic states of the ion. The matrix element in (4) is the probability amplitude that Ψ_f is obtained by annihilating an electron of momentum $\mathbf{p}$ and spin coordinate σ in the target state Ψ_0. We use second-quantized notation, in which the operator that annihilates such an electron is

$$\sum_{j,\alpha,m} \phi^j_{\alpha m}(\mathbf{p},\sigma)a^j_{\alpha m}. \tag{5}$$

The function $\phi^j_{\alpha m}$ is the single-particle orbital in momentum-spin space for total angular momentum j with projection m. In a many-body calculation the effect of certain complete manifolds is often represented by replacing the manifolds with pseudostates (usually denoted by the use of a bar, e.g., $\bar{6}s$). The notation $\phi^j_{\alpha m}$ includes pseudostates. The quantum number j and the label α together uniquely specify the orbital, apart from the projection number m. The operator $a^j_{\alpha m}$ is the usual fermion annihilation operator with quantum labels j, α, and m. The orbital $\phi^j_{\alpha m}(\mathbf{p}\sigma)$ is defined by

$$\phi^j_{\alpha m}(\mathbf{p}\sigma) = \phi^j_\alpha(p)y^{jl}_m(\hat{\mathbf{p}},\sigma), \tag{6}$$

where y^{jl}_m is a normalized spin-angle harmonic. The spin-angle integration in the matrix element of (4) is done by the Wigner-Eckart theorem to give [5,8]

$$G_f(p) = (2J_0 + 1)^{-1}$$
$$\times \sum_{j,\alpha,\beta} \phi^j_\alpha(p)\phi^j_\beta(p) < \Psi_0||(a^j_\alpha)^\dagger||\Psi_f><\Psi_f||a^j_\beta||\Psi_0>. \tag{7}$$

We now consider only the final states $|\Psi_f : J_f M_f>$ that contain the one-hole configuration resulting from the annihilation of a characteristic orbital $\phi^i_\alpha(p)$ in Ψ_0. These ion states all belong to a symmetry manifold characterized by $j\alpha$ and obey the manifold-closure relation

$$\sum_{f \in j\alpha} |\Psi_f : J_f M_f><\Psi_f : J_f M_f| = 1. \tag{8}$$

The closure property may be used in conjunction with Eq. (7) (in unreduced form) to determine the sum rule for the structure factors belonging to a particular manifold,

$$W_{j\alpha}(p) = \sum_{f \in j\alpha} G_f(p)$$

$$= (2J_0 + 1)^{-1} \sum_{j,\alpha,\beta}{}' \phi^j_\alpha(p)\phi^j_\beta(p) < \Psi_0|(a^j_\alpha)^\dagger \cdot a^j_\beta|\Psi_0>, \tag{9}$$

where the last factor in (9) is just the density matrix for the ground state of the target. The prime over the summation is used to indicate that the sum over orbitals is restricted to those which obey the vector coupling selection rules. We may now define the generalized spectroscopic factor for a particular symmetry manifold by

$$G_f(p) = S_{f,j\alpha}(p)W_{j\alpha}(p).$$ (10)

It is obvious from (9) that the spectroscopic factors obey the sum rule

$$\Sigma_{f\epsilon j\alpha}S_{f,j\alpha}(p) = 1.$$ (11)

The Target Hartree-Fock Approximation

In general, $S_f(p)$ and $W_{j\alpha}(p)$ are calculated by many-body methods such as configuration-interaction or Green's-function methods. However, for many closed-shell targets, nearly 100% of the spectroscopic strength of the many-body wave function Ψ_0 is attributed to the Hartree-Fock (or Dirac-Fock) configuration Φ_0. It is reasonable to make the target Hartree-Fock (or Dirac-Fock) approximation THFA (or TDFA) in which

$$|\Psi_0 >= |\Phi_0 > .$$ (12)

This effects an enormous simplification in $G_f(p)$ and $W_{j\alpha}(p)$. Once again we are interested only in the shell $j\alpha$, so we have $\beta = \alpha$. For a closed-shell target $J_0 = 0$. We express $G_f(p)$ in terms of a momentum-independent spectroscopic factor $S_{f,j\alpha}$ [5,8]

$$G_f(p) = [\phi_\alpha^j(p)]^2| < \Psi_f||(a_\alpha^j)^\dagger||\Phi_o > |^2$$
$$= (2j+1)S_{f,j\alpha}[\phi_\alpha^j(p)]^2.$$ (13)

The structure factor in the THFA is simply the square of the radial momentum-space orbital multiplied by the electron multiplicity and the spectroscopic factor

$$S_{f,j\alpha} = (2j+1)^{-1}| < \Psi_f||(a_\alpha^j)^\dagger||\Phi_0 > |^2.$$ (14)

In this approximation the spectroscopic factor is indepent of p. It is equal to 1 in the approximation that the ion structure is given by a hole in Φ_0. We may consider it as the probability that Ψ_f consists of a hole in Φ_0 with quantum numbers j,α. The THFA manifold structure factor is

$$W_{j\alpha}(p) = [\phi_\alpha^j(p)]^2 < \Phi_0 : 00|a_\alpha^{j\dagger} \cdot a_\alpha^j|\Phi_0 : 00 >$$
$$= (2j+1)[\phi_\alpha^j(p)]^2.$$ (15)

Equation (15) gives an experimental definition of the orbital $\phi_\alpha^j(p)$, valid in the case where electron correlation splits the single-particle ion state into fragments f in the manifold $j\alpha$. The EMS experiment may be used to provide an experimental measurement of the HF single-particle energy,

$$\epsilon_{ja} =< \Phi_0 : 00|(a_{\alpha m}^j)^\dagger H a_{\alpha m}^j|\Phi_0 : 00 >$$
$$= \sum_{f\epsilon j\alpha,m} < \Phi_0 : 00|(a_{\alpha m}^j)^\dagger|\Psi_f : jm > \epsilon_f < \Psi_f : jm|a_{\alpha m}^j|\Phi_0 : 00 >$$
$$= \sum_{f\epsilon j\alpha} S_{f,j\alpha}\epsilon_f,$$ (16)

where ϵ_f is the experimentally determined separation energy of the state f.

Initial-state Configuration Interaction.

The PWIA structure factor is given in general by (7), which contains the effects of both initial- and final-state configuration interaction. A detailed calculation of the target-ion overlap sometimes reveals sensitivity to initial-state configuration interaction. However in nearly every case the target ground state Ψ_0 is dominated by the Hartree-Fock configuration Φ_0 so that the manifold structure factor (9) is dominated by the THFA term (15) if the orbital ϕ_α^j is occupied in Φ_0.

The situation changes completely if ϕ_α^j is unoccupied in Φ_0. If there are no occupied orbitals of symmetry $j\alpha$ then the manifold structure factor is strongly dependent on the coefficients of the configurations containing ϕ_α^j. Such a case is the (e,2e) reaction on He, leaving the ion He^+ in a 2p state. The n=2 helium ion states provide a sensitive test of initial-state CI, highly accurate correlated wavefunctions are required to fit the measured cross sections(see Refs.3,5,9).

Distorted-wave Calculations for Atoms.

In the distorted wave impulse approximation (DWIA) we must replace the structure factor (4) by the distorted structure factor. In the THFA (or TDFA) approximation this is for an atomic target simply given by[5]

$$G_f(p) = S_{f,ja} \sum_m | < \chi^{(-)}(\mathbf{p}_A)\chi^{(-)}(\mathbf{p}_B)|\phi_{\alpha m}^j \chi^{(+)}(\mathbf{p}_0) > |^2, \qquad (17)$$

where $\chi^{(\pm)}(\mathbf{p})$ are elastic-scattering wave functions for electrons in the appropriate equivalent local static exchange potentials.

RESULTS FOR SOME ATOMS

The remarkably-detailed agreement of the PWIA with the noncoplanar-symmetric experiment of Lohmann and Weigold[4] for hydrogen over a wide range of total energies has been mentioned earlier (fig. 2). It confirms the validity of the orbital mapping in the case where we know the orbital exactly. Some of the work on the noble gases He, Ar, Kr, and Xe, has also been discussed[3,5-9]. We will examine the case of argon in some more detail, since it contains all the complications of final-state CI in a case where the final states with larger spectroscopic factors can be experimentally resolved. In addition there are two occupied valence states, 3p and 3s, in the target Hartree-Fock configuration, so it is possible to test the consistency of the spectroscopic factor determination using two methods of normalisation, the spectroscopic sum rule and comparison of the structure factor with the manifold structure factor for a different manifold.

For the argon ion the 3p manifold has essentially only one state in the LS coupling representation. Its momentum profile shape at 1500 eV is described within experimental error by the DWIA in fig. 3. The PWIA-THFA gives an undistorted mapping for p $\leq$ 1.5 a.u. as in the other noble gas atoms[5]. The normalisation for the whole analysis is carried out for the $3p^{-1}$ state at $\phi = 10°$. The experiment is normalised to the DWIA and the PWIA is multiplied by 0.874 to agree with the DWIA at this point.

The 3s manifold has several states with spectroscopic factors greater than 0.01. The shapes of the different momentum profiles for the strongly excited states at large separation energy are independent of energy and have the characteristic 3s orbital momentum distributions up to a maximum momentum $\sim$ 1 a.u.[5]. The $3s^{-1}$ spectroscopic factors, listed in table 1, are independent of energy and momentum. In fact the DWIA-THFA describes the shape and normalization of the 3s manifold

Table 1. Spectroscopic factors for observed states (or unresolved clusters of states) in the 3s manifold of Ar+. The experimental error in the last figure is shown in parentheses

ε_f(eV)	Dominant Configuration	E = 1000eV(Ref.7)		E = 1500eV(Ref.10)		
		p=0.43	p=0.92	p=0.1	p=0.55	p=0.92
29.3	$3s3p^6$	0.55(2)	0.53(4)	0.55(2)	0.55(2)	0.54(3)
35.2				0.01(1)	0.01(1)	0.01(1)
36.7	$3s^2 3p^4 4s$	0.03(1)	0.05(2)	0.03(1)	0.04(1)	0.04(2)
38.6	$3s^2 3p^4 3d$	0.17(2)	0.14(4)	0.15(1)	0.14(2)	0.15(2)
41.2	$3s^2 3p^4 4d$	0.07(1)	0.06(2)	0.07(1)	0.07(2)	0.06(2)
42.7	$3s^2 3p^4 5d$			0.01(1)	0.03(1)	0.01(2)
		0.06(2)	0.08(2)			
43.4				0.06(1)	0.05(1)	0.06(2)
43.5-55	$Ar^{++}+e$	0.12(1)	0.14(3)	0.12(1)	0.12(1)	0.13(3)

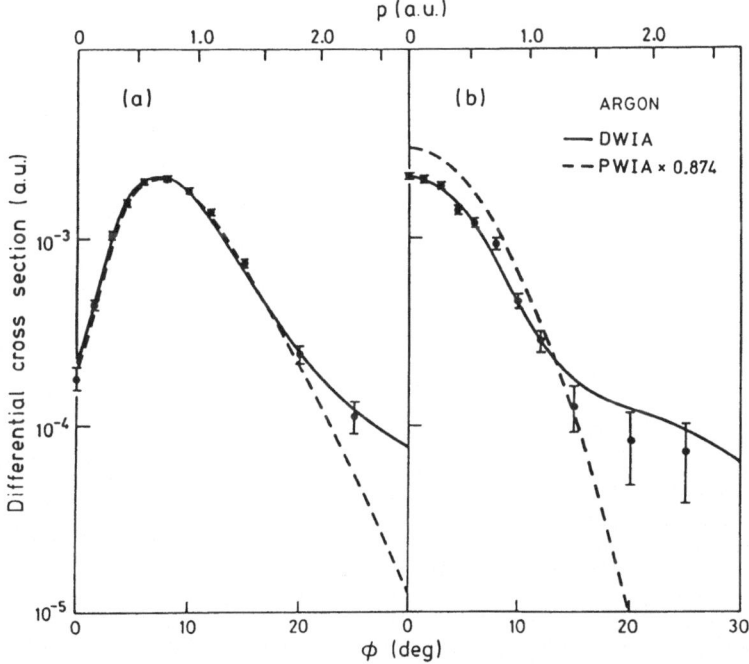

Fig. 3
The 1500eV (e,2e) reaction for the (a) 3p and (b) 3s manifolds of argon. The whole experiment is normalised to the DWIA theory at $\phi = 10°$ for the 3p manifold. Angular acceptance (full width) $\Delta\theta = 0.8°$, $\Delta\theta = 1°$ has been folded into the calculations. From Ref. 10.

Table 2. Eigenvalues and spectroscopic factors for observed states of
Pb⁺. The spectroscopic factors for the 6p manifold are evaluated at
p = 0.35 (upper) and p = 0.55 (lower)

Experiment		Theory	
ε_f(eV)	S_f	ε_f(eV)	S_f
6p manifold			
7.42	0.944±0.01	7.10	0.907
	0.933±0.01		0.901
9.16	0.056±0.006	8.76	0.062
	0.067±0.007		0.065
6s manifold			
14.59	0.762±0.008	13.79	0.729
18.35	0.227±0.005	18.34	0.214
20.34	0.011±0.003	19.97	0.036

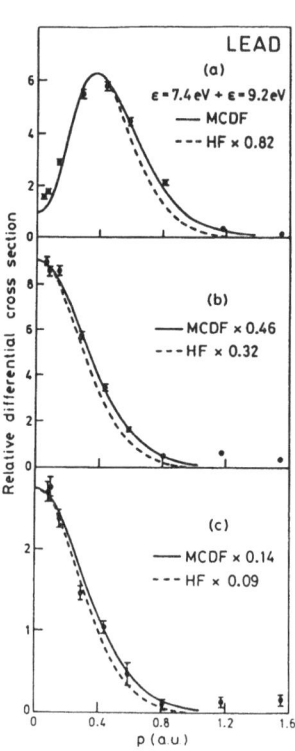

Fig. 4
Momentum profiles for the 1000eV
(e,2e) reaction on lead (Ref. 11).
Curves are calculated in the PWIA.
The whole experiment is normalised
to the peak for the 6p manifold (a).
The 14.6 and 18.4eV states of the
6s manifold are shown in (b) and (c)
respectively.

structure factor for all momenta at 1500eV (fig. 3) and at 1000eV (Ref. 7). The PWIA underestimates the 3s cross section relative to the 3p manifold by a factor of 0.7. This is due to neglect of absorption of the electron waves.

The orbital energy defined by (17) as the weighted centroid of the 3s manifold is at 35.2 ± 0.2eV, in excellent agreement with the Hartree-Fock value of 34.76eV. For atomic lead relativistic orbitals and configuration interaction in initial and final states must be used. Experiments at 1000eV and their analysis are described by Frost, Mitroy and Weigold[11].

The valence electrons are, in terms of the Hartree-Fock determinant, two in the $6p_{1/2}$ orbital and two in the $6s_{1/2}$ orbital. For such a large atom relativistic effects (essentially spin-orbit coupling) are important in the structure, so it is sensible to base the discussion on jj coupling with relativistic orbitals[12] (Multi-configurational Dirac-Fock Optimal level, MCDF-OL).

The ion states at 7.4eV and 9.2eV are identified by the PWIA (which is expected to give an undistorted mapping up to about p=1) as belonging to the 6p manifold. The manifold sum is shown in fig. 4(a) in which the cross section is well described by the relativistic orbitals (full curve) and significantly less well by the nonrelativistic orbital (dashed curve). Note that the long-range orbitals are compressed in momentum space so that a large part is mapped by the PWIA. With no interaction of target jj-coupling configurations only one ion state would be expected in the 6p manifold in the absence of final-state CI. The simplest realistic description of a two-state ion manifold requires a two-configuration target wave function

$$|\Psi_0> = a|6p_{1/2}^2 > +b|6p_{3/2}^2 >, \qquad (18)$$

since we expect small final-state CI for the topmost valence states on the basis of experience with other atoms. The coefficients a and b were determined by the MCDF-OL calculation to be respectively 0.9625 and -0.2711. The branching ratio for the two ion states is determined by the experiment as an average for the whole angular range and for $\phi \leq 7.75°$. The experimental values are respectively 0.072 ± 0.007 and 0.059 ± 0.006. The MCDF-OL value is about 0.08 with a slight momentum dependence in the same sense as the experiment. Inclusion of final-state CI in the complete overlap calculation (7) reduces the theoretical branching ratio to about 0.07 with a similar momentum dependence.

The ion states at 14.6eV, 18.4eV and 20.3eV are identified by the PWIA as belonging to the 6s manifold. The relativistic orbital again gives a better description of the momentum dependence than the nonrelativistic orbital (fig. 4(b) and (c)). The validity of the mapping concept is confirmed by the fact that the momentum distribution shapes are identical for the three states, whose splitting must therefore be due to final-state CI. The experimental data for the first two states (fig. 4(b) and (c)) are normalized to the 6p manifold sum at $p = 0.6$, but the spectroscopic factors so determined (0.46 and 0.14) are nearly a factor of two smaller than those determined from the spectroscopic sum rule for the 6s manifold. This is in conformity with the failure of the PWIA to give the correct relative normalisation for manifold sums found for neon, argon and xenon[5]. In those cases the DWIA corrected this deficiency. The sum rule is expected to give correct spectroscopic factors, which are given in table 2 for both the 6p and 6s manifolds, together with the results of the calculation of $G(p)$ (eq. 4) using initial- and final-state CI. The spectroscopic factors for the 6p manifold, as expected for initial-state CI, are momentum dependent. Experimental spectroscopic factors are given for the angular averages $\phi \leq 7.75°$ and all ϕ respectively. Theoretical spectroscopic factors are calculated for the weighted means of p, 0.35 and 0.55 in the respective cases. Spectroscopic factors for the 6s manifold are essentially momentum-independent, confirming the extension of the THFA to the one-hole linear combination for the case of significant initial-state CI. Here the 6s-hole linear combination is essentially $a|6s^{-1}6p_{1/2}^2 > +b|6s^{-1}6p_{1/2}6p_{3/2} > +c|6s^{-1}6p_{3/2}^2 >$, where a, b and c are in essentially the same proportion for the corresponding target and ion configuration.

Note that if initial-state CI is insignificant the spectroscopic factors for the two states in the 6p manifold would be 1 and 0 respectively, instead of the experimental values of approximately 0.94 and 0.06. Although initial-state CI is again small the near absence of final-state CI for upper valence states makes it observable as in the case of helium.

Comparison of PES with EMS

PES data are often mistakenly compared directly with EMS. The two probe quite different regions of momentum space. Ignoring the small momentum of the photon, high energy PES probes details of the wavefunction at $p \geq 10$, far in excess to that probed in EMS. At such high momenta the valence momentum density has become very small and the structure factor may be sensitive to correlations in the target state. Momentum dependent spectroscopic factors have been measured for helium[3,5,9] and for lead[11]. Experience with EMS has also shown that electron distortion effects become important at high momentum (≥ 1.5 a.u.) even for total energies of over 1000eV. This is not surprising since the high momentum components of the wavefunction arise predominantly from the regions close to the nucleus, where the potential is very strong. It is just these regions which are probed in the XPS and in the higher energy synchrotron radiation PES experiments. Care should thus be taken in making direct comparisons between EMS and PES data. Spectroscopic information must be independent of the energy of the probe, as is demonstrated in the EMS experiments.

EMS MEASUREMENTS FOR MOLECULES

For molecular targets distorted wave calculations are much more difficult and have not yet been done. The PWIA has therefore been employed using either the complete overlap or the THFA. This requires the spherically-averaged square of the momentum-space orbital $\phi^\kappa(\mathbf{p})$, where κ labels the symmetry manifold and the orientation of the orbital. Molecular orbitals are generally calculated as linear combinations of coordinate-space functions centred at the atomic sites. They depend on the radial coordinate $\mathbf{r}$ of the electron and the ground-state equilibrium coordinates $\mathbf{R}_s$ of the atomic sites s.

$$\psi^\kappa(\mathbf{r}) = \Sigma_s \Omega_s^\kappa(\mathbf{r} - \mathbf{R}_s), \tag{19}$$

The atomic site function Ω_s^κ has angular properties given by spherical harmonics and is expanded in terms of basis functions, usually primitive Gaussians or Slater functions.

The momentum space orbital is

$$\phi^\kappa(\mathbf{p}) = \Sigma_s \Lambda_s^\kappa(\mathbf{p}) exp(-i\mathbf{p} \cdot \mathbf{R}_s), \tag{20}$$

where

$$\Lambda_s^\kappa(\mathbf{p}) = (2\pi)^{-3/2} \int d^3r exp(i\mathbf{p} \cdot \mathbf{r})\Omega_s^\kappa(\mathbf{r}) \tag{21}$$

Thus in momentum space all of the atomic orbitals are centred at the origin and the information about the nuclear positions is contained in the phase factors exp(-i$\mathbf{p} \cdot \mathbf{R}_s$) in the linear combination (20). These phases have a small effect for $p << 1/R_{st}$, where R_{st} is the distance between atoms s and t. The bond oscillations are therefore more noticeable in the inner valence and core orbitals, these being extended in p-space whereas they are highly localized in r-space.

The PWIA has been thoroughly checked for molecules, the momentum profile shapes are found to be independent of energy for momenta $p < 1/\bar{r}$, where $\bar{r}$ is roughly the rms radius of the atom-centred functions used as a basis for describing the molecular wave function. For high momenta ($p \geq 1.5$) distortion effects become

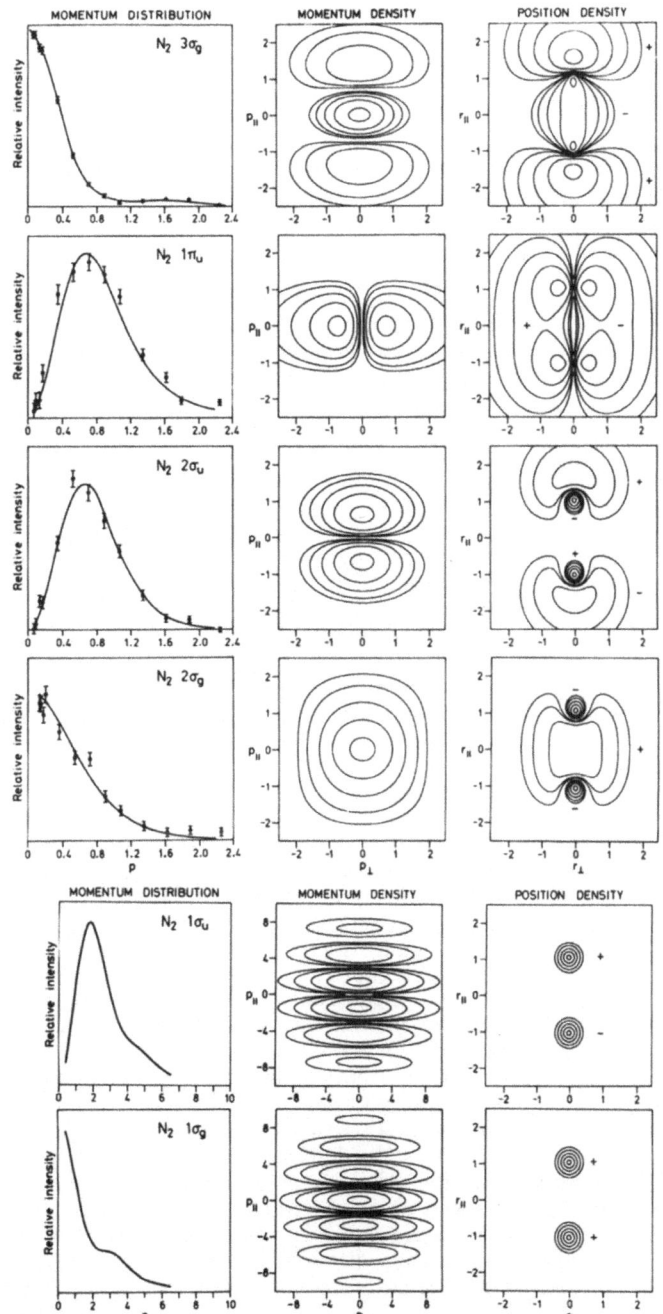

Fig. 5
Momentum distributions and momentum and position density maps for N_2 given by the orbital wave functions of Snyder and Basch (Ref. 14). The sign of the position space wave function is indicated. Contours are at 1%, 3%, 10%, 30% and 80% of the maximum intensity. The scales are in a.u. The data are from Ref. 13.

noticeable at the lower energies. The fact that EMS experiments are most sensitive and accurate in the low momentum region is advantageous since this region largely corresponds to the outer regions in coordinate space (as can be seen from the inverse weighting of the Fourier-transform). The cross sections for the outer valence orbitals are well described as a function of momentum even at 400eV, while the spectroscopic factor for the inner valence orbitals is underestimated by typically 20%. This discrepancy is removed at a total enery of 1200eV, and the spectroscopic sum rule is satisfied for all the valence-orbital manifolds.

Figure 5 shows the recent 1500eV data of Cook, Pascual and Weigold[13] for the valence $3\sigma_g$, $1\pi_u$, $2\sigma_u$ and $2\sigma_g$ orbitals of N_2 compared with the orbital wave functions of Snyder and Basch[14]. The agreement is very good, both in shape and in magnitude for all four valence orbitals.

Calculated position and momentum density maps are shown in figure 5 as well as the spherically averaged momentum distributions. No measurements exist for the two core orbitals, but the calculations are included in the figure to highlight the differences between the momentum and position space pictures for these orbitals. The momentum density maps highlight the interference effects due to the nuclear positions. In momentum space core orbitals are rich in information on molecular geometry and orientations. This promises to be a fruitful area of research, in particular in studying the properties of molecules on surfaces.

The absence of density in the charge density map along the $r_{\parallel} = 0$ plane in the $2\sigma_u$ orbital indicates the strong antibonding character of this orbital. This nodal plane also appears in the momentum density maps since symmetry is conserved. The Snyder and Basch wave function overestimates the spherically averaged momentum density in the region from 0.9 to 1.4 a.u., probably because it does not give enough contraction of the momentum density in the bond perpendicular ($p_{\perp}$) direction, which is typical of an antibonding orbital. The π_u orbital on the other hand is bonding, which in momentum spce is evidenced by a contraction of the momentum density in the $p_{\parallel}$ direction.

The outermost $3\sigma_g$ orbital is non-bonding. The two nodal planes at p = ± 0.75 lead to a narrow spherically averaged momentum distribution. The small amount of density in the lobes at large $p_{\parallel}$ gives rise to the small spherically averaged density at large p. The wave function of Snyder and Basch underestimates the density in these lobes.

Figure 6 shows the 1000eV separation energy spectrum for water obtained by Cambi et al[15] at $\phi = 0°$ and $\theta = 47°$. The three outer valence orbitals ($1b_1$, $3a_1$, and $1b_2$) have small cross sections at low momentum. The inner valence $2a_1$ orbital is significantly split due to final state correlation effects. The inner valence region has been fitted using calculated ionisation energies and spectroscopic factors obtained from an extended 2ph-TDA Green's function calculation (fig. 6(a)) and by a large CI calculation (fig. 6(b)). The CI calculation gives a very good description of the data up to nearly 40eV, but it considerably underestimates the spectroscopic strength above about 37eV. The $2a_1$ momentum distribution is in good agreement with that given by good MO wavefunctions.

The measured momentum distributions for the outer valence orbitals are shown in figure 7, together with calculated momentum distributions given by several MO calculations as well as two complete CI overlap calculations (eq. 4). The data are from Cambi et al[15] and Bawagan et al[16]. The calculations have been normalized to the measured $1b_2$ momentum profile. The very precise SCF calculations of Bawagan et al (1987) are effectively converged at the Hartree-Fock limit for total energy, dipole moment and momentum distribution. The observed significant discrepancies in calculated and observed momentum distributions are independent of the basis set. They were, however, largely removed by CI calculations of the full target-ion state overlap amplitude. The CI wave functions for the final ion and target ground states, generated from the accurate Hartree-Fock limit MO basis sets, accounted for up to 88% of the correlation energy. This work, and the earlier work of Dixon et al[17] and of Williams, McCarthy and Weigold[18], clearly shows the need for adequate

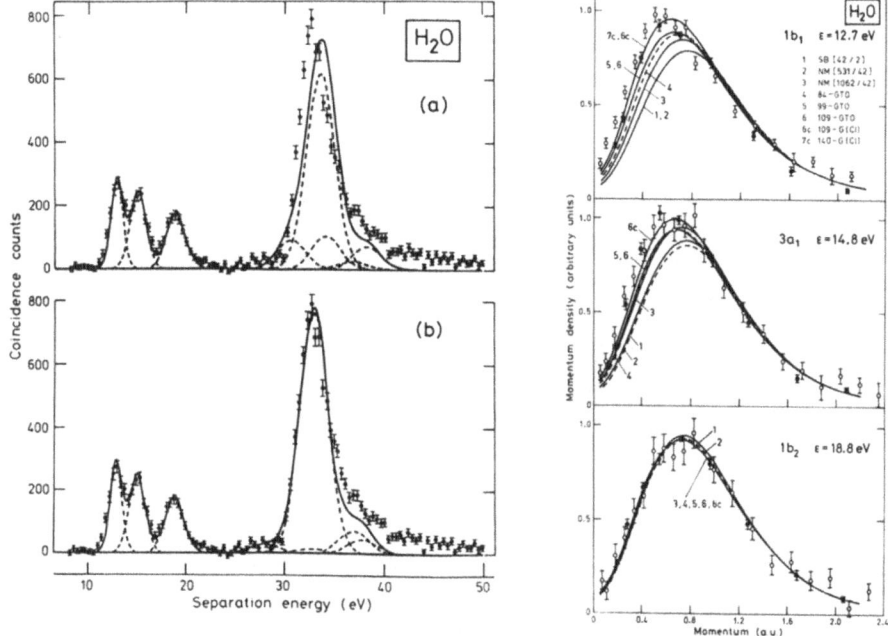

Fig. 6
The 1000eV separation energy spectrum for water at $\phi = 0°$ and $\theta = 47°$ compared with simulated spectra (Ref. 15). The inner valence region ($\epsilon \geq 25eV$) is fitted using the calculated $2a_1$ ionisation energies and spectroscopic factors obtained from (a) an extended 2ph-TDA Green's function calculation and (b) a MRD CI calculation using a gaussian with fwhm of 3.3eV for each pole.

Fig. 7
Comparison of the experimental momentum distributions of the outer valence $1b_1$, $3a_1$ and $1b_1$ orbitals of H_2O compared with spherically averaged momentum distributions calculated using (a) the molecular orbital wave functions of (1) Snyder and Basch (Ref. 14); (2) and (3) Neumann and Moskowitz (1968); (4)-(6) 84, 99 and 109 basis function sets used by Bawagan et al (Ref. 16); and (b) the full overlap amplitude calculated with CI wave functions generated by the indicated basis function sets for the initial ground state and final ion states (6c) and (7c) (Ref. 16). The open circles are the data of Bawagan et al and the solid circles of Cambi et al.

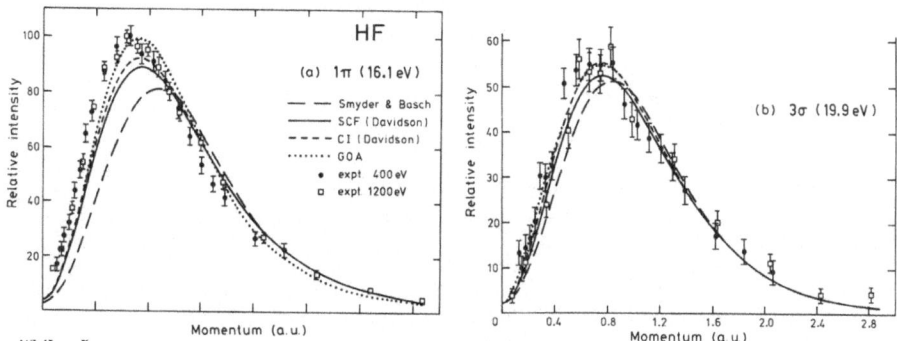

Fig. 8
The momentum distributions obtained for the outer valence orbitals of HF using 400eV and 1200eV noncoplanar symmetric kinematics compared with several calculated momentum distributions (Ref. 19).

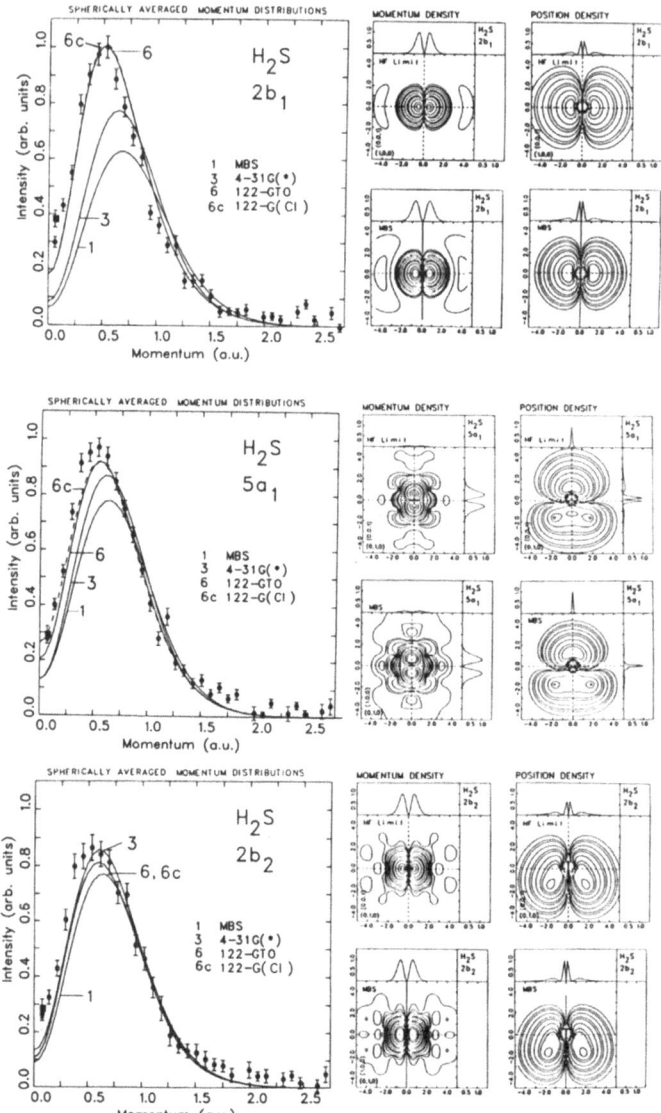

Fig. 9

The momentum distributions for the outer valence orbitals of H_2S obtained at 1200eV compared with those given by various SCF orbital wavefunctions as well as the complete overlap function including both initial and final state CI (Ref. 21). The calculations are placed on a common intensity scale by normalizing the 122-GTD calculation to the maximum of the $2b_1$ distribution. Also shown are orbital density maps generated by the MBS and 122-GTO wavefunctions.

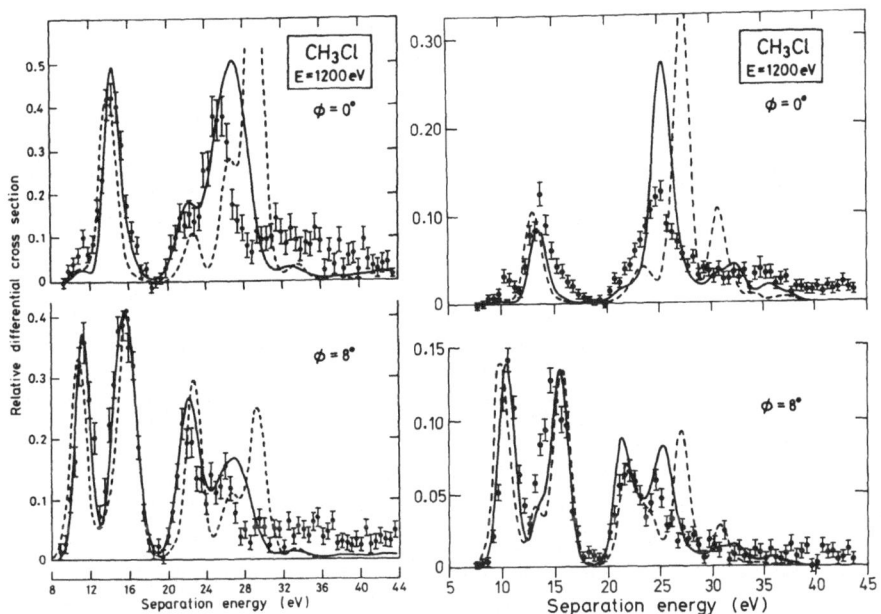

Fig. 10
Comparison of measured and calculted separation energy spectra for CH_3Cl (Ref. 22) and CH_3Br (Ref. 23). The dashed curves refer to the many-body Green's function calculations with the unpolarised basis sets, the full curves to the ones with the polarised basis sets. The data were taken at 1200eV and $\phi = 0°$ corresponds to $p \sim 0.06$, and $\phi = 8°$ to $p \sim 0.67$.

consideration of electron correlation effects in describing the low momentum parts of the $1b_1$ and $3a_1$ electron distributions.

A similar effect can be seen in the outer valence 1π and 3σ orbitals of HF. Fig. 8 shows the data of Brion et al[19] compared with the orbital wavefunctions of Snyder and Basch[14] and of Davidson[20], as well as with the full overlap using both the CI technique [20] and the Green's function many body method[19]. Again, the better the calculation the better the agreement with the data.

The recent EMS data and momentum profile calculations of French, Brion and Davidson[21] for the outer valence orbitals of H_2S are shown in fig. 9. Also included in the figure are the momentum and position density maps obtained with the minimal basis set (MBS) and Hartree-Fock limit SCF orbital wavefunctions. Relative normalization between the different orbitals has been maintained. The sensitivity of the calculated momentum profiles to the quality of the wavefunctions is again evident.

The Inner Valence Region

The inner valence region often shows complex structure, generally due to final state correlation effects. The momentum profiles for individual transitions are usually in very good agreement with the calculated inner valence HF orbital momentum distributions, but the strength is split among many final ion states. The explanation of this structure provides a considerable challenge to quantum chemists.

The halomethanes offer an example. Fig. 10 shows the 1200eV valence separation energy spectra for CH_3Cl and CH_3Br at two angles ϕ compared with simulated spectra (Refs. 22 and 23). In the simulated spectra the ionisation energies and spectroscopic strengths are given by extended 2ph-TDA Green's function calculations, and the angular dependence by the corresponding SCF orbitals of the polarised basis set MO calculations. In the calculated spectra the widths of the peaks include only the experimental energy resolution and not any natural widths or splittings of the states. The calculations are normalised to the observed intensity for the outermost (2e) orbital at $\phi = 8°$. The full curves refer to the calculations based on polarised MO basis, and the dashed curve the calculations with the unpolarised basis, the former having d-functions centred on the carbon and halogen sites, and a p-function on the hydrogen atom. The basis set effects are quite marked, with the better basis giving a much improved result.

The inner valence region for the halogen molecules Cl_2 and Br_2 are shown in fig. 11 (Ref. 24 and 25). The inner valence σ_u and σ_g orbital manifolds are severely split, being dominated by final state electron correlations. Frost et al compared their measured separation enrgy spectra with the results of several detailed many-body calculations, using both Green's function and CI methods based on both unpolarised and polarised SCF wave functions. The CI calculation and the polarised Green's function calculations (based on the algebraic diagrammatic construction method accurate up to fourth order in perturbation theory) are able to predict the gross features and origins of the inner valence ionisation spectra better than the unpolarised Green's function calculation. This demonstrates again the necessity for adequate basis sets in the many-body calculations. Although the higher energy region is still not adequately explained by these calculations, the CI (for Cl_2) and ADC(4) calculations give quite accurate energies, intensities and origins of the ionisation strength in the region of the main σ_u and σ_g strength. This demonstrates the importance of double excitations for the description of ionic states in the inner valence region.

Fig .12 shows the valence separation energy spectra of CCl_4 at three angles (momenta) compared with simulted spectra based on a Green's function calculation (Ref. 26). The inadequacy of this calculation in the inner valence region can again be seen. Fig. 13 shows momentum profile for the inner valence region ($\epsilon = 23 - 47.5eV$) where the $1t_2$ and $1a_1$ orbitals should dominate. The calculated summed $1t_2$ and $1a_1$ momentum distributions, normalized to the outer valence cross

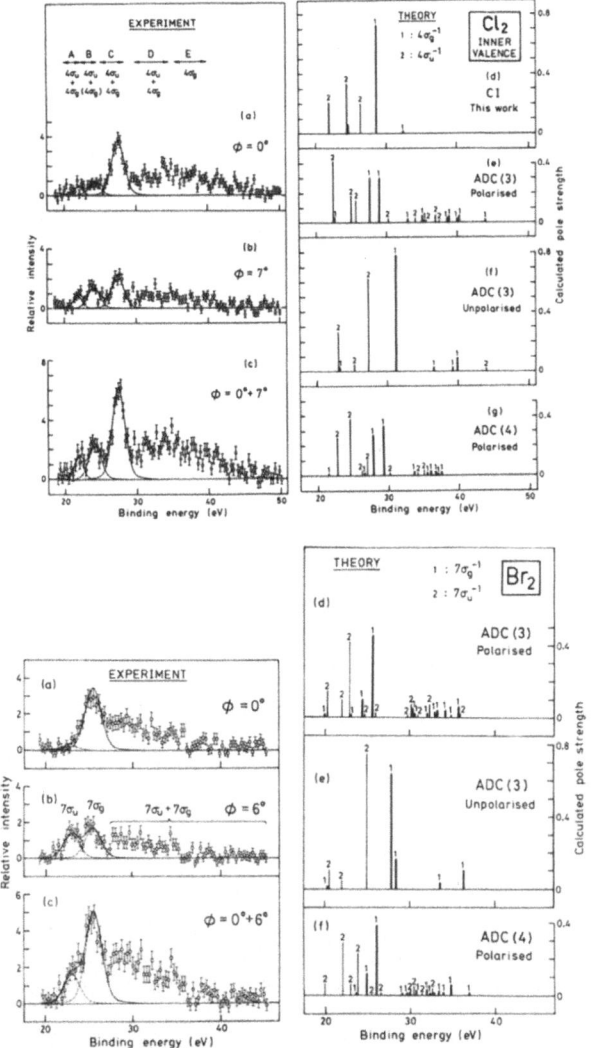

Fig. 11
Experimental and calculated binding energy spectra for the inner valence regions of Cl_2 (Ref. 24) and Br_2 (Ref. 25). The heights of the lines in the theory panels are proportional to the calculated pole strengths using the indicated many-body methods. The ADC(3) many-body Green's function calculations were carried out using both polarized and unpolarized basis sets.

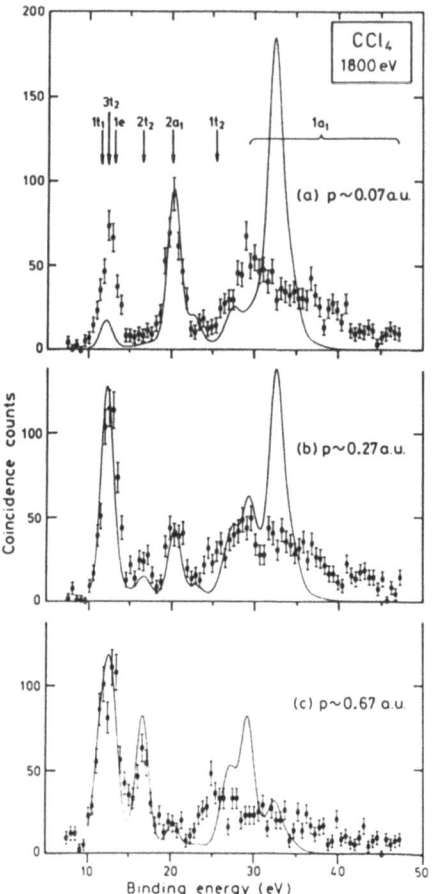

Fig. 12
The measured separation energy spectra for CCl_4 compared with simulated spectra based on ionization energies and spectroscopic factors given by a 2ph-TDA many-body Green's function calculation (Ref. 26). The calculation is normalized to the p~0.67a.u. data for the outer valence band.

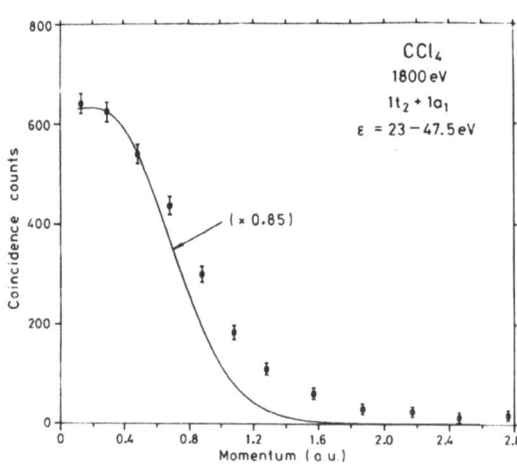

Fig. 13
The observed summed inner valence momentum distribution for CCl_4 compared with that given by the combined $1t_2$ and $1a_i$ orbital wavefunctions. The theory has been normalized to the data as discussed in caption for fig. 12 (Ref. 26).

sections, are significantly narrower than the measured profile. This is found to be independent of incident energy over the measured range 1000eV-1800eV[26]. The broadening of the momentum profile implies that the inner valence electrons must be significantly more localized than predicted by the SCF wavefunctions. Similar effects are also observed by Grisogono, Weigold and von Niessen for the inner valence $1a_1$ orbitals of $CHCl_3$ and CH_2Cl_2, although for CH_3Cl there is good agreement between theory and experiment.

SUMMARY

The application of the EMS technique to molecules has been greatly aided by the creative partnership between theoretical quantum chemists and electron momentum spectroscopists. In general the measured spherically averaged momentum distributions are well described by independent particle SCF orbital wave functions at or close to the Hartree-Fock limit. The plane wave impulse approximation appears to be quite adequate above about 1000eV for both momentum distribution shapes (for $p \leq 1$) and comparative manifold sums of spectroscopic factors. In most cases the target Hartree-Fock approximation is also applicable. In many cases the measured momentum distributions can easily differentiate between wave functions of different quality.

In some cases Hartree-Fock wave functions and the target Hartree-Fock approximation are inadequate to describe the momentum distributions, and many-body effects in both the initial and final states must be allowed for. This seems to be particularly true for the outermost orbitals of the second row hydrides. In these cases the measured momentum distributions generally contain more low momentum components than given by the variationally derived molecular orbitals. The largest contribution to the low momentum region comes from the diffuse part of the position space wave function. The experimental mapping of wave functions can therefore provide much needed additional and complementary information to the usual variational optimisation technique (which emphasises the large-p components) used in molecular wave function calculations and modelling. This is particularly important for optimising wave functions used for calculating physical and chemical properties such as dipole moments, bonding and reactivity, which depend on the longer range behaviour of the charge density.

The splitting of the spectroscopic strength in the outer valence orbital manifolds is generally negligible, the transitions being dominantly single particle (hole) in nature, the independent particle model working well for both initial neutral ground state and final ionic state. This is not the case for the inner valence region where significant splitting of the spectroscopic strength is generally observed. This is mainly due to many-body effects in the final ionic states. Correlations in the initial state are often negligible. It is thus still usually possible to use the target Hartree-Fock approximation and to measure single particle orbital momentum distributions. In many cases the inner valence structure is so rich and complicated that many-body calculations and experiments must go hand in hand in unravelling the structure information. The many-body calculations, based on both CI and Green's function techniques, are often qualitatively rather than quantitively correct. Fortunately, however, the better the calculation the better the results are as compared to experiment.

ACKNOWLEDGEMENT

Financial support for much of the work was provided by the Australian Research Grants Scheme. I am also indebted to my many colleagues and collaborators, in particular C.E. Brion, J.P.D. Cook, L. Frost, A.M. Grisogono, J. Mitroy, I.E. McCarthy, R. Pascual and W. von Niessen.

1. E. Weigold, S.T. Hood and P.J.O. Teubner, Phys.Rev.Lett. **30** (1983) 475.
2. R. Camilloni, A. Giardini-Guidoni, R. Tirribelli and G. Stefani, Phys.Rev.Lett **29** (1972) 618.
3. I.E. McCarthy and E. Weigold, Phys.Rep. **27C**(1976)275; E. Weigold and I.E McCarthy, Adv.At.Molec.Phys. **14** (1978) 127.
4. B. Lohmann and E. Weigold, Phys.Lett. **86A** (1981) 139.
5. I.E. McCarthy and E. Weigold, Rep. on Prog. in Phys. 1987 (in press).
6. A.J. Dixon, I.E. McCarthy, C.J. Noble and E. Weigold, Phys.Rev.A **17** (1978 597.
7. I.E. McCarthy and E. Weigold, Phys.Rev.A **31** (1985) 160.
8. J.P.D. Cook, I.E. McCarthy, J. Mitroy and E. Weigold, Phys.Rev.A **33** (1986 211.
9. J.P.D. Cook, I.E. McCarthy, A.T. Stelbovics and E. Weigold, J.Phys.B **1** (1984) 2339.
10. R. Pascual, I.E. McCarthy and E. Weigold (to be published).
11. L. Frost, J. Mitroy and E. Weigold, J.Phys.B **19** (1986) 4063.
12. I.P. Grant, B.J. McKenzie, P.H. Norrington, D.F. Mayers and H.C. Pyper Comput.Phys.Comm. **21** (1980) 2093.
13. P.J.O. Cook, R. Pascual and E. Weigold (to be published).
14. L.C. Snyder and H. Basch, Molecular Wave Functions and Properties (Wile N.Y., 1972).
15. R. Cambi, A.G. Ciullo, A. Sgamellotti, C.E. Brion, J.P.D. Cook, I.E. McCarth and E. Weigold, Chem.Phys. **91** (1984) 373.
16. A.O. Bawagan, C.E. Brion, E.R. Davidson and D. Feller, Chem.Phys. **11** (1987) 19.
17. A.J. Dixon, S. Dey, I.E. McCarthy and G.R.J. Williams, Chem.Phys. **21** (1977 81.
18. G.R.J. Williams, I.E. McCarthy and E. Weigold, Chem.Phys. **22** (1977) 281
19. C.E. Brion, S.F. Hood, I.H. Suzuki, E. Weigold and G.R.J. William J.Elec.Spectrosc. **21** (1980) 71.
20. A.O. Bawagan, C.E. Brion and E.R. Davidson (to be published).
21. C.L. French, C.E. Brion and E.R. Davidson (to be published).
22. A. Minchinton, J.P.D. Cook, E. Weigold and W. von Niessen, Chem.Phys. **11** (1987) 251.
23. A. Minchinton, J.P.D. Cook, E. Weigold and W. von Niessen, Chem.Phys. **9** (1985) 21.
24. L. Frost, A.M. Grisogono, I.E. McCarthy, E. Weigold, C.E. Brion, A.O. Bawa gan, P.K. Mukherjee, W. von Niessen, M. Rosi and A. Sgamellotti, Chem.Phys **113** (1987) 1.
25. L. Frost, A.M. Grisogono, E. Weigold, C.E. Brion, A.O. Bawagan and W. vo Niessen (to be published).
26. A.M. Grisogono, E. Weigold and W. von Niessen (to be published).

AUTODETACHMENT SPECTROSCOPY OF NEGATIVE IONS

W. C. Lineberger

Department of Chemistry
Joint Institute for Laboratory Astrophysics
University of Colorado
Boulder, Colorado 80309

INTRODUCTION

One of the exciting developments in ionic spectroscopy in recent years has been the application of very high resolution spectroscopic techniques to the study of molecular negative ions. These studies provide an excellent opportunity to investigate the interaction between the weakly bound outermost electron and the vibrational and rotational degrees of freedom of the negative ion. Such studies also provide an important test of electron-molecule scattering theories, in that the photodetachment process can be viewed via detailed balancing as an electron scattering event with one (or a few) partial waves present.

Prior to 1983, rotationally resolved spectra had been obtained for only two negative ions[1,2]. More recently, the techniques of velocity modulated infrared absorption[3], photodetachment in ion cyclotron resonance traps[4,5] and auto detachment spectroscopy[6-11] have substantially expanded the scope of such studies. Autodetachment spectroscopy, the subject of this review, is particularly suited for a probe of electron-molecule interactions.

In autodetachment spectroscopy, a tunable laser excites transitions between levels of the ground electronic state of the anion and an excited vibrational or electronic state of that ion. The upper state level must lie above the detachment threshold so that it can autodetach to the (neutral + e$^-$) continuum. The autodetachment events are seen as resonances in the electron photodetachment signal wherever a autodetaching state is excited. The location of the structure provides direct information on the structure of the excited state of the ion and in addition, the line widths of the transitions can be related to the lifetime of the resonance, thereby providing direct information on the dynamics of the autodetachment process and electron-molecule scattering. In this paper, we provide examples of such studies of CH_2CN^- and PtN^- ions, probing the dynamics of the interaction of a slow electron with a strong dipolar field. The lifetimes of the autodetaching levels, together with information on the final state provide direct information on the mechanisms of rotational autodetachment and indicate the most important propensity rules.

EXPERIMENTAL

The coaxial laser-ion beam spectrometer used for the present work has been previously been described in detail.[7] Negative ions are extracted from a hot cathode discharge source, mass-selected and accelerated to 3000 eV. An electrostatic quadrupolar field merges the ion beam with the output of a single mode dye laser. After traveling some 30 cm coaxially, the ions and photodetached neutrals are separated by a second quadrupolar field. The electrons that are detached in the interaction region are collected by a weak solenoidal magnetic field and counted with an electron multiplier. Data are obtained by scanning the laser frequency while monitoring the electrons formed, normalized to photon flux and ion current. The resolution of the spectrometer is less than 20 MHz, limited by the Doppler spread in the kinematically compressed ion beam.[7,12]

RESULTS AND DISCUSSION

In, CH_2CN^-, the excited electronic state is a dipole bound state lying just below the photodetachment threshold. This state arises from the long-range interaction of an electron with the dipole field of the molecular core. While these states can be viewed as a negative ion analog of Rydberg states, important rotation-electronic coupling effects are present owing to the anisotropic electron dipole interaction, absent in Rydberg states.

Figure 1 shows the total photodetachment cross-section for CH_2CN^- in the region near the photodetachment threshold. Throughout this region, photodetachment is dominated by auto-ionizing resonances, with direct photodetachment playing only a relatively minor role. There are two vibronic bands in the spectrum. The autodetachment structure in the vicinity of 12,400 cm^{-1} is due to transitions between the vibrationless levels of the two electronic states of CH_2CN^- and the structure near 12,000 cm^{-1} is associated with a hot-band transition involving two quanta of the hydrogen out-of-plane bending mode. The sharp structure arises from unresolved Q-branch transitions and can be recognized to be characteristic of a perpendicular electronic transition associated with a slightly asymmetric prolate rotor.[13] A high resolution scan of a single Q-branch reveals individual rotational transitions, as shown in Fig. 2. At this resolution, high J states are broader and definitely shorter lived than low J states. In yet higher resolution, the widths of the individual rotational transitions can be measured, providing direct information on autodetachment lifetimes.

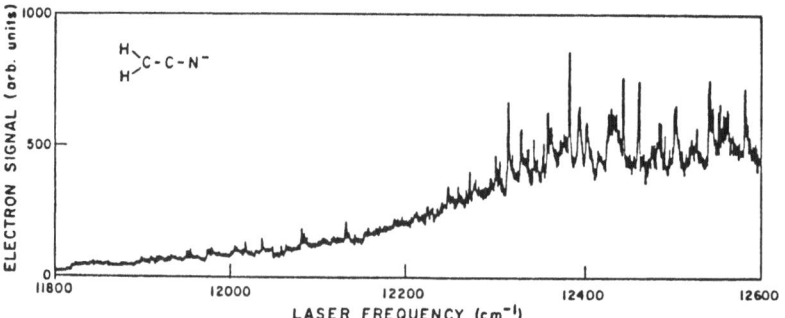

Fig. 1. CH_2CN^- Threshold Photodetachent at 1 cm^{-1} Resolution.

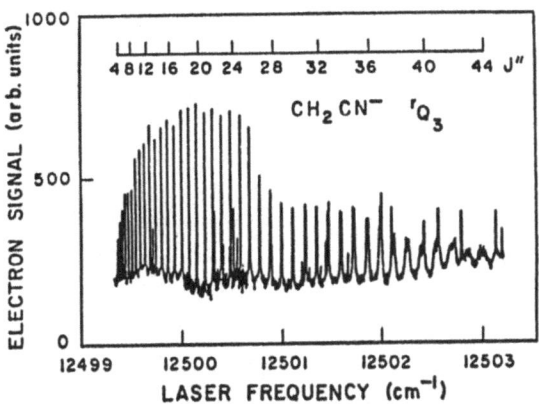

Fig. 2. High Resolution (< 20 MHz) Scan of the $^rQ(3\leftarrow2)$ Branch in CH_2CN^-.
Note the broadening of lines toward high J.

Approximately 5000 individual transitions were recorded and fit to
an asymmetric top Hamiltonian. The rotational constants obtained from
this fit shows that in both upper and lower states the ion is nearly a
symmetric prolate rotor whereby each rotational level can be labeled by
J, the total rotational angular momentum, and the nearly good quantum
number K, the projection of J along the $C-C\equiv N$ axis of the ion. The ion
is near planar and the outermost electron is bound by 65 cm^{-1} in a very
diffuse (30 Å) orbital on the positive (H_2C) end of the molecule.

The line widths of the upper state are plotted as a function of J
and K in Fig. 3. The most striking feature in this plot is the abrupt
rise in autodetachment rate near J=33, and the relative K independence of
the autodetachment rate. This effect is not due to the total rotational
energy since, for example, K=8 rotational levels lie about 600 cm^{-1} above
K=1 levels with the same J. Rather, the widths depend primarily on
autodetachment selection rules and propensity rules which couple the
negative ion state to the (neutral + e^-) continua.

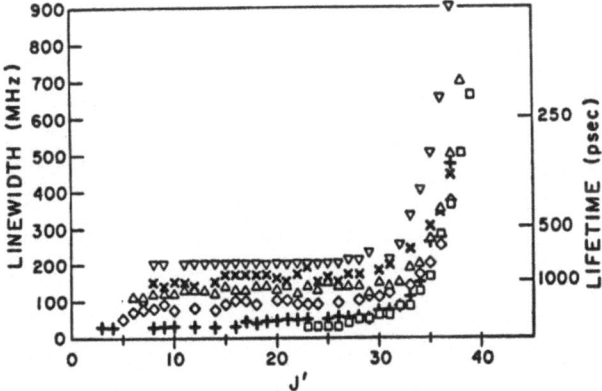

Fig. 3. Experimental Autodetachment Linewidths for the CH_2CN^-
Dipole-bound state as a function of the J and K rotational
quantum numbers. □, K=1; +, K=3; ◊, K=5; Δ, K=6; ×, K=7;
∇, K=8.

73

The upper state can only decay by rotation-electronic coupling, transferring at least 60 cm^{-1} of rotational energy to the weakly bound electron. Detailed analysis show that transitions involving changes in K do not exert substantial torque on the electron, making energy associated with rotation about the C-C≡N axis relatively ineffective. Rather, end over end rotations do produce substantial forces on the electron bound at the positive end of the molecule and lead to autodetachment. The rapid onset at J=33 corresponds to the ΔJ=-2 or -3, ΔK=0 process becoming energetically accessible. This same propensity is seen even more graphically in the case of PtN^{-}.

REFERENCES

1. P. L. Jones, R. D. Mead, B. E. Kohler, S. E. Rosner, and W. C. Lineberger, J. Chem. Phys. 73, 4419 (1980); W. C. Lineberger and T. A. Patterson, Chem. Phys. Lett. 13, 40 (1972); G. Herzberg and A. Lagerqvist, Can. J. Phys. 46, 2363 (1968).
2. P. A. Schulz, R. D. Mead, P. L. Jones, and W. C. Lineberger, J. Chem. Phys. 77, 1153 (1984).
3. L. M. Tack, N. H. Rosenbaum, J. C. Owrutsky, and R. J. Saykally, J. Chem. Phys. 84, 7056 (1986); K. Kawaguchi and E. Hirota, J. Chem. Phys. 84, 2953 (1986); B. D. Rehfuss, M. W. Crofton, and T. Oka, J. Chem. Phys. 85, 1785 (1986).
4. J. Marks, D. M. Wetzel, P. B. Comita, and J. I. Brauman, J. Chem. Phys. 84, 284 (1986).
5. R. C. Stoneman and D. J. Larson, J. Phys. B. 19, L405, (1986).
6. D. M. Neumark, K. R. Lykke, T. Andersen, and W. C. Lineberger, J. Chem. Phys. 83, 4364 (1985); M. Al-Za'al, H. C. Miller, and J. W. Farley, Chem. Phys. Lett. 131, 56 (1986).
7. R. D. Mead, K. R. Lykke, W. C. Lineberger, J. Marks, and J. I. Brauman, J. Chem. Phys. 81, 4883 (1984).
8. J. Marks, J. I. Brauman, R. D. Mead, K. R. Lykke, and W. C. Lineberger, J. Chem. Phys. (submitted).
9. T. Andersen, K. R. Lykke, D. M. Neumark, and W. C. Lineberger, J. Chem. Phys. 86, 858 (1987).
10. K. R. Lykke, K. M. Neumark, T. Andersen, V. J. Trapa, and W. C. Lineberger, in Laser Spectroscopy VII, edited by Y. R. Shen and T.W. Hänsch (Springer Verlag, Berlin 1985), pp. 130-133.
11. K. R. Lykke, K. K. Murray, and W. C. Lineberger, Phys. Rev. A, in press.
12. B. A. Huber, T. M. Miller, P. C. Cosby, H. D. Zeman, R. L. Leon, J. T. Moseley,, and J. R. Peterson, Rev. Sci. Instrum. 48, 1306 (1977).
13. G. Herzberg, Electronic Spectra of Polyatomic Molecules (Van Nostrand, New York, 1967) pp. 247-261.
14. K. R. Lykke, D. M. Neumark, T. Andersen, V. J. Trapa and W. C. Lineberger, J. Chem. Phys. (submitted).

RECENT EXPERIMENTAL RESULTS RELATED TO SHAPE RESONANCES

Richard Hall

Groupe de Spectroscopie par Impact Electronique et Ionique[*]
Université Pierre et Marie Curie, 4 Place Jussieu T12-E5
75252 Paris, France

INTRODUCTION

Resonances in electron-molecule scattering have been with us for many years and the subject has been reviewed on numerous occasions (see, for example, the account of resonance phenomena given by Herzenberg[1]). They play a particularly important role in vibrational excitation producing dramatic enhancement of the cross sections at energies of a few eV which can take on various forms depending on their lifetime. Long lifetime resonances give sharp structures (O_2, NO) whereas short lifetimes produce broad structureless peaks (H_2, HCl), for lifetimes between these two extremes one can observe broad structure with superimposed oscillations (N_2, CO_2). The important point to be retained here about resonances is that they enable a light fast projectile to transfer large amounts of energy to the slow heavy nuclei of a molecule. Such a large exchange of energy is very improbable in a direct collision. The electron can even go from a positive to a negative energy in the dissociative attachment process which leads to the production of stable negative ion fragments.

An account of resonant processes in binary electron-molecule collisions is not the goal of this paper. Nevertheless it is of interest here to present recent observations on the prototype resonance of N_2 and H_2 obtained by Allan[2,3] which represent the experimental state-of-the-art with regard to instrumental sensitivity. Allan has based his instrument on the trochoïdal electron spectrometer which uses crossed electric and magnetic fields as an electron energy filter. One spectrometer produces a low dispersion electron beam and a further two, in tandem, analyse the forward scattered electrons[4] (fig. 1). This instrument gets its sensitivity from a relatively long path length in the target gas and collection over large solid angles near threshold. This instrument is capable of analysing cross sections

---

[*]In association with the "Centre National de la Recherche Scientifique" (U A 774).

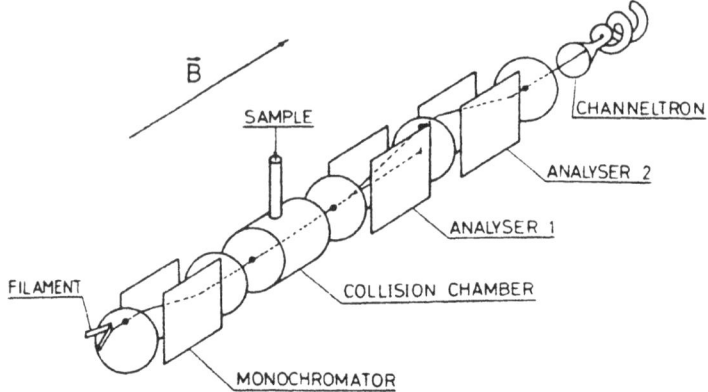

Fig. 1. Schematic diagram of the electron spectrometer of Allan[4].

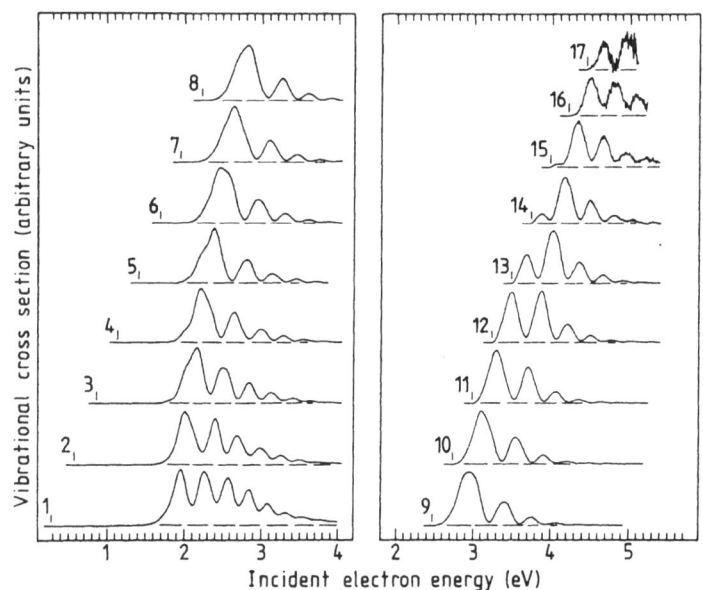

Fig. 2. Vibrational excitation cross sections for the 0→v transitions in e- N_2 scattering. The excitation thresholds are indicated by vertical lines. The heights of the curves were chosen arbitrarily for clarity of presentation.

in the threshold region as low as 10^{-22} cm^2. This is demonstrated in figure 2 which shows vibrational cross sections of N_2 up to v=17, the intensity of this latter level being more than six orders of magnitude weaker than that of v=1 (6.10^{-16} cm^2). As indicated above the $^2\Pi_g$ state of N_2^- has an intermediate lifetime and the spectacular shifting oscillations result from an interference phenomenon known as the "boomerang effect"[1]. Note that there is no vibrational excitation outside the resonance zone; all the vibrational excitation transits via the intermediate negative ion state.

The vibrational cross sections measured by Allan[3] in H_2 are shown in figure 3 for levels up to v=6. Here the cross section profile evolves from a smooth hump for v=1 to a strongly structured broad peak for v=6. The presence of this structure in the cross sections of high levels was predicted by Mündel et al[6] and in a resonance picture would be due to the onset of the "boomerang effect" as the nuclei explore larger internuclear separations where the resonance lifetime gets longer. However there is some debate as to whether the resonance picture is applicable in H_2 as the lifetime (10^{-16} sec) would be about the same as the time the electron takes to fly past the molecule, moreover non-resonant theory gives a good description of the lowest vibrational cross sections (see the review of Lane[7])· Furthermore scattering models where a resonance is not an explicit assumption are obtaining good agreement with observations for the phenomena associated with dissociative attachment in the 4 eV region[8].

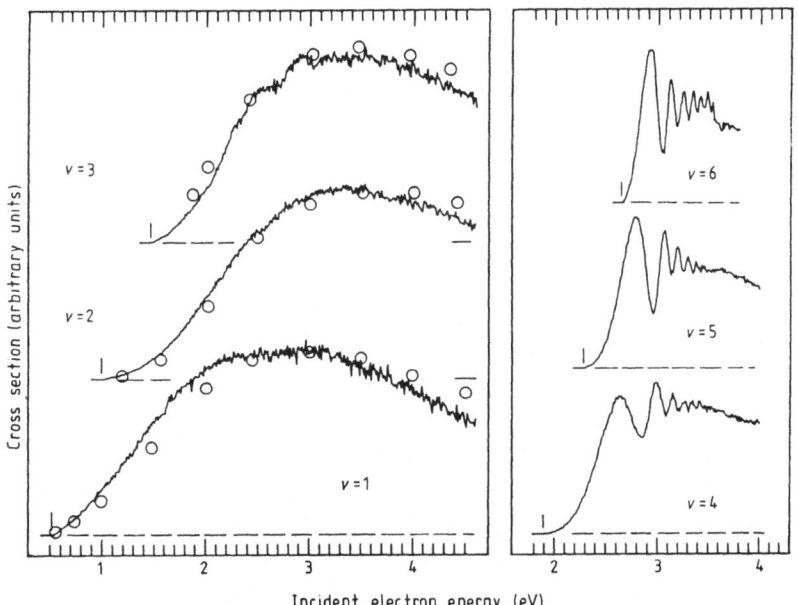

Incident electron energy (eV)

Fig. 3. Vibrational excitation cross sections in H_2 . The vertical scales of the individual curves were chosen arbitrarily for clarity of presentation. The short vertical bars indicate the threshold energies. The open circles show the integral cross sections of Ehrhardt et al[5], normalised to the present data at the maximum of the v= 1 curve.

Even though some caution must be taken when the frontier between a resonant and a non-resonant mechanism is ill-defined, the resonant picture representing the formation and the decay of an intermediate negative ion is nevertheless a most useful one for interpreting and understanding many salient features of low-energy electron-molecule scattering. In what follows it will be shown how general ideas associated with resonances lead to a broad understanding of several physical phenomena observed when one moves away from the pure binary collision system and considers situations where either the electron or the molecule or both are correlated with one or more partners. It will be seen how resonances still reveal themselves in electron collisions with condensed molecules (section 1) in negative ion and Rydberg atom collisions with molecules (section 2) and in molecular collisions with metal surfaces (section 3). New information has been obtained on all these subjects because we already have accumulated a large amount of knowledge on resonances but these studies reveal that all is far from known and more work both experimental and theoretical remains to be done.

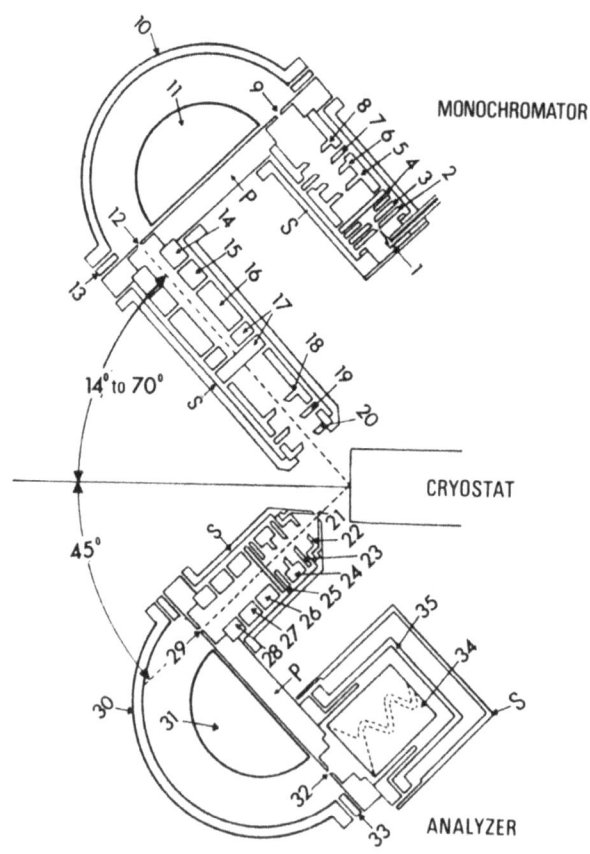

Fig. 4. Schematic diagram of hemispherical electron-energy-loss spectrometer used for surface studies by Sanche and Michaud[9].

1. ELECTRON SCATTERING FROM MOLECULES ON SURFACES

Our knowledge on the interaction of low-energy electrons with matter has been restricted for a large part to electron scattering from uncorrelated particles in the gas phase. However new fields are opening up with the aim to study more complex states of matter such as molecules on surfaces where a change of phase has taken place. The experimental requirement are more constraining than for electron scattering off gases. Where surface phenomena are concerned cleanliness is paramount and pressures in the 10^{-11} Torr range have to be attained. Furthermore the surface under investigation has to be cooled to low temperatures (< 20K) in order to condense the gas. Otherwise the electron spectrometers used to study the inelastic processes are identical to those developed for gas phase studies except the materials have to be carefully chosen to withstand bake-out to temperature near 400 °C. A drawing of a typical instrument used in surface studies built by Sanche and Michaud [9] is shown in figure 4.

Similarly in interpreting the observations on electron scattering from adsorbed molecules one has drawn heavily on knowledge aquired in gas phase studies as it so happens that the phenomena are closely related. When vibrational excitation of condensed molecules is studied as a function of incident electron energy, enhancement is found over restricted energy ranges due to resonances. Vibrational excitation functions observed by Demuth et al.[10] are shown in figure 5 for N_2, CO, O_2 and H_2 on a silver surface at 20K. The full lines are for monolayers or less and the dashed lines (N_2, CO) are for condensed multilayers while the dotted lines are the gas phase excitation functions. In each case except for perhaps H_2, the structure corresponds to the gas phase phenomena but shifted to lower energies and the effect of increasing the number of layers is to reduce this shift for N_2 whereas the opposite occurs for CO, the shift increasing. Also notice that for these surface molecules the secondary oscillations are absent.

These resonances allowed excitation of many vibrational levels of the ground state to be detected and thus deduce the fundamental vibrational constants. These constants were found to be little different from those of the gas phase molecules. Similarly electron energy loss spectroscopy near 3 eV has shown that H_2 on a surface can rotate without hindrance and that the internuclear separation in the adsorbed phase varies by less than 2%[11-13]. These facts then indicate that the molecules are only very weakly bound to the surface (physisorbed) and are hardly perturbed by the presence of the metal. However we have seen above that for the molecular negative ion states of surface molecules the energies are shifted and they lose their structure. When the N_2^- resonance was observed under high resolution for the v=1 level the oscillatory structure was seen to reappear as the number of layers increased from 2 to 8 [9](figure 6) and remained for thick films[14]. The spacing between the "vibrational peaks" of the resonance is again very near that observed in the gas phase whereas the resonance lifetime of the condensed molecules was estimated to be 3.10^{-15} sec as opposed to 8.10^{-15} sec for free N_2. At monolayer coverage the strctureless width would correspond to an even shorter lifetime of about 5.10^{-16} sec.

The lifetime of the N_2 resonance is determined by the symmetry of the resonance and the interaction potential. The symmetry imposes the partial waves which can form the resonance (l=2 for the $^2\Pi_g$ state of N_2) and thus dictates the centrifugal potential. The potential interaction is made up of the polarization and exchange potentials which when combined with the centrifugal term generates a barrier through which the extra electron can tunnel. The degree of ease with which this

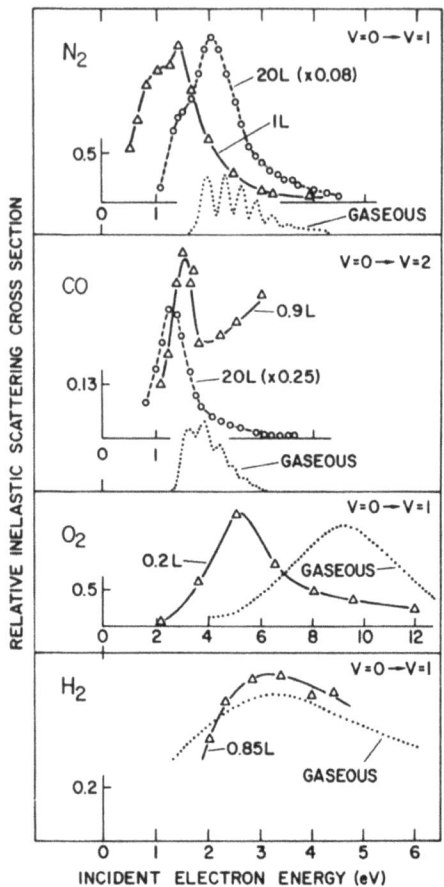

Fig. 5. Relative inelastic scattering cross sections for N_2, CO, O_2, and H_2 adsorbed on silver (solid lines). The dashed lines represent condensed multilayers of N_2 and CO while the dotted lines are the gas-phase spectra.

tunneling takes place determines the resonance lifetime. Above we have seen that the lifetime of N_2^- in thick films is near that of a free molecule. This would imply that the trapped electron is localised on the N_2 molecule and the symmetry of the resonance is well defined by $^2\Pi_g$. Thus the wavefunction of the extra-electron does not appreciably overlap with the surrounding molecules to reduce its symmetry or lose its identity. When the N_2 film thickness is reduced to a monolayer or less the lifetime is reduced which in this picture would imply that the symmetry is lowered by the presence of the metal substrate. The lower symmetry allows lower partial waves to contribute to the process with their associated lower barrier and greater ease of tunneling. This reduced symmetry will at the same time have the effect of lowering the state energy. As the symmetry is conserved, at least for many monolayers, then the general reduction of the resonance energy which attains about 0.7 eV for N_2 on a surface can be attributed to two factors: firstly, the attractive potential between the negative ion and its image in the metal and secondly the polarization of the surrounding molecules. A final point should be made here and concerns the case of H_2. As figure 5 shows there is little or no shift between the structure observed in the two phases, and this probably reflects the non-resonant character of the process as mentioned above.

Up to this point the observed features in the condensed phase are similar to those in the gas phase, the neighbouring molecules and metal substrate only acting as a perturbation. However there are two gas phase resonant processes which are not observed for surface molecules. The first is excitation via Feshbach or core-excited resonances. The reason for the absence of these resonances is no doubt due to the size of the Rydberg orbitals of which they are composed. These orbitals will have a strong interaction with the surrounding molecules and are no longer localized on any single one. The second unobserved process is resonant scattering in the elastic channel. A proposed explanation for this has been put forward and concerns energy loss to intermolecular motion caused by the polarization discussed above[15].

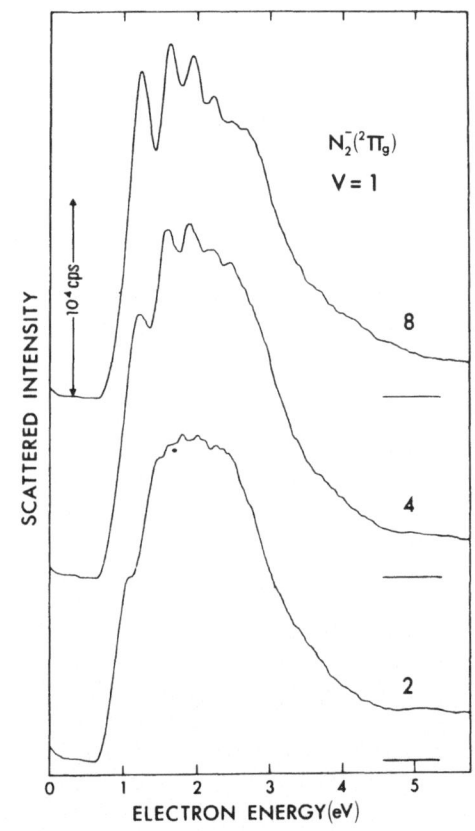

Fig. 6. Vibrational structure of the N_2^- temporary negative ion for films composed of 2, 4, and 8 monolayers. The horizontal base lines on the right-hand side represent the zero intensity for each curve.

Another gas phase resonant process which also exists for condensed gases is dissociative attachment[16,17]. This has been observed for O_2 by Sanche[16] and the O^- yield as a function of energy is shown in figure 7. The lower curve represents the O^- cross section for gaseous O_2. Note that here the resonance location would appear to be the same for both the gaseous and the condensed phases. We have seen above that polarization and image effects serve to lower the resonant energy but then in dissociative attachment the O^- ion thus formed must overcome this same field in order to escape from the surface. Thus the higher the O^- formation energy i.e. the higher the electron energy, more probable will be its escape, and one can imagine that these two effects cancel each other out giving the impression that the resonance does not move. One can also notice in figure 7 that the resonance intensity compared to the polar dissociation process which occurs above 15 eV is much lower for adsorbed molecules than for free ones. This would indicate a lower cross section for O^- from O_2 on the surface and can be accounted for by a shortening of the resonant lifetime, to which the dissociative attachment process is particularly sensitive, by the mechanism discussed above for the case of N_2^-.

In an experiment described below[18] in section 3 where electrons impinge both on a gas beam and on a surface (see figure 16) we were able, by detecting near zero energy H^- ions, to study simultaneously dissociative attachment to H_2 for both phases. Figure 8 shows the H^- yield with hydrogen in the system (upper curve) in which case the H^- ions come both from the gas and the surface. When the hydrogen gas is

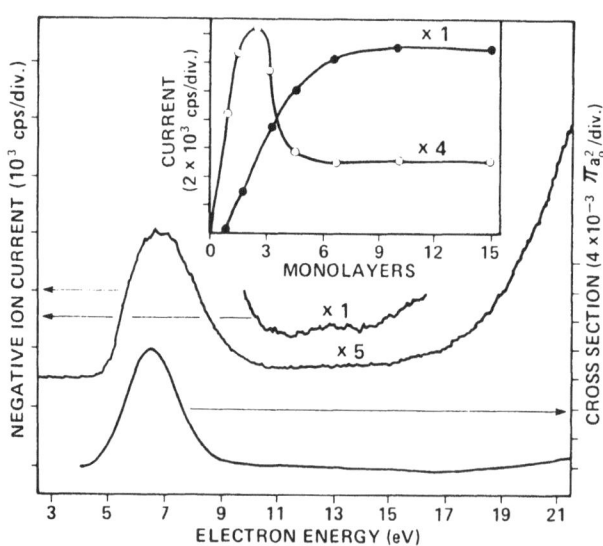

Fig. 7. Energy dependence of the O^- yields produced by electron impact on a six-"monolayer"-thick film of condensed O_2 and on gaseous O_2 (bottom curve). The inset shows the film-thickness dependence of the O^- signal produced by 7 eV (full circles) and 21 eV (open circles) electrons. The intensity of the signal is obtained by multiplying the left scale by the factor which appears on top of the upper curves.

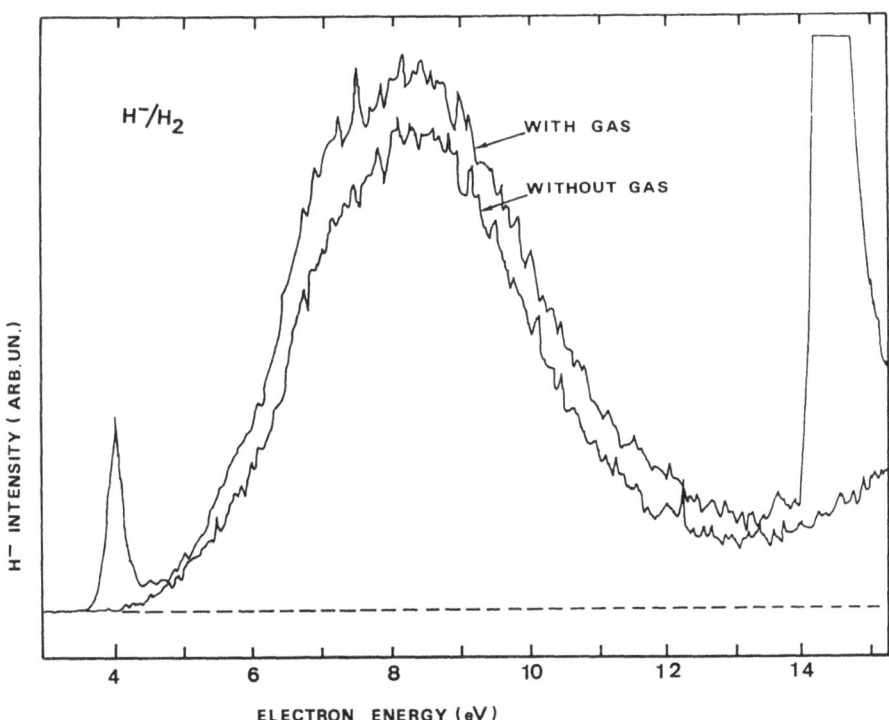

H^-/H_2

WITH GAS

WITHOUT GAS

H⁻ INTENSITY (ARB. UN.)

ELECTRON ENERGY (eV)

Fig. 8. Yield of H⁻ ions with near zero energy against
incident electron energy. The 4 eV and 14 eV peaks
result from dissociative attachment to the first and
second dissociation limits, respectively. The central
peak is produced by dissociative attachment to surface
molecules.

removed (lower curve) the ions come solely from the surface. This
experiment was not performed under the "clean" conditions usually
encountered in surfaces studies, nevertheless it shows the trends one
would expect from the above discussion. First of all the 3.7 eV and 14
eV processes are not seen on the surface, this would be attributed to
the fact that the H⁻ ions are formed at their process thresholds with
little or no energy and are unable to overcome the attractive
polarization and image potentials. This same effect was also observed
for the O⁻ from CO process at threshold[16], which was not detected for
adsorbed molecules. In the energy region around 8 eV the dissociative
attachment process leads to energetic H⁻ ions which we do not detect
from the gas phase as the instrument is tuned for near zero energy
ions. These ions are however detected from the surface which would
indicate that they have sufficient energy to escape from the surface but
their low final energy would imply that they are also partialy relaxed
perhaps due to collisions with surface molecules.

In this section we have seen that high resolution spectrometers
developed for scattering studies in the gas phase are now being used
for molecular targets condensed on surfaces. These studies reveal
that the shape resonances which are now becoming well known in the
gas phase are also present with some modifications in the condensed

phase. In fact electron transmission observations have shown that, at least for N_2, the N_2^- resonance can still be detected when N_2 is buried in a condensed rare gas matrix[19], indicating that an electron even under these circumstances can be localised on one molecule during the resonance lifetime. This result indicates a technique which could provide information on resonances in radicals which are   not easily amenable to study in gaseous form.

2.ELECTRON DETACHMENT IN NEGATIVE ION MOLECULE COLLISIONS

Amongst atomic particles negative ions have special properties. The interaction between an electron and an atom is weak and short ranged consequently the electron is only weakly bound in a negative ion and generally has no excited states. The structure can be considered as an atom with an external electron, for example, the electrons of the simplest negative ion must be represented as 1s1s'. The binding energy of H^- is   0.75eV and that of the simplest negative ion with an outer p electron, F^-, is 3.4eV. Negative ion binding energies are all lower than the lowest atomic ionization potential(Cs with 3.9 eV).

An electron is readily detached from a negative ion in collisions with other particles and this reaction:

$$A^- + B \longrightarrow A + B + e$$

represents a particularly simple bound–free process. A knowledge of the energy distribution of the detached electron is an important factor for the comprehension of the detachment mechanisms. One way of achieving this goal is by measuring the energy of the neutral resulting

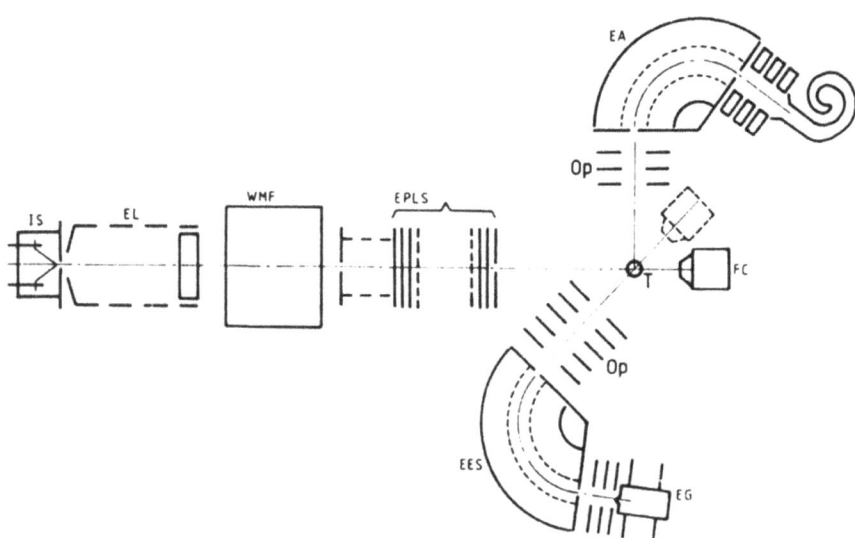

Fig. 9. Schematic diagram of the experimental set up:
IS, ion source;EL, einzel lens;WMF, Wien mass filter;
EPLS,exponential potential lens stack;Op,   optics;
FC, Faraday cage;EA, 127° electrostatic analyser;
EES, 127° electron energy selector;EG, electron gun;
T, target gas beam.

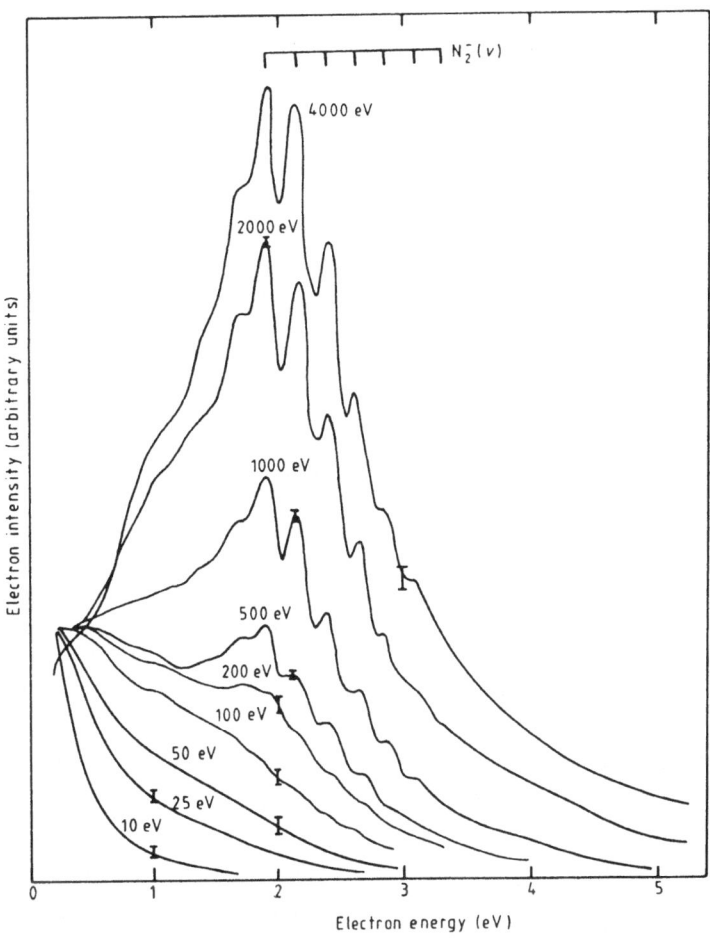

Fig. 10. Detached electron energy spectra for $H^- - N_2$ collisions.

from the collision and such observations have been particularly rich[20]. The other way consists of the direct observation of the detached electron and a schematic diagram of such an experiment is shown in figure 9. The negative ion beam is generated by a discharge ion source and crosses a target gas beam effusing from a hollow needle. The electron energy is measured with a cylindrical electrostatic filter. A second filter produces an electron beam which is used for calibration purposes[21]. Here again the experiment represents an extension of techniques developed in electron scattering to the study of other collision processes.

In the case of H^- collisions with rare gases the electron spectra all show a broad peak with a maximum below 1 eV followed by a monotonic decay[22]. When molecular targets are studied new structure appears as can be seen for N_2 in figure 10. This structure is the result of charge exchange to shape resonance i.e. the reaction:

$$H^- + N_2 \longrightarrow N_2^- + H$$
$$\longrightarrow N_2 + e$$

The energy location of the structure due to N_2^- seen in elastic electron scattering is indicated in the figure and shows that in charge exchange the structure is more complex. In the latter process, the N_2^- "levels" are all populated and the populations will be determined by Franck-Condon factors and the energy defect between H^- and N_2^-. The energy defect depends on the internuclear separation of N_2 and would thus favour the lowest "levels" of N_2^-. Hence the structure in the figure results primarily from the decay of "v=0" of N_2^- to the same level of N_2 and the additional structure at lower electron energies is the result of decay to excited N_2 levels.

Information on the way the "levels" of N_2^- are populated by H^- can be obtained from the binary collision model. In this model there is no interaction between the core of the negative ion (the H atom) and the target. The collision system then reduces to the scattering of a free electron by the target with the electron having an energy corresponding to the negative ion velocity. For example, H^- at 2 keV would be the equivalent of about a 1 eV electron. This model has been developed and refined by Vu Ngoc Tuan et al.[23] to take into account the electron affinity and the orbital velocity distribution of the outer electron of H^-. The only input into this model is the electron-target

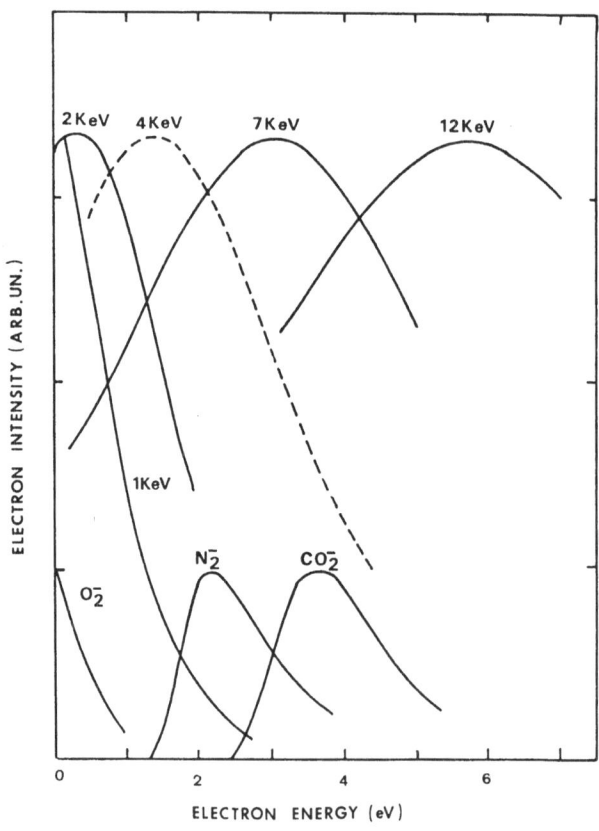

Fig.11. Energy distribution of the electron wave packet at the incident H^- energies. The envelopes of the indicated resonances are displayed in the lower part.

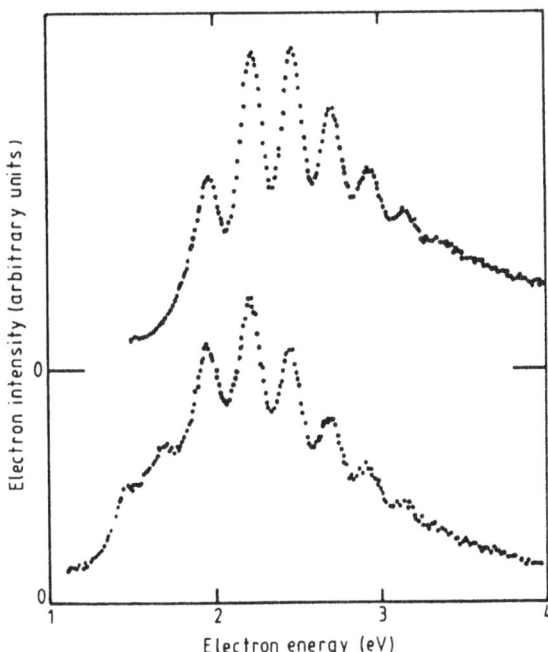

Fig.12. Upper curve, electron intensity elastically scattered
at 20° on N_2 as a function of incident energy; lower
curve, energy spectrum of electrons scattered at 20°
on N_2 when a 'white' incident beam is used.
The electron energy is that of the electron after the
collision.

scattering amplitude which can be experimental or theoretical. The
equivalent electron energy distribution for H^- energies between 1 keV
and 12 keV is shown in figure 11. Also shown are the envelopes of the
O_2^-, N_2^- and CO_2^- resonances. The maximum probability for charge
exchange to the resonance will occur when the peak in the electron
distribution coincides with the resonance, i.e. at about 5 keV for N_2 and
twice this for CO_2 It can also be seen from the figure that for H^- at
1keV colliding with N_2 will favour population of the lowest "levels" of
N_2^- to produce a non-Franck-Condon behaviour. These phenomena have
been observed by Vu Ngoc Tuan et al.[24] for N_2 and CO_2 targets.

The above model indicates that the H^- beam can be likened to an
electron beam with a very broad energy distribution. Such a beam can
be simulated using the apparatus of figure 9 by rapidly modulating the
incident electron beam energy with a saw-tooth wave-form whose
amplitude is a few eV wide. The scattered electron spectrum in N_2 with
the mean incident electron energy of the modulated beam near 2 eV is
shown in the lower part of figure 12 and as can be seen is very much
like the spectrum obtained in detachment. The upper part of the figure
shows the elastic cross section for electron scattering. The spectrum
using the "white" electron source is a sum over all entrance and exit
channels and the effect of the resonance decaying to levels of N_2
other than v=0 is to make the main structure less prominent and add
auxiliary peaks below 2 eV.

In the above binary collision model the charge exchange process to a shape resonance depends on the ion velocity. Thus this process will be less prominent for F^- with a mass of 19 than for H^- at the same energy as its velocity will be over 4 times slower. This is indeed the case as can be observed in figure 13 which shows detached electron spectra for F^- on N_2. The resonance structure is hardly visible even at 4 keV compared to that of H^- on N_2 at 750 eV. The threshold for charge exchange with an F^- projectile is about 1 keV compared to 50 eV for H^-.

Molecular oxygen stands apart from N_2 as O_2^- in the v=0 level is a stable ion. This state only becomes a resonance for v=4 and above and moreover has a very long lifetime corresponding to a width of about 1meV. From figure 11 we can see that O_2^- levels will be populated at the lowest energies and in fact they are observed[22] at 4.5 eV incident energy for H^-. If the detached electrons are observed at 4 keV under high resolution the structure becomes quite complex due to the resonance coupling to excited vibrational levels of O_2. Notice that the peaks are double due to spin-orbit splitting (20 meV) and the relative intensities are compatible with the statistical weights.

Another situation where electron transfer to resonances can strongly infuence a collision process is destruction of atoms in high Rydberg states. For these species the outer electron is much more weakly bound than for typical negative ions and the binary collision picture should be an even better description of the process. D^+ formation in collision of D atoms in Rydberg states up to n=71 with N_2 has been studied by Koch[25] and the results are shown in figure 15. Also

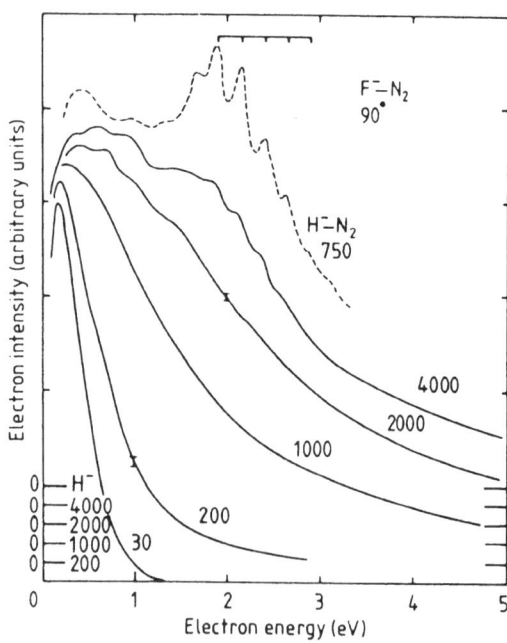

Fig.13. Detached electron spectra obtained in F^- - N_2 collisions at indicated energies (in eV).The upper spectrum (broken curve) corresponds to a H^- - N_2 collision at 750 eV. The N_2 vibrational levels are also indicated.

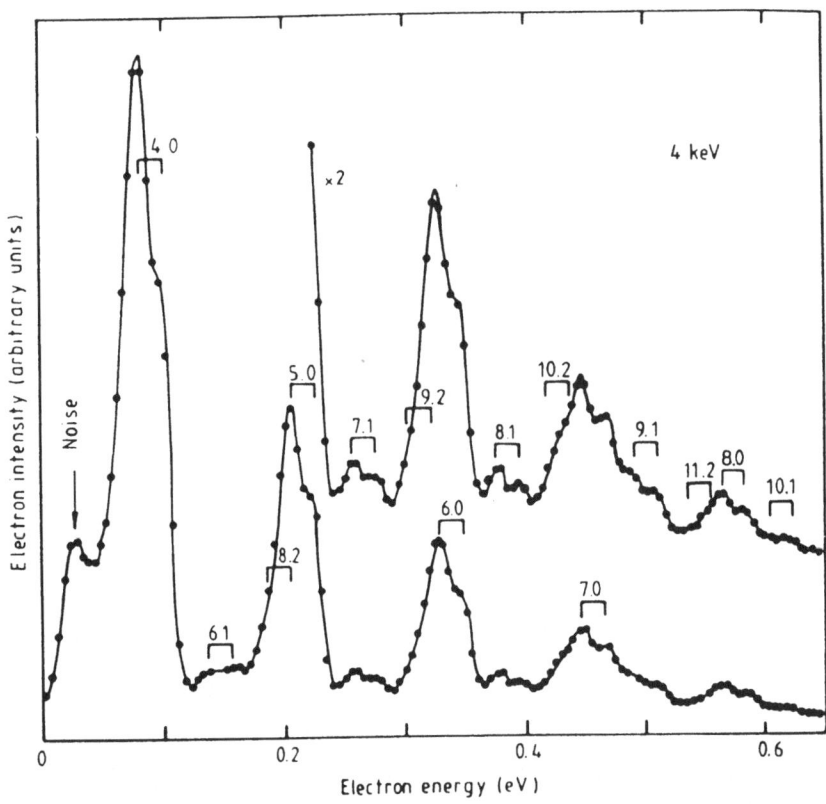

Fig.14. Detached electron energy spectra for $H^- - O_2$ collisions: High resolution details of the O_2^- resonance structure. The numbers above the brackets correspond to the initial O_2^- and final O_2 vibrational levels respectively.

shown is the total destruction cross section as well as the $e-N_2$ cross section on the electron energy scale corresponding to the atom velocity . The N_2 resonance and D^+ formation envelopes are clearly very similar and would indicate that electron transfer to a temporary negative ion is an important process in Rydberg atom collisions with molecules when the energy of the excited electron corresponding to the atom velocity coincides with a resonance of the target particle. Note that the low electron binding energy of the Rydberg electron leads to electron wave packets with energy spreads much smaller than those of figure 11 for H^- and the widths are now sufficiently low to be able to resolve the structure of the N_2^- resonance.

In this section it has been shown how a knowledge of resonances in electron molecule collisions has led to an understanding of phenomena observed in experiments on electron detachment from negative ions and from atoms in high Rydberg states. Furthermore these experiments can also yield information on the resonances themselves. Charge exchange from H^- to O_2 and observation of the autodetaching electrons allowed spectroscopic informations to be obtained on the ground state of O_2^- .

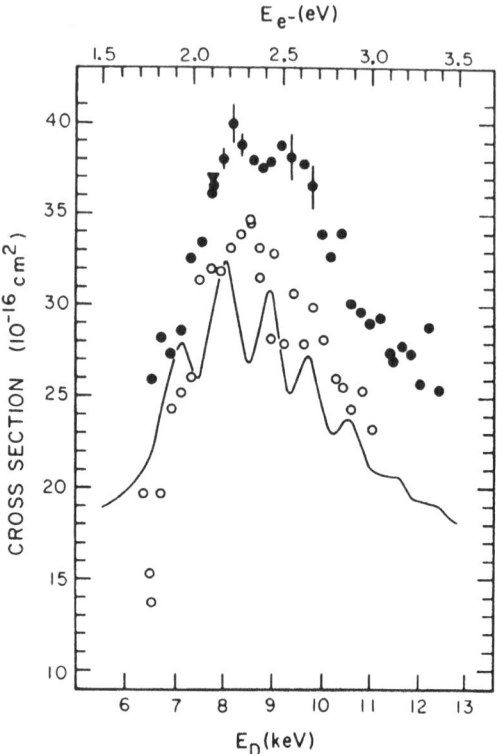

Fig. 15. Collisions of D with n=46 and N₂. Open circles :
ionization cross section; full circles : total
destruction cross section. The full line curve is
the electron scattering cross section.

3. RECOMBINATIVE DESORPTION OF H ATOMS ON METAL SURFACES

Here is described an application of a resonant process
(dissociative attachment) which allows, via theoretically determined
cross sections, the evaluation of vibrational populations of H_2
molecules prepared in a beam. This beam is then used to study the
effect of vibrational energy on the 14 eV dissociative attachment
process of H_2. However the most important result of this study is the
identification of the process by which the vibrational levels of H_2 are
populated. They are populated by recombinative desorption of atomic
hydrogen on metal walls i.e.:

$$H + H(wall) \longrightarrow H_2(v)$$

and the precise mechanism which would lead to the formation of $H_2(v)$ is
thought to proceed by the formation of an intermediate negative ion.

The method employed for detecting vibrationally excited H_2 is
based on the dissociative attachment process which goes to the first
limit at 3.72 eV. i.e.:

$$e + H_2(v) \longrightarrow H^- + H .$$

The cross sections for this process have been determined at threshold by Allan and Wong[26] up to v=4. Subsequently theoretical models were adjusted to these observations and cross sections have been published for all vibrational and rotational levels[27,28]. These cross sections are characterized by sharp peaks at threshold whose magnitude increases rapidly from 10^{-21}cm^2 for v=0 to a few times 10^{-16}cm^2 for v=6 and remains roughly constant for higher levels. Thus, crossing an electron beam of variable energy with a molecular beam of hydrogen and observing the H$^-$ yield allows relative vibrational populations to be determined.

The experimental set-up[18] is shown in figure 16. The electron beam is generated by a cylindrical 127° electrostatic selector and has an energy width of about 100 meV at half maximum. The H$^-$ ions are detected by means of a quadrupole mass spectrometer placed at an angle of 90° with respect to the incident electron beam and is preceded by electrostatic optics which are tuned to be particularly sensitive to zero energy particles. When the filament in the gas cell is heated to temperatures between 1700K and 2500K the H$^-$ yield as a function of incident electron energy are as shown in figure 17. The peaks correspond to dissociative attachment to molecules with different vibrational energies: as the internal energy increases less electron energy is required to produce dissociation, consequently the threshold moves to lower impact energies. The processes appear as peaks because the cross sections have peaks at threshold and this is accentuated by the tuning of the H$^-$ collection optics. Notice that levels up to the maximum of v=9 are detected. Dissociative attachment to higher levels is exothermic and the H$^-$ ions are no longer formed with zero energy. Notice also that there is no appreciable rotational excitation associated with the vibration.

The relative vibrational populations can be determined from the observed peak intensities using the theoretical dissociative attachment cross sections. The populations derived from the spectrum of figure 17 are shown in figure 18 on a semi-log scale. Two series of results are displayed and correspond to different theories. These populations show a double distribution with a break at v=3. The theories are in reasonable agreement for v > 3 but differ somewhat for lower levels. The populations of the first three excited levels are well approximated by a straight line for both theories which corresponds to Boltzmann temperatures near 2500K. The populations of the higher levels are well in excess of those corresponding to this temperature and cannot be readily approximated by a straight line.

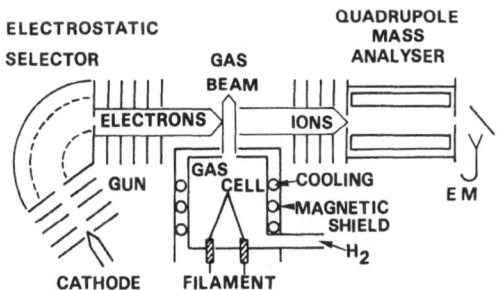

Fig.16. Schematic diagram of the experimental apparatus of Hall et al[18].

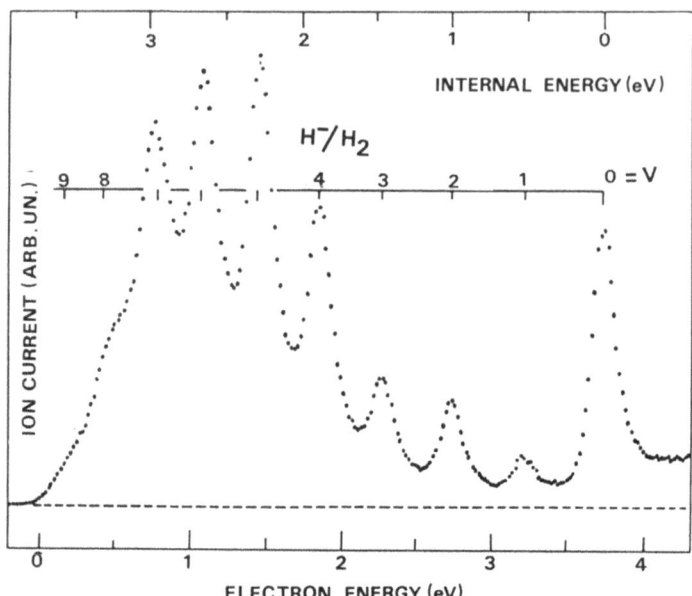

Fig.17. H⁻ yield as a function of electron energy. The peaks
correspond to dissociative attachment at threshold
to H_2 in the indicated vibrational levels.

The vibrational excitation results from recombinative desorption
of atomic hydrogen on the cell surfaces, i.e. recombination of atomic
hydrogen from the gas phase with adsorbed hydrogen atoms on the
surface followed by desorption of the resulting molecules. The atomic
hydrogen is generated by dissociation of H_2 on the incandescent
filament. The evidence in support of this mechanism was obtained when
H_2 and D_2 were mixed together in the cell. The H⁻ spectrum then
revealed additional peaks from dissociative attachment to HD and the D⁻
spectrum originated essentially from HD , as the D⁻ from HD cross
sections are larger than those from D_2.

If the straight line in figure 18 through v=1 to 3 is extrapolated to
v=0, then assuming that this line represents the populations of the
molecules formed by wall recombination , the proportion of recombined
molecules in the beam is 4% or 13% depending on the theory. If we
compare this to the probability of a molecule colliding with the filament
(1.0) then we derive that the probability of a molecule being atomized
on the filament is 0.04 with Gauyacq's theory[28] and 0.13 with the
theory of Bardsley and Wadehra[27]. The former value would be more in
keeping with the known sticking coefficient of 0.07 for H_2 on
tungsten[29].The molecules must first stick to the filament before being
atomized.

The dynamics of the recombinative desorption process can be
rationalized using the potential energy curves of the interaction of
hydrogen with a metal surface and a schematic representation of such
curves is shown in figure 19. In our gas cell with a pressure of 10^{-3}
torr the surface will be covered by adsorbed hydrogen in both atomic
and molecular form. The atomic hydrogen has 2.24 eV available at the

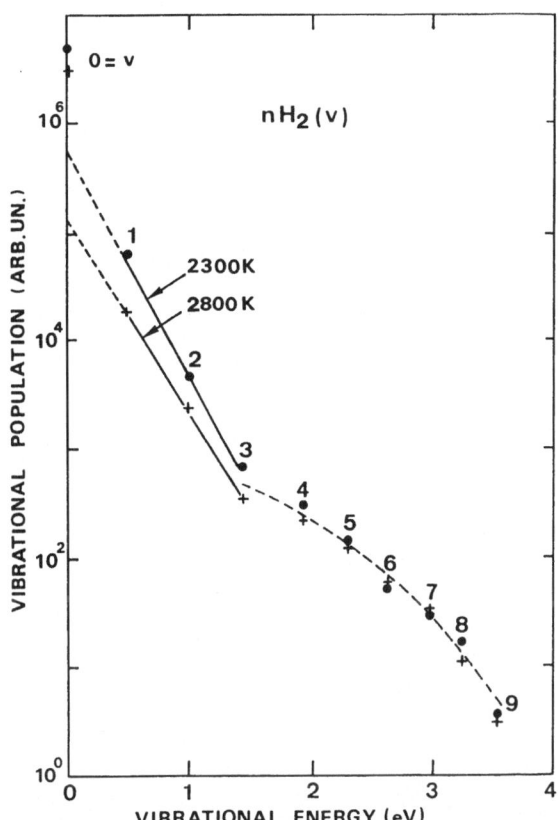

Fig.18. Vibrational population as a function of vibrational energy on a semi-log scale. The full circles correspond to the theoretical dissociative attachment cross sections of Bardsley and Wadehra[27] and the crosses to those of Gauyacq[28].

most for recombination with an adsorbed atom with respect to the H_2 molecule limit. Thus one would expect levels of H_2 up to v=4 (1.88 eV) to be populated but not levels v=5 (2.28 eV) and above. In fact our observations show that molecules in vibrational levels up to at least v=9 (3.56 eV) are populated, and it would appear that recombination occurs to atoms in the bulk in order for sufficient energy to be available to populate these high levels, the bulk states being populated by the same atomic hydrogen flux from vacuum.

The precise mechanism for vibrational excitation is not clear but a model presented by Gadzuk[30] may be relevant. This model is proposed to account for vibrational excitation of molecules resulting from surface collisions and invoques the formation of a temporary negative ion. As a molecule (H_2, N_2 etc...) approaches the surface its electron affinity increases due to image-potential effects until a point is reached where the affinity level becomes degenerate with a Fermi level of the metal and an electron can tunnel to the molecule to form a negative ion. The subsequent motion of the molecule then continues on the negative ion curve up to the classical turning point where the motion is reversed. On the way out the molecule will then lose the electron to a vacant level of the metal when the affinity level returns above the Fermi level energy. As in the gas phase, during the negative ion lifetime the internuclear separation will be modified leading to vibrational excitation which one can imagine to depend on Franck-Condon factors between the negative ion state at the instant the decay takes place and the neutral molecule on condition that the decay is fast. Hence this process leads to the transfer of translational energy to vibrational energy via the formation of an intermediate negative ion. The recombinative desorption process would then be the half-collision analog of the above model. In this case the H atoms, which are thought to be partially charged on th surface, recombine to form H_2^- which subsequently moves away from the surface along the H_2^- potential before decaying and leaving H_2 in an excited level.

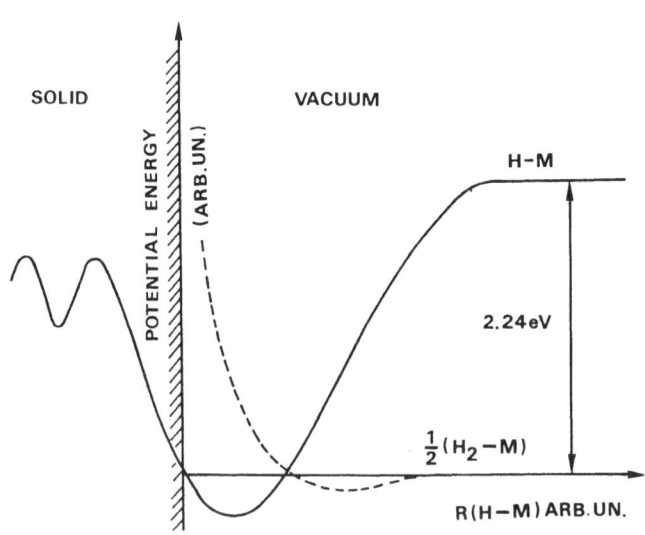

Fig.19. Schematic potential energy curves for hydrogen on a metal.

The results of these observations are of interest to interstellar chemistry. Hydrogen is the most abundant molecule in space and is believed to be formed by recombinative desorption on grains[31]. Recombinative desorption can also contribute to the vibrational populations in a hydrogen discharge. Vibrational excitation plays a key role as it allows H^- ions to be produced efficiently by dissociative attachment: a process which is used for generating intense H^- beams[32].

An immediate application of the H_2 beam containing known populations of excited molecules has been to study of the dependence of the 14 eV dissociative attachment process on the internal energy [33]. Figure 20 shows the H^- yield as a function of energy for the two dissociation limits. The two curves are superimposed so that the dissociation limits coincide and the main peaks are normalized to the same intensity. These experiments were performed with a different cell geometry and at a higher pressure compared to those of figure 17. Under these conditions rotational levels are also populated as can be seen in the figure. Further structure can be seen above 14 eV and results from predissociation of resonances[34]. The cross section enhancement with vibrational level at 14 eV is 3.4 and 4.3 for v=1 and 2

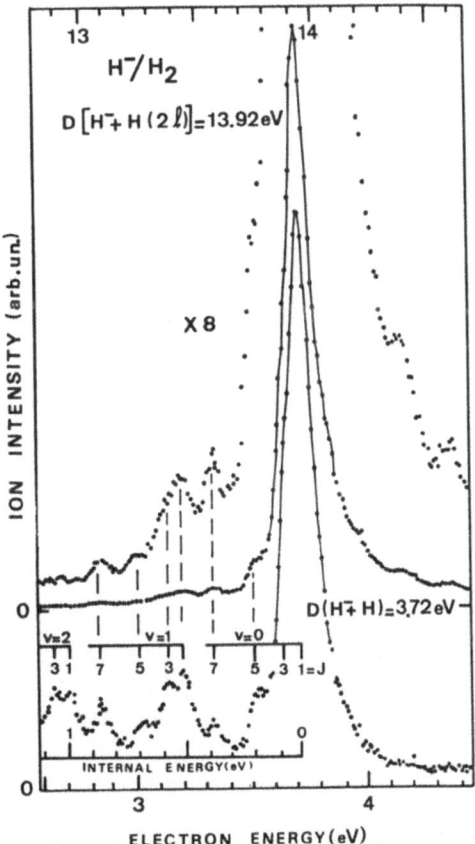

Fig.20. H^- yield as a function of electron energy in the region of the first (lower curve) and the second dissociation limit (upper curve). The initial ro-vibrational levels of H_2 are indicated.

respectively, much lower than the values of 30 and 500 at 3.7 eV. The cross section was observed to be insensitive to rotational excitation. The 14 eV process is considered to proceed via a Feshbach resonance whose exact idendity is still an open question[35]. Feshbach resonances have long lifetimes and as such the cross sections are expected to be relatively insensitive to internal energy. In fact the observed enhancement can be simply explained by the Franck-Condon factors between the ground state level and the resonant state. A good agreement with the observations was obtained using the potential curve parameters of the "a" series resonance determined by Comer and Read[36].

The determination of the vibrational populations described above relies on theoretical dissociative attachment cross sections and demonstrates the importance of having available precise values. The two sets of theoretical results which were used, while being similar for high vibrational levels, vary by a factor of about 2 for levels below v=4. This difference has important consequences and leads to a factor of 3 in the determination of the number of recombined molecules. Both these semi-empirical theories were adjusted to observations[26] which were normalized to total dissociative attachment cross sections[37] and isotope effects[38] which date back to the sixties. At the time the experimental techniques did not benefit from high resolution electron beams and mass selection of the charged fragments, consequently a re-evaluation of these cross sections using more sophisticated techniques would be most desirable.

CONCLUSION

The observations presented here show how resonances, originally discovered in electron – molecule binary collisions, can play an important role in other physical situations. Our knowledge of resonances has been used to interpret the new phenomena and these analyses demonstrate the importance of precise theoretical and experimental data. As an example two sets of theoretical dissociative attachment cross sections were employed to determine vibrational populations in H_2 and gave differing results. Both calculations were semi-empirical and relied upon experimental cross sections which were measured twenty years ago and since then have not been supported by other observations. This example goes to show that even for the prototype electron – H_2 system the data are still partially uncertain. Thus although resonances were discovered a generation ago there is still a need for more accurate information even for the simplest systems.

Acknowledgements:

I wish to express my gratitude to the following people who during collaborations past and present contributed to much of the work presented in this paper: Iztok Čadež, Vladimir Esaulov, Jean-Pierre Grouard, Michel Landau, Jean-Louis Montmagnon, Francis Penent, Françoise Pichou and Catherine Schermann.

REFERENCES

1. A. Herzenberg in "Electron-Molecule Collisions", Plenum (1984)
2. M. Allan, J. Phys. B 18,4511 (1985)
3. M. Allan, J. Phys. B 18,L451 (1985)
4. M. Allan, Helvet. Chim. Acta. 65,2008 (1982)
5. H. Ehrhardt, D.L. Langhans, F. Linder and H.S. Taylor , Phys. Rev. 173,222 (1968)

6. C. Mündel, M. Berman and W. Domcke , Phys. Rev. A 32,181 (1985)
7. N.F. Lane, Rev. Mod. Phys. 52,29 (1980)
8. J.P. Gauyacq, Europhys. Lett. 1,287(1986)
9. L. Sanche and M. Michaud, Phys. Rev. B 30,6078 (1984)
10. J.E. Demuth, D. Schmeisser and P. Avouris, Phys. Rev. Lett. 47,1166 (1981)
11. P. Avouris, D. Schmeisser and J.E. Demuth, Phys. Rev. Lett. 48,199 (1982)
12. S. Anderson and J. Harris, Phys. Rev. Lett. 48,545 (1982)
13. R.E. Palmer and R.F. Willis, Surf. Sci. 179, L1 (1987)
14. L. Sanche and M. Michaud, Phys. Rev. B 27,3856 (1983)
15. U. Fano, J.A. Stephens and M. Inokuti, J. Chem. Phys. 85,6239 (1986)
16. L. Sanche, Phys. Rev. Lett. 53, 1638 (1984)
17. L. Sanche and L. Parenteau, J. Vac. Sci. Technol. A4, 1240 (1986)
18. R.I. Hall, I. Čadež, M. Landau, F. Pichou and C. Schermann, to be published
19. L. Sanche, G. Perluzzo and M. Michaud, J. Chem. Phys. 83,3837 (1985)
20. V.A. Esaulov, Ann. Phys. Fr. 11, 493 (1986)
21. J.L. Montmagnon, V. Esaulov, J.P. Grouard, R.I. Hall, M.Landau and C. Schermann, J. Phys. B 16, L143, (1983)
22. V.A Esaulov, J.P. Grouard, R.I. Hall, M. Landau, J.L. Montmagnon, F. Pichou and C. Schermann, J. Phys. B 17,1855 (1984)
23. Vu Ngoc Tuan, V. Esaulov, J.P. Gauyacq and A. Herzenberg, J. Phys. B 18, 721 (1985)
24. Vu Ngoc Tuan, V. Esaulov and J.P. Gauyacq, J. Phys. B 17,L133 (19 84)
25. P.M. Koch, Phys. Rev. Lett. 43,432 (1979)
26. M. Allan and S.F. Wong, Phys. Rev. Lett. 41, 1791 (1978)
27. J.N. Bardsley and J.M. Wadehra, Phys. Rev. A 20,1398 (1979)
28. J.P. Gauyacq, J. Phys. B 18, 1859 (1985)
29. P.W. Tamm and L.D. Schmidt, J. Chem. Phys. 55,4253 (1971)
30. J.W. Gadzuk, J. Chem. Phys. 79,6341 (1983)
31. W.W. Duley and D.A. Williams, Interstellar Chemistry. Academic Press. London (1984)
32. J.R. Hiskes, Comments At. Mol. Phys. 19,59 (1987)
33. I. Čadež, R.I. Hall, M. Landau, F. Pichou and C. Schermann, XV ICPEAC, Brighton 1987 and to be published
34. M. Tronc, R.I. Hall, C. Schermann and H.S. Taylor, J. Phys. B 12, L279 (1979)
35. A. Huetz and J. Mazeau, J. Phys. B 16, 2577 (1983)
36. J. Comer and F.H. Read, J. Phys. B 4,368 (1971)
37 D. Rapp, T.E. Sharp and D.D. Briglia, Phys. Rev. Lett. 14,533 (1965)
38. G.J. Schulz and R.K. Asundi, Phys. Rev. 158,25 (1967)

THRESHOLD MEASUREMENTS OF ELECTRON AND PHOTON COLLISIONS

WITH ATOMS AND SMALL MOLECULES

George C King

Department of Physics, Schuster Laboratory
Manchester University
Manchester   M13 9PL

INTRODUCTION

Something interesting usually happens when you excite a collision process close to its threshold. The energy of the incident particle (in the present case, an electron or a photon) is relatively small, ie just greater than the threshold energy of the reaction. Further, one or more electrons of very small energy ($\leqslant$.01eV) can be produced. These conditions can have a profound influence on the collision. For example electron exchange can take place and electron-electron correlation effects can play a dominant role. Threshold excitation often leads to new and unexpected results. The present review deals with some of the characteristics of threshold excitation by electrons and photons and some of the aspects of the experimental techniques involved. As will be seen there are similarities betwen excitation by electrons and photons. For example the resonances observed in electron atom scattering have some similarities to the doubly excited states induced by photon excitation since both atomic systems consist of two electrons clinging to a positively charged core.

The present review will concentrate on work that has been performed recently at Manchester University and Daresbury Laboratory Synchrotron Radiation Source. Several recent studies using threshold excitation will be described together with the information and the physical insight that can be gained from them. The review will also try to identify areas for future study.

Theoretically, threshold studies are interesting because different approaches are often required especially when two low energy electrons are involved. Experimentally, threshold measurements are interesting because they involve the collection and energy analysis of very low energy electrons and this makes the experiments exacting and demanding.

Characteristics of Threshold Excitation

Typical reactions of interest may be represented by:

$$e^-(E_i) + A \rightarrow A^* + e^-_{scatt}$$

$$h\nu(E_i) + A \rightarrow A^+ + e^-_{ejected}$$

where e^-_{scatt} or $e^-_{ejected}$ has an energy that is much smaller than the excitation energy, E_i, and may be ~ few meV. It is these very low energy electrons that are detected in the present studies.

A threshold spectrum is obtained by sweeping the energy of the incident electron or photon beam while collecting only those scattered or ejected electrons with almost zero energy. The relationship between the shapes of inelastic cross sections and the threshold spectrum is illustrated in figure 1. The upper three curves represent cross sections of three neighbouring excited states of the target. As the energy of the incident beam passes through the threshold for the process, a very low energy electron is produced. The shaded parts of the curves represent the range of electron energies that can be transmitted by the threshold analyser. The width of these shaded parts corresponds to the resolution of the threshold analyser, which typically can be a few meV. Thus the yield of detected electrons as a function of incident energy contains peaks that map out the energy levels of the target. This produces the threshold spectrum shown in the lowest curve of figure 1. Further, as can be seen, the intensity of a threshold peak is a sensitive function of the shape of the cross section near to threshold.

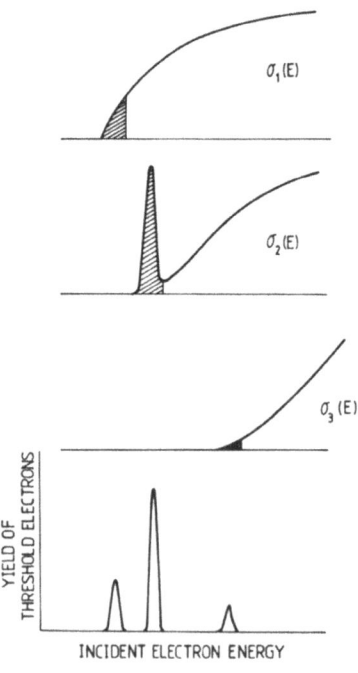

Figure 1   Schematic diagram of inelastic cross sections and the associated threshold electron spectrum.

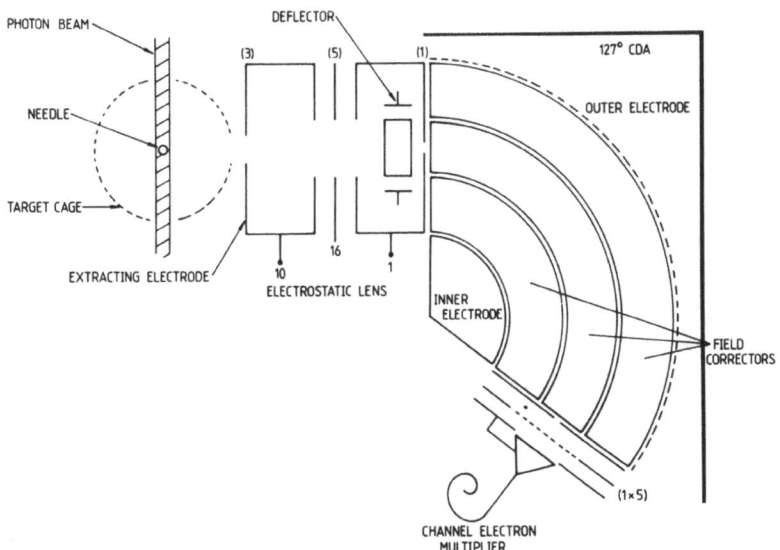

Figure 2 Schematic diagram of a threshold electron analyser. From King et al (1987)

There are several reasons for studying threshold excitation. Firstly, it provides a powerful probe for studying the dynamics of the collision. The energy of the electrons are low and if two or more electrons are involved, they move slowly away from the positive core allowing plenty of time for electron-electron interactions to become dominant. Post collision interactions are a dramatic demonstration of such interactions. Secondly, in the case of electron impact, dipole selection rules are considerably relaxed when the energy of the incident electron is close to threshold. For example, the target and incident electrons may exchange and so singlet-triplet transitions may be excited. Threshold excitation is thus a powerful spectroscopic technique. Thirdly, as illustrated in figure 1, the yield of threshold electrons is very sensitive to the shape of the cross sections close to threshold, which are often dominated by the presence of resonances. Threshold excitation can provide valuable information about the shape of these threshold cross sections.

EXPERIMENTAL CONSIDERATIONS

The apparatus for threshold studies consists of crossed beams of the target atoms or molecules and the excitation source, photons or electrons, and a threshold electron analyser. In the studies described here the electrons were provided by a high resolution electron monochromator (eg Brunt et al 1977) while the photons were provided by the Daresbury Synchrotron Radiation Source used in conjunction with a toroidal grating monochromator. The basic requirement of the threshold analyser is that low energy electrons ($\leqslant 10$meV) should be transmitted with high efficiency (~100%) whilst more energetic electrons are discriminated against.

A schematic diagram of the threshold analyser that was used in the threshold photoionization studies at Daresbury Laboratory, (King et al 1987), is shown in figure 2.

The essential parts of the analyser are: a cage surrounding the interaction region, an extracting electrode outside the exit aperture of the cage, an electrostatic lens, a 127° cylindrical deflection analyser (CDA) and an electron detector. The cage surrounding the interaction region is a crucial part of the analyser. It allows penetration of the extracting field and passage of the photon beam and yet being highly transparent allows energetic photoelectrons to leave the region without producing low energy secondary electrons at solid surfaces. The target gas beam emanates from a gold-plated platinum-iridium hypodermic needle of bore 0.5mm. The extracting electrode which has an aperture of 3mm is situated 2mm outside one of the exit holes of the cage. This electrode produces the extracting field that draws out the low energy photoelectrons and directs them towards the analyser. The effect of the extracting field on the threshold photoelectrons is illustrated in figure 3 which shows the computed trajectories of electrons that are produced at the centre of the interaction region with an initial kinetic energy of 2meV.

As can be seen, electrons having an energy as low as this are collected over 4π sr, and form a crossover near the exit hole of the target cage. The beam crossover region is imaged onto the entrance plane of the CDA by a three-aperture asymmetric-voltage lens. The CDA has a mean radius of 25.4mm and its purpose is to prevent transmission of energetic photoelectrons that are ejected in the direction of the threshold analyser. A channel electron multiplier with a conical input is used to detect transmitted electrons.

The threshold analyser was used in conjunction with a toroidal grating monochromator attached to the Daresbury Laboratory SRS. This provided photons over the energy range 10 to 100eV with an energy bandwidth as small as 0.015eV.

The high performance of the threshold analyser is demonstrated by the threshold photoelectron spectrum of argon in figure 4. Threshold electrons are produced as the photon energy crosses the energies of the $^2P_{3/2}$ and $^2P_{1/2}$ states of Ar$^+$. The observed positions of the peak maxima in the spectrum of figure 4 are 15.760 and 15.937eV respectively in agreement with the spectroscopic values. The fact that the maxima of the threshold peaks occur at the spectroscopic values indicates that the widths of the peaks are determined by the energy spread of the photon beam and that the energy resolution of the threshold

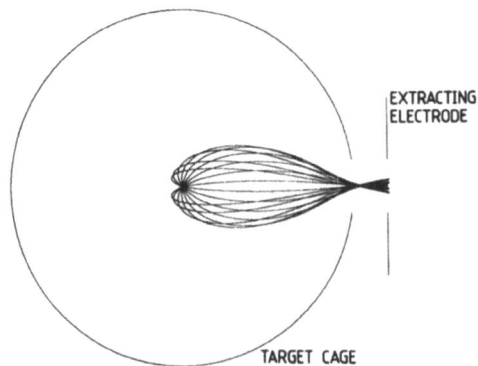

Figure 3   Computed trajectories of low energy (2meV) photoelectrons produced at the interaction region.

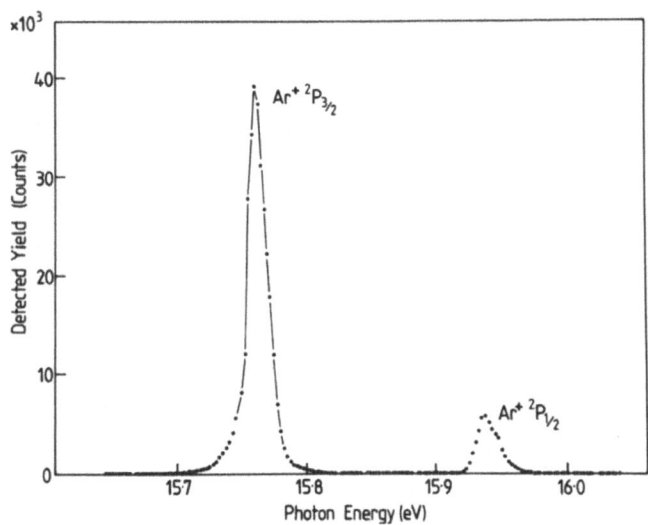

Figure 4   Threshold photoelectron spectrum of argon.   From King et al
          (1987)

spectrometer is much smaller than the observed widths of the peaks.
The count rate in the $^2P_{3/2}$ peak was 4×10^4 cps, while the signal to
noise ratio was approximately 10^3:1, as is clearly demonstrated by the
almost complete absence of noise in the energy region between the two
peaks.   The 11s' autoionizing state of argon lies 3meV above the $^2P_{3/2}$
ion core state and the resulting ejected electrons contribute to the
intensity of the peak corresponding to the $^2P_{3/2}$ ion state which
explains why the ratio of the heights of the two peaks is not the ratio
of statistical weights, 2:1.   The observed ratio of the peaks is 6.7:1
which indicates that the energy resolution of the spectrometer is
approximately 3meV (Peatman et al 1975).

THRESHOLD ELECTRON IMPACT STUDIES

Threshold Electron Spectroscopy

     In threshold electron spectroscopy, states are excited by electron
impact at incident energies that are within a few hundredths of an
electron volt above their excitation energies.   The selection rules
that govern the excitation are therefore considerably relaxed allowing
the excitation and observation of a wide range of optically forbidden
transitions (Hall and Read 1984).

     An example of a threshold electron spectrum in CO, over the energy
range 5.8 to 11eV, (Hammond et al 1985) is shown in figure 5, where the
wealth of structure shown by the technique is evident.

Of particular note is the fact that the more intense peaks correspond
to optically forbidden, triplet transitions.   In this respect threshold
spectroscopy is a complementary technique to optical absorption
spectroscopy or high incident energy electron energy loss spectroscopy.
Figure 5 also demonstrates the high sensitivity of the technique.   This
enables high vibrational levels ($\nu > 30$) to be observed.   Clearly the

spectra give information about: energy positions, peak intensities, vibrational spacings and hence potential curves of valence and Rydberg states.

The threshold spectra also give information about resonance mechanisms at threshold. For example the sequence of spectra shown in figure 6 was obtained by altering the detection band of the threshold analyser. Figure 6 (a), (b) and (c) correspond to the detection of scattered electrons of energy 0 to 5meV, 5 to 15meV and 70 to 90meV respectively. A comparison of the three spectra in figure 6 reveals that in moving from near-threshold detection conditions (a) and (b) to an above-threshold detection condition figure 6(c), there are distinct changes in the spectrum. The most obvious change is in the intensity of the $B^1\Sigma^+$ $v=0$ peak at 10.777eV relative to the $b^3\Sigma^+$ $v=0$ state at 10.399eV. This is interpreted as being due to a resonance situated close to the threshold of the $B^1\Sigma^+$ $v=0$ state, probably a resonance of configuration $(CO^+X^2\Sigma^+)(3s\sigma3p\pi)^2\Pi$ in agreement with the earlier observations of Mazeau et al (1972). This interpretation is supported by the metastable excitation function measurements of Newman et al (1983). It is interesting to note that the two features corresponding to the $B^1\Sigma^+$ and $b^3\Sigma^+$ states exhibit similar behaviour as threshold is approached to the equivalent states (the 2^1S and 2^3S) in helium (Cvejanović and Read 1974b, Nesbet 1975).

Also of note, in the spectrum of figure 5, is the broad bump at 9.66eV. This bump is due to the detection of low energy stable O^- ions which result from the process:

$$e + CO(^1\Sigma^+) \rightarrow CO^{-*} \rightarrow O^-(^2P) + C(^3P)$$

Detailed measurements of this process have been made by Hall et al (1977) who used a magnetic filter to separate the O^- ions from the

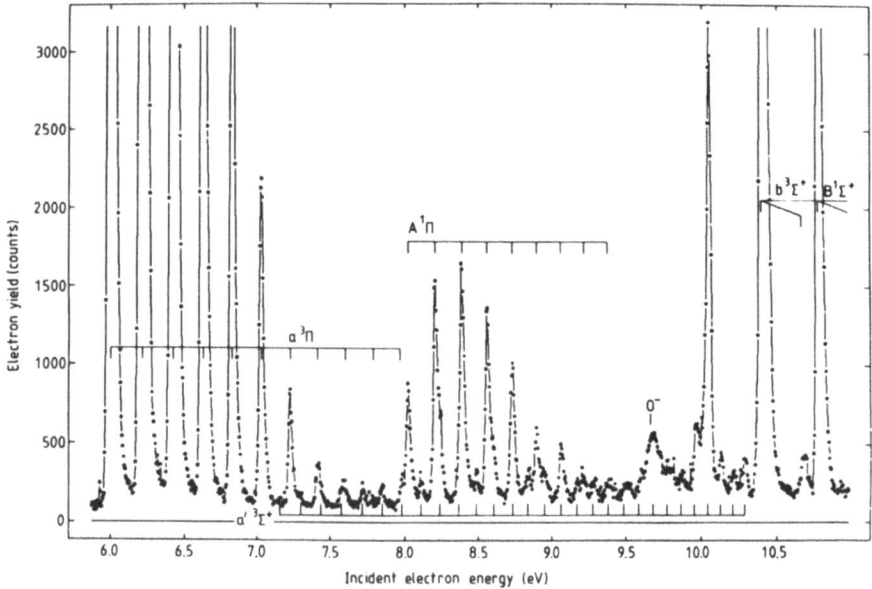

Figure 5. The threshold electron spectrum of CO. From Hammond et al (1985).

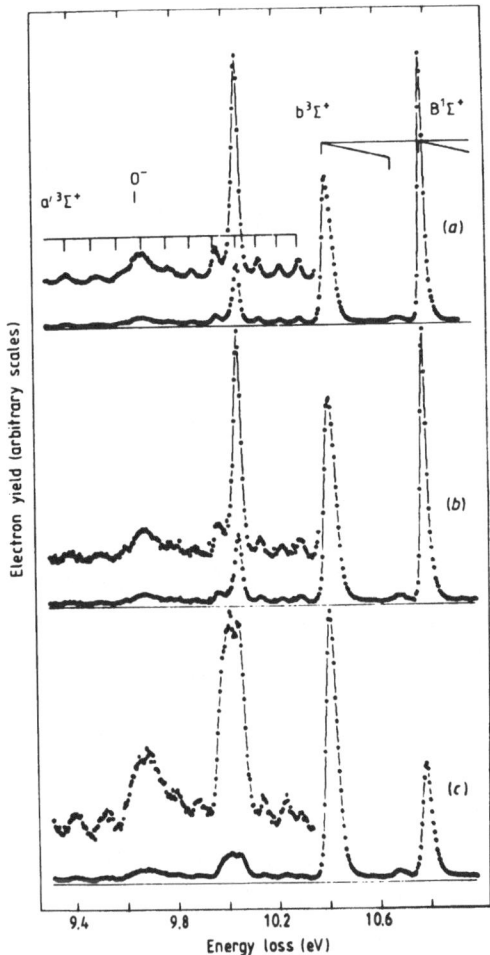

Figure 6   Sequence of threshold spectra of CO under various pass-band conditions.   The pass bands in the three spectra were: (a) 0 to 10meV, (b) 5 to 15meV and (c) 70 to 90meV, see text.   From Hammond et al (1985).

electrons in the scattered beam.   This feature illustrates the point that charged particles follow the same trajectories in electrostatic systems regardless of their mass or charge.   The particles can be separated, for example in mass, by placing a magnetic field or an RF field just before the detector.   This suggests that the technique of threshold detection could be usefully used in processes which result in the production of very low energy stable negative ions, for example associative detachment of triatomic molecules.

Threshold electron spectroscopy provides a different kind of information above the ionization potential of the target species, because then there are two electrons which proceed from the excited target.   In this region post collision interaction (Read 1976, Niehaus 1978) can cause peaks to apparently shift to higher excitation energies.   Thus, an incident electron excites an autoionizing neutral state of the target and is scattered inelastically.   When the neutral

state autoionizes the scattered electron finds itself in the Coulomb
fields of the ejected electron and, more importantly, of the positive
ion. The faster the process of autoionization and the slower the
velocity of the scattered electron, the closer the scattered electron
is to the ion at the time of autoionization. To escape from the ion,
the electron has to have sufficient kinetic energy to overcome the
Coulomb potential. In threshold experiments, where the electrons are
detected when they have a very low energy, the scattered electrons are
observed only when they are just able to overcome the Coulomb
attraction and escape from the ion. Hence the observed peaks have an
energy higher than that of the autoionizing state. The measured shift
ΔE in the energy of the peak gives an estimate of the lifetime, τ, of
the autoionizing state (Hammond 1985)

$$\tau \approx \hbar R^{\frac{1}{2}} \Delta E^{-3/2}$$

where $\hbar$ is Planck's constant and R the Rydberg constant.

An example of a threshold spectrum obtained in Xenon in the region
12-14eV, near to the two ionization potentials is shown in figure 7.
There are three particular points of note in this spectrum. The
horizontal lines with bars indicate the well known spectroscopic
energies of the states of Xe^+. Clearly the threshold peaks belonging
to these excited states are shifted in energy, where the shifts appear
to depend on the quantum numbers n and ℓ. Also, peaks of high values
of ℓ tend to have a larger intensity than expected. These large
intensities of high ℓ states is again a result of long range
interactions and correlations which result in the exchange of large
values of angular momentum. Finally, there may be seen a cusp-like
feature at the higher ionization potential, again due to the effects of
electron-electron correlation (cf Cvejanović and Read 1974a).

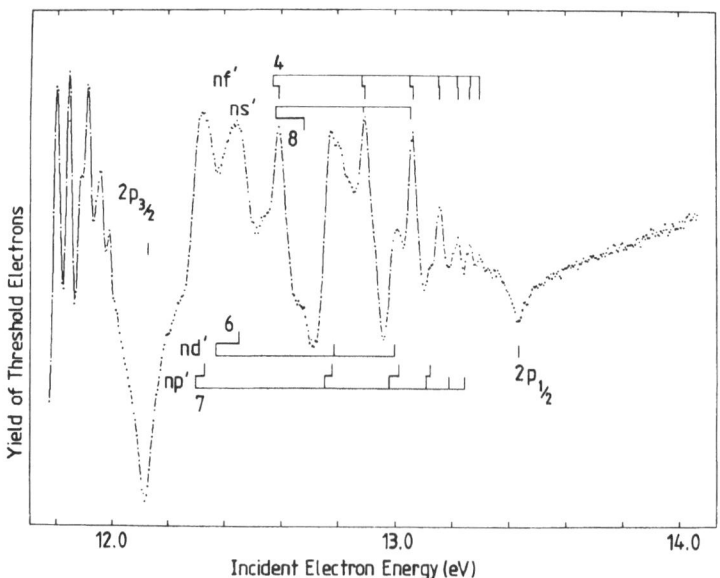

Figure 7   The threshold electron spectrum of xenon over the energy
range including the $^2P_{3/2,1/2}$ ionization potentials. A
sloping background has been subtracted from the data. From
Hammond et al (1988).

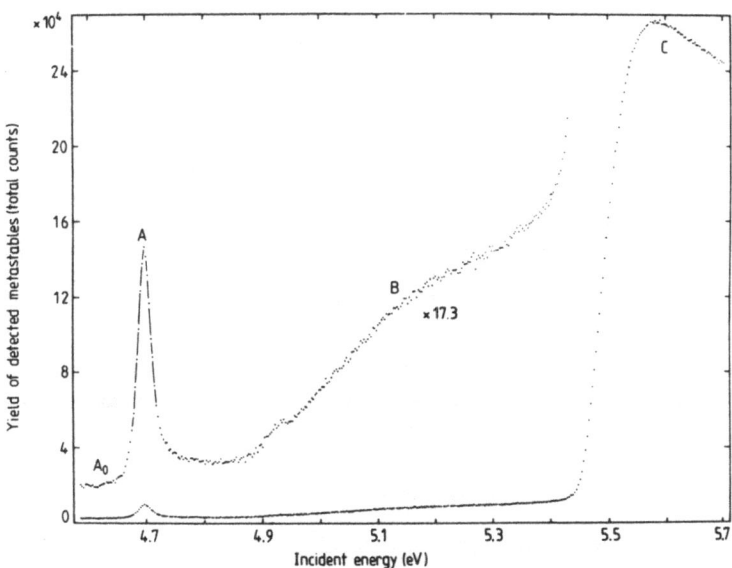

Figure 8   Metastable-atom excitation function in mercury.   From Newman et al (1985).

Metastable Excitation Function Measurements

Metastable excitation function measurements provide a way of studying excitation cross sections close to threshold that avoids the detection of very low energy electrons. It is particularly valuable in the study of resonance enhancement of these cross sections that occurs near the threshold. In this technique the electron beam is again crossed with the target atomic or molecular beam, but here a detector is placed downstream of the interaction region, to intercept the atoms that are excited to metastable states. The yield of atoms in metastable states is measured as a function of the incident electron energy to provide an excitation function of the metastable state which contains structure due to resonances that decay into the metastable state. Nearly all of the metastable atoms produced can be intercepted by the detector, and detected with relatively high efficiency (~10-50%). Hence the measured excitation function is close to a total cross section measurement and further the technique is very sensitive. This high sensitivity of the method makes it particularly valuable for studying the higher lying and weaker intensity resonance structures.

An example of a metastable excitation function, obtained in mercury, is shown in figure 8.

It corresponds to the sum of the excitation functions of the $6^3P_{0,2}$ states of mercury in the region of their excitation thresholds. The differences in the shape of the excitation functions of the two states is evident and it is seen that the technique allows the observation of the resonance enhancement of the 6^3P_0 state near its threshold. This region and the higher energy region of the metastable excitation function of mercury which is very rich in resonance structure has been discussed in detail by Newman et al (1985).

A further example of a metastable excitation function, obtained in Ne, is shown in figure 9 (Buckman et al 1983).

Of particular note here are the features labelled P which occur at neutral state thresholds. These features are thought to be due to s-wave virtual states or to non-valence resonances consisting of a Rydberg electron weakly bound to the neutral excited state (cf Nesbet 1975).

So far metastable excitation function measurements have usually corresponded to a sum of metastable excitation measurements. Recently (Zubek 1986, Zubek and King 1987), it has been possible to isolate individual metastable excitation functions. When the metastable atom strikes the detection surface it ejects an electron whose energy depends on the energy of the metastable state and the work function of the detection surface. By analysing the kinetic energy of the ejected electrons, it is possible to isolate an individual metastable excitation function. Zubek and King (1987) have used this technique to measure the excitation function of an individual metastable state in mercury.

THRESHOLD PHOTOELECTRON SPECTROSCOPY

Threshold photoelectron spectroscopy has been shown to be a powerful technique in the study of photoionization of atoms and molecules (see for example Peatman et al 1969, 1978, Delwiche et al 1981). The technique allows the direct and accurate determination of ion excitation energies and the measurement of photoionization cross-sections close to their thresholds rather than far above the threshold as in conventional photoelectron spectroscopy, which it therefore complements. Important advantages of threshold photoelectron spectroscopy over conventional spectroscopy at fixed wavelength are that very high transmission can be achieved for electrons of near-zero kinetic energy without loss of energy resolution and that Doppler broadening becomes very small for such electrons.

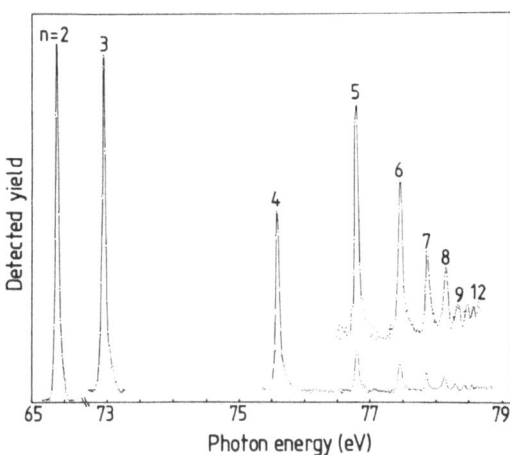

Figure 9   Metastable-atom excitation function in neon.   From Buckman et al (1983).

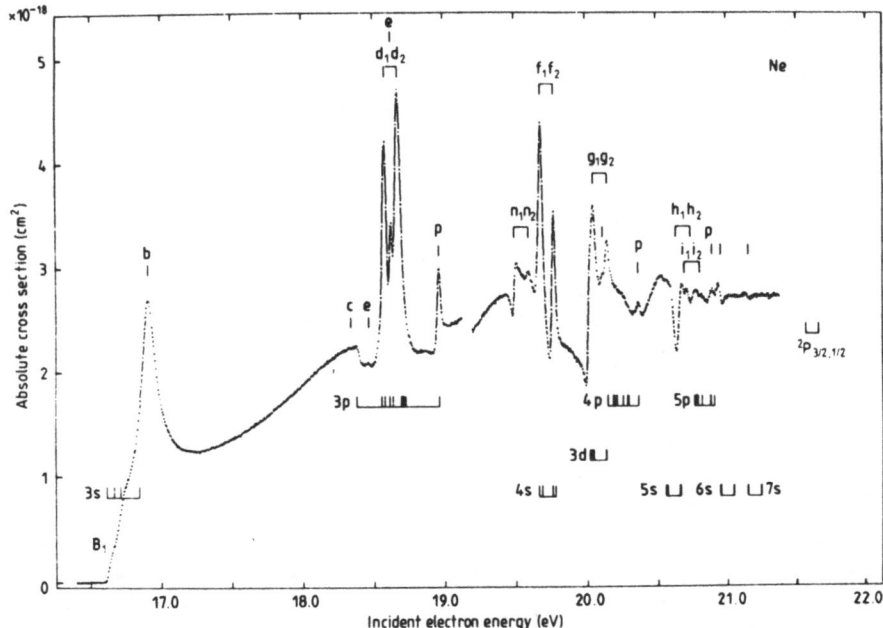

Figure 10   Threshold photoelectron spectrum of heliun.   From King et al (1988).

An example of a threshold photoelectron spectrum, obtained in helium is shown in figure 10.

There are two particular points of note about this spectrum. Firstly, although He$^+$ is hydrogenic, the observed intensities of the threshold peaks do not vary as $^1/_n{}^{*3}$. Secondly, the intensities do not vary smoothly with n, with peaks of even values of n having an enhanced intensity. One possible explanation for this second observation is the occurrence of Wannier ridge resonances (see for example Buckman et al 1983). These consist of atomic states which have two electrons, with high values of n, lying on opposite sides of the positive ion core. Such resonances would be expected to lie just below ionization thresholds of helium, ie He** (n_1=6, n_2=6) would lie just below the He$^+$(n=6) ion state, and they could lead to the observed enhanced intensities. Their energy spacing, (Buckman et al 1983) could explain why only ion states with certain values of n are enhanced.

A second example of a threshold photoelectron spectrum, obtained in xenon is shown in figure 11.

The region of the spectrum includes the (4d)$^{-1}_{5/2, 3/2}$ ionic states of Xe at 67.55 and 69.54eV respectively. Threshold peaks corresponding to these ion states would be expected and are in fact observed. What is, at first sight, surprising is the occurrence of threshold peaks at the energies of the inner-shell neutral states of Xe and the similarity of the spectrum to the inner-shell energy loss spectra of xenon obtained by King et al 1977. This suggests a decay mechanism for the inner-shell states involving more than one electron with electron interactions causing one of the electrons to have a very low energy.

Threshold Photo-double-ionization Studies

In the photo-double-ionization process two low energy electrons are produced, and these recede slowly from the positive ion core. In this system the electron-electron correlations are so strong that the two electrons cease to move independently of one another. Then the independent electron model fails to provide an adequate starting basis for classification purposes (see for example Herrick and Kellman 1980). The reaction in helium may be written,

$$He + h\nu \rightarrow He^{++} + e^- + e^-$$

where the two outgoing electrons have only a small amount of energy to share ($\leq 1eV$). If the two electrons proceed on opposite sides of the ion core at the same radial distance and with the same velocity, ie along the so called Wannier Ridge, then ionization can occur. If however, one of the electrons moves further away from the core than the other, the innermost electron will tend to screen it. The outer electron will gain energy at the expense of the inner electron which may then be recaptured by the ion and ionization will be frustrated.

A threshold photoelectron spectrum obtained in helium (King et al 1988) near the double ionization potential is shown in figure 12. The cusp like feature at the double ionization potential clearly demonstrates the effects of electron-electron correlations.

Doubly Excited States of Helium

The doubly excited states of neutral helium where both electrons have high values of principal quantum number offer an ideal system in which to study electron-electron correlations. The core is

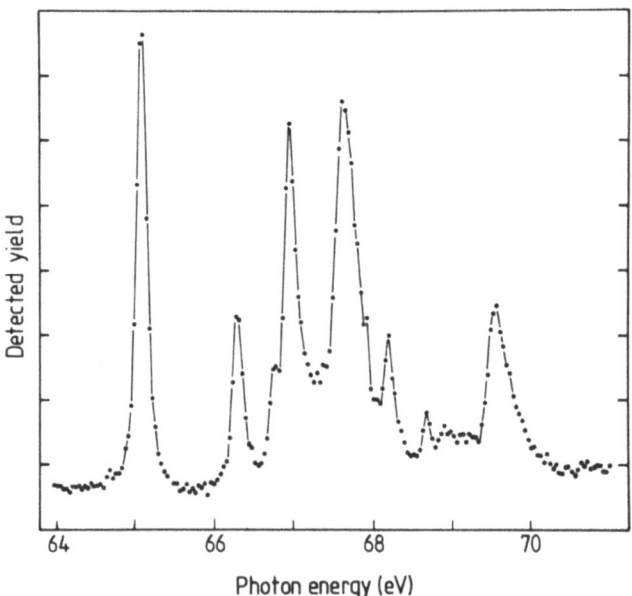

Figure 11   Threshold photoelectron spectrum of zenon.   From Zubek et al (1988a).

110

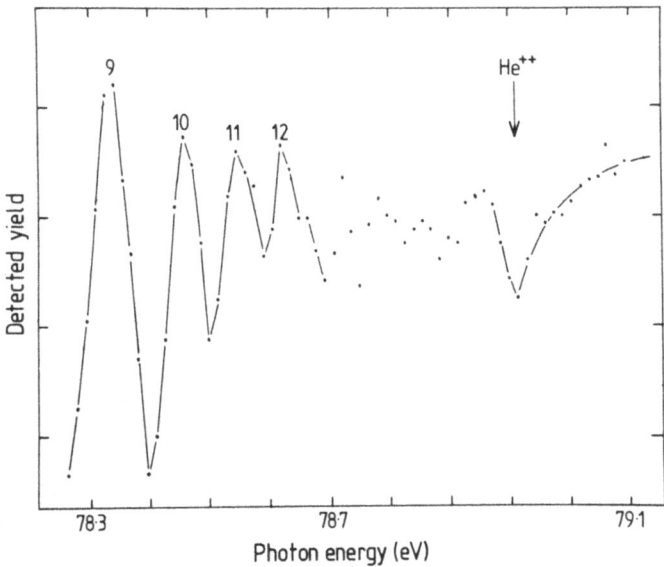

Figure 12   Threshold photoelectron spectrum of helium.   From King et al
(1988).

particularly simple and if photon impact is used, the angular momentum
and parity is well defined.   In the work of Madden and Codling (1963)
the doubly excited states were observed for the first time, using
synchrotron radiation in a conventional photoabsorption spectrum.
Below the n=2 level of He$^+$ there are three ^{1}P series, the strongest of
which was seen in the initial work of Madden and Codling.   An important
point about these and other doubly excited states, as pointed out by
Cooper et al (1963) is that the wavefunctions represent strong
admixtures of single particle configurations and that they therefore
cannot be classified in general by single-particle quantum numbers,
Lipsky et al (1977), Herrick and Kellman (1980).   Although the ±
classification scheme (Cooper et al (1963)) provides some guidance, as
do the various other classification models (eg Klar and Klar (1980))
the nature of the quantum numbers of these and other doubly excited
atomic states has still not been firmly established.   Two recent group
theoretical approaches, which may result in the closing of this gap in
knowledge are those of Herrick and Kellman (1980) and Iachello and Rau
(1981).

In photoabsorption studies the doubly excited states show up as
relatively small structures on top of a background due to direct
ionization.   This large background could make it difficult to
distinguish the higher lying and smaller intensity states.   It is these
higher lying states where both electrons have large values of principal
quantum number, that are of particular interest in correlation studies.
Zubek et al (1988b) have used a technique in which this large
background is avoided by detecting the electron produced in the
autoionizing decay of the doubly excited state.   As described below
this necessitates ramping the photoelectron energy analyser in
synchronism with the incident photon energy.

For these measurements the apparatus shown in figure 2 was employed but the photoelectron energy analyser was operated in a conventional mode and photoelectrons that were emitted in a direction at right angles to both photon and gas beams were energy analysed. The photon beam, with an energy spread of typically 50meV, was ramped across the region containing the double excited states of interest. These states decay by autoionization to an ion state of helium and it is these photoelectrons that were selected by the analyser. In conventional photoelectron spectroscopy, the energy of the photon beam is kept fixed, while the energies of the emitted photoelectrons are analysed. Here the photon energy was ramped in synchronism with the collection energy of the analyser which was tuned to an energy equal to the photon energy minus the energy of the final ion state. The resulting yield of photoelectrons is an excitation function of the ion state which contains sharp structures due to the doubly excited states that decay into it. There are two ways in which the ion state may be excited, either by direct excitation or through autoionization via the doubly excited state. This results in an interference effect and the doubly excited states appear as Fano profiles.

An excitation function of the $He^+(1s)$ state is shown in figure 13. The energy region includes doubly excited states that converge to the $He^+(n=2)$ ion state. No correction has been made for the transmission function of the analyser, which was quite flat in this energy region, and no background has been subtracted. The excellent signal to noise ratio shows the very high sensitivity of the analyser and a number of doubly excited states may be seen. The energies of the observed features are in good agreement with the results of Madden and Codling (1963).

The sensitivity of the technique makes it particuarly suitable for studying the higher lying and smaller intensity doubly excited states. It is for these states, that both electrons have high values of principal quantum number where the effects of electron-electron correlations will be the most dominant. Excitation functions of the $He^+(n=2)$ and $He^+(n=3)$ Zubek et al (1988b) states are shown in figures 14 and 15.

Structure due to doubly excited states, that decay into these ion states is very evident in these spectra. The technique allows different ion states to be isolated and thus the decay of a doubly excited state into various final ion states to be studied. This in turn gives decay routes, and partial cross-sections and hence should elucidate classification schemes for the doubly excited states.

Future Developments

Future developments that suggest themselves are the application of coincidence techniques to the study of the photodouble ionization process. Such studies would enable the angular correlations of the two outgoing electrons to be measured and also the way in which the two electrons share the available energy. It would be essential to have photon beams of high flux for such coincidence studies and it seems likely that such sources will become available with the use of insertion devices on synchrotron radiation sources.

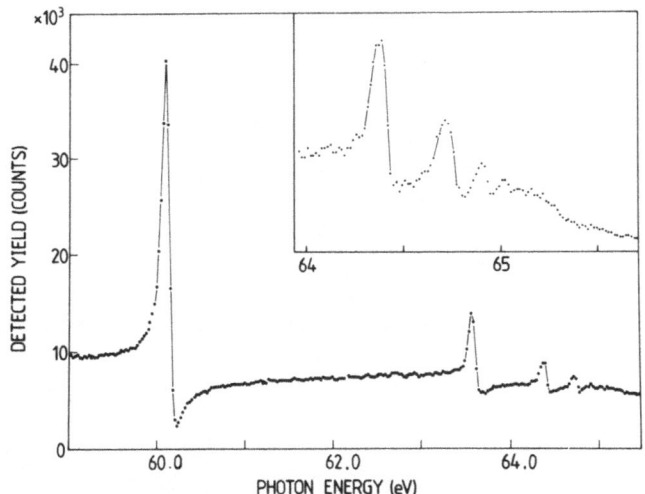

Figure 13   Excitation function of the He$^+$ (1s) state.   From Zubek et al (1988b)

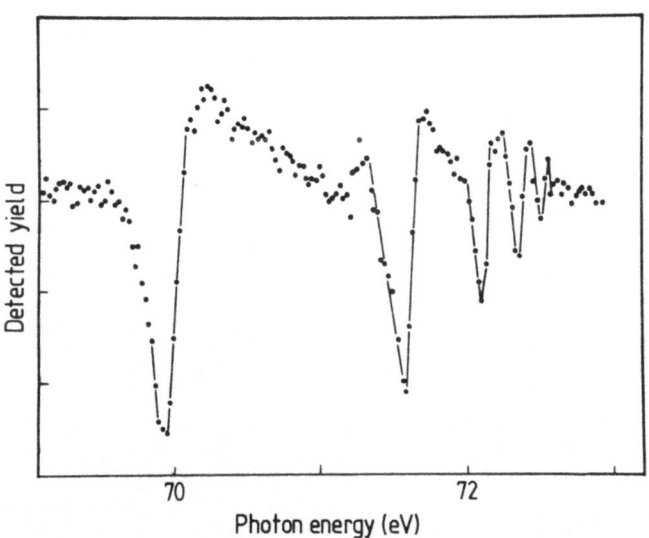

Figure 14   Excitation function of the He$^+$(n=2) states over the energy region just below the He$^+$(n=3) states.   From Zubek et al (1988b).

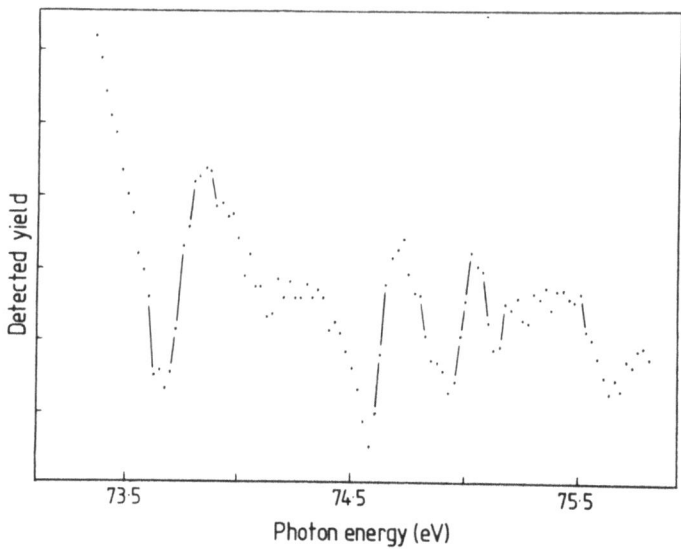

Figure 15   Excitation function of the He$^+$(n=3) states over the energy
region   just   below   the   He$^+$ (n=4)   states.   Zubek   et   al
(1988b).

Acknowledgements

It is a pleasure to acknowledge the workers involved in these
studies, including D and S Cvejanonić, S Buckman, P Hammond, D Holland,
A MacDowell, D S Newman, F H Read, P M Rutter, J B West and M Zubek.

References

Brunt J N H, Read F H and King G C, 1977 The realisation of high energy
resolution using the hemispherical electrostatic energy selector in
electron impact spectroscopy, J Phys E: Sci Instrum 10 134

Buckman S J, Hammond P, King G C and Read F H 1983, High resolution
electron impact excitation functions of metastable states of neon,
argon, krypton, and xenon, J Phys B: Atom Molec Phys 16 4219

Cooper J N, Fano U and Prats F 1963, Classification for two-electron
excitation levels of helium, Phys Rev Letts 10 518

Cvejanović S and Read F H 1974a,   Studies of threshold electron impact
excitation of helium, J Phys B: Atom Molec Phys 7 1841

Cvejanović S and Read F H 1974b, A new technique for threshold
excitation spectroscopy, J Phys B: Atom Molec Phys 7 1180

Delwiche J, Hubin-Franskin M J, Guyon P M and Nenner I 1981,
Autoionization of OCS by threshold photoelectron spectroscopy, J Chem
Phys 74 4219

Hall R I and Read F H 1984, Electron-Molecule Collisions, Ed I
Shimamura and K Takayanagi (New York: Plenum) 35

Hall R I, Cadez I, Schermann C and Tronc M 1977, Differential crossections for dissociative attachemtn in CO, Phys Rev A 15 599

Hammond P, King G C, Jureta J and Read F H 1985, The threshold spectrum of carbon monoxide, J Phys B: Atom Molec Phys 18 2057

Hammond P, Read F H and King G C 1988, J Phys B: Atom Molec Phys, to be published.

Herrick D R and Kellman M E 1980, Novel supermultiplet energy levels for doubly excited He, Phys Rev A 21 481

Iachello F and Rau ARP 1981, Group theoretical approach to two-electron correlations in atoms, Phys Rev Lets 47 501

King G C, Tronc M, Read F H and Bradford R C 1977, An investigation of the structure near the $L_{2,3}$ edges of argon, the $M_{4,5}$ edges of krypton and the $N_{4,5}$ edges of xenon using electron impact with high resolution, J Phys B: Atom Molec Phys 10 2479

King G C, Zubek M, Rutter P M and Read F H 1987, A high resolution threshold elctron spectrometer for use in photoionization studies, J Phys E: Sci Instrum 20 440

King G C, Zubek M, Rutter P M, Read F H, MacDowell A A, West J and Holland D M P 1988, A study of the threshold photoelectron spectrum of helium from the $He^+(n=2)$ state to the double ionization potential, J Phys B: Atom Molec Phys, to be published

Klar H and Klar M 1980, An accurate treatment of two-electron systems using hyperspherical co-ordinates, J Phys B: Atom Molec Phys 13 1057

Lipsky L, Anaria R and Connealy M J 1977, AE Data Nucl Data Tables 20 127

Madden R P and Codling K 1963, New autionizing atomic energy levels in He, Ne and Ar, Phys Rev Letts 10 516

Mazeau J, Greteau F, Joyez G, Reinhardt J and Hall R I 1972, Resonances in electron scattering from CO, J Phys B: Atom Molec Phys 5 1890

Nesbet R K 1975, Vibrational calcualtions of electron impact excitation of He near the 2^3S and 2^1S thresholds, Phys Rev A 12 444

Newman D S, Zubek M and King G C 1985, A study of resonance structure in mercury using metastable excitation by electron impact with high resolution, J Phys B: Atom Molec Phys 18 985

Newman D S, Zubek M and King G C 1983, Metastable excitation measurements in CO and N_2 by high resolution electron impact using a low work function detector, J Phys B: Atom Molec Phys 16 2247

Niehaus A 1978, Electronic and Atomic Collisions, Ed G Watel (Amsterdam: North Holland) 185

Peatman W B, Borne T B and Schlag E W 1969, Photoionization resonance spectra 1. Nitric oxide and benzene, Chem Phys Letts 3 492

Peatman W B , Kasting G B and Wilson D 1975, The origin and elimination of spurious peaks in threshold electron photoionization spectra, J Elec Spec Rel Phen 7 233

Peatman W B, Gotcher B, Gurtler P, Koch E E and Saile V 1978, Transition probabilities at threshold for the photoionization of molecular nitrogen, J Chem Phys 69 2089

Read F H 1976, The Physics of Electronic and Atomic Collisions, Ed J S Risley and R Geballe (Seattle: University of Washington Press) 176

Zubek M, King G C, Rutter P M and Read F H 1988, Photoionization function measurements in Helium, J Phys B: Atom Molec Phys, to be published

Zubek M 1986, A detector for metastable molecules with an adjustable detection threshold, J Phys E Sci Instrum 19 463.

Zubek M and King G C 1987, The observation of a new metastable state in mercury and the measurement of its excitation function by high resolution electron impact, J Phys B: Atom Molec Phys 20 1135.

Zubek M, King G C, Rutter P M and Read F H 1986a, Threshold photoionization studies in Xe and N_2, J Phys B: Atom Molec Phys: to be published.

ELECTRON SCATTERING IN MOLECULAR OXYGEN: A HIGH RESOLUTION STUDY USING

A SYNCHROTRON RADIATION PHOTOIONIZATION SOURCE

D. Field[*], G. Mrotzek[*], D.W. Knight[†], S. Lunt[‡] and
J.P. Ziesel[‡]

[*] School of Chemistry, University of Bristol, Bristol, U.K.
[†] Present address: Department of Science, Bristol Polytechnic
   Bristol, U.K.
[‡] SERC, Daresbury Laboratory, Warrington, U.K.
[‡] Bât.351, LCAM, Université Paris-Sud, Orsay, France

INTRODUCTION

The work described here is concerned with the low energy scattering of
electrons by molecular oxygen in the kinetic energy range zero to 1.3 eV.
Recent work by other groups in this area has been concerned with electron
transmission through oxygen (Stephen and Burrow 1986; Zecca et al.1986).
The present paper reports electron scattering by a supersonic beam of
oxygen and appear to be the first crossed-beam study since that of Linder
and Schmidt (1971).

Oxygen provides a very good test of the resolution of the incident
electron beam at low energies in our new apparatus since resonances in the
zero to one eV range are known to be narrow and each resonance exhibits
spin-orbit fine structure with a splitting previously measured to be
20±2 meV (Land and Raith 1974; Stephen and Burrow 1986). Theoretical
studies have predicted that the lifetimes of the scattering resonances may
be as long as 10ps or more and that rovibrational resonances may have widths
of less than 1 meV (Herzenberg 1969; Koike and Watanabe 1973; Parlant and
Fiquet-Fayard 1976). Of course we cannot expect to observe resonance linewidths of less than 1 meV experimentally since the incident electron beam
is certain to be of very much poorer energy definition. Experimental
resonance widths are therefore very likely to be instrumental in origin and
we may expect only to observe an envelope of overlapping rovibrational
lines. Nevertheless our goal has been to observe rotational phenomena in
our. spectra and we have found that the resolution of the photoionisation
source described below has been sufficient for this purpose.

EXPERIMENTAL

Electrons are produced in our system by using synchrotron radiation
to photoionise Ar. The apparatus was sited at the Daresbury Synchrotron
Radiation Source (SRS) at SERC Daresbury Laboratory. The electron source
is closely similar in design to that described in Field et al. 1984. We
chose to use a photoionisation source since conventional analysers have
proved to have a limiting resolution of the order of 10 meV. We have
coupled our photoionisation source with a supersonic molecular beam target

and it is a crucial feature of the experiment that this target gas has a
low rotational temperature in order that a comparatively small number of
rovibrationally inelastic resonances contribute to our scattering signals.

II(i) The Electron Source

A schematic diagram of the apparatus is shown in figure 1. Synchrotron
radiation (hν) enters the photoionisation cell containing argon at a typical
pressure of 10 mtorr. The radiation itself emerges from the exit slit of
a 5-metre MacPherson monochromator fitted with a 1200 l/mm holographic
grating. The radiation was focussed into the photoionisation cell using
a 4:1 demagnifying ellipsoidal mirror. In order to obtain a maximum signal
of photoelectrons it is necessary to straddle the Ar ionisation threshold
at 786.72 Å (Yoshino 1970; Moore 1971). Electrons were withdrawn from the
photoionisation cell using an electric field of 0.4 or 0.2 V/cm and
accelerated to an energy of 1 eV at element 4. Beyond this point electrons
entered a four-element lens (L1 to L4) (Martinez et al. 1983) which acted
to focus the electrons onto the supersonic molecular beam. Electrons
scattered from the molecular beam were detected over an angular range 60°
to 120° using carefully shielded channel plates sited 18 mm from the
scattering centre. In the experiments described there was no energy
analysis of the scattered electrons and we do not discriminate between
elastic and inelastic events. The apparatus has now been upgraded to
include energy analysis of the scattered electrons using an electron energy
analyser.

For high resolution we require that the interaction region where the
electrons strike the molecular beam is subject to an electric field of less
than 2 to 3 meV/cm. To achieve this we surrounded the scattering centre
with an Aerodag-coated aluminium box. We found no noticeable deterioration
of the resolution below the theoretical value (see section V) and therefore
none which could be attributed to the presence of electric fields in the
interaction region. With regard to applied potentials the two critical
values are those of the region of photoionisation (at zero V) and the
scattering centre. 16-bit stability voltage supplies (HYTEC 650) were used
for the front and back plates of the photoionisation cell (S1 and S3) and
for the fourth element of the lens (L4), the box structure and various other
close-by elements. Other voltages in the system were set up using 12-14
bit stability supplies (GEC 1216V). The apparatus was magnetically shielded
with a double mu-metal shield.

The molecular beam source (Campargue 1984) was built under the direction of Professor Campargue. We have not attempted to measure the beam
kinetic or rotational temperatures in our present work. We know however
from experiments in the laboratory of Professor Campargue that beams of
this sort have kinetic temperature of 15-20K for unseeded oxygen and around
5K if, as in some of our experiments, we seed oxygen into helium.

II(ii) Experimental Operating Conditions

With a beam energy of 2 GeV collecting up to 5 mrad vertically and 5
mrad horizontally the SRS yields 2.3×10^{10} photon s^{-1} at 786.7 Å measured
using a calibrated Al diode with 331 mA in the ring and 125 μm slitwidths
on the 5 m MacPherson monochromator (0.2 Å resolution). Placing a 2 mm
diameter disc at the scattering centre we measured an electron current of
1 pA at the scattering centre for an electron kinetic energy of 500 meV.
The efficiency of the source in terms of the ratio of number of electrons
produced in the source and those perfectly focussed with unit magnification
at the molecular beam is at present ∿10% at 500 meV energy. The low

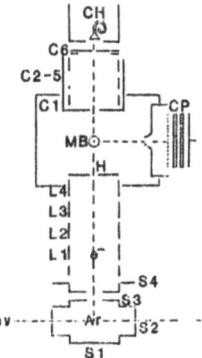

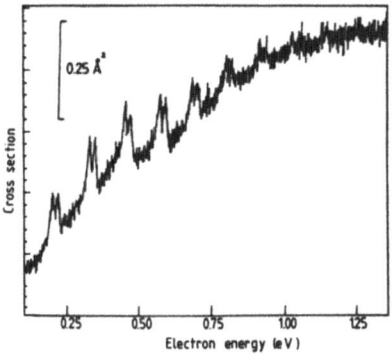

Fig.1. The apparatus: hν = synchrotron radiation, S1,S2,S3 = elements of the photoionisation cell, L1-4 electron lens, H = 5 mm collimating hole, MB = molecular beam, CP = channel plates, C1-C6 = electron collection region, CH = channeltron.

Fig.2. The variation of the scattered electron count rate with electron kinetic energy. Photon bondwidth $\Delta(h\nu)$ = 0.2 Å, field across photoionisation cell (E) = 0.4 V/cm, beam stagnation pressure (P_o) = 1300 mbar.

currents are offset by the high beam densities which typically approach and may exceed 10^{13} cm^{-3} in the scattering region.

III Experimental Results

The variation of the scattered electron count rate with electron kinetic energy over the range of angles 60-120° is shown in fig.2, for the energy range 0.1 to 1.3 eV. Resonant structure arises through the capture by O_2 of an incoming electron at specific energies. The resulting O_2^- decays to yield rotationally and vibrationally excited O_2. The resonances are designated by the vibrational state of the O_2^- intermediate formed in the collision. Levels of O_2^- with $v' \leqslant 3$ are bound (Celotta et al. 1972) and the first scattering resonance should therefore appear for $v' = 4$. This lies at 91 meV (Land and Raith) but we do not record this resonance since it lies a little below the practical limit of the apparatus. The relative heights of the resonances in fig.2 agree well with those given by other groups with a 10-15% error introduced at low energies (<350 meV) due to masking at a 5 mm collimation hole in L4. From data in fig.2 we find values for ω_e and $\omega_e x_e$ for O_2^- to be 136±0.5 meV and 1.0±0.5 meV in agreement with the values given in Stephen and Burrow and Linder and Schmidt.

Each of the resonances in fig.2 is split by spin-orbit coupling since the O_2^- intermediate is in a $^2\Pi_g$ electronic state. This spin-orbit splitting is well-known from the time-of-flight transmission studies of Land and Raith and from the work of Stephen and Burrow. However these splittings have never been observed before in the scattered electron signal and their identification was an encouraging sign that high resolution had been achieved in the incident electron beam. In the course of our experiments we have observed some remarkable effects due to differing rotational populations in the target beam. For example we find in fig.3 two traces of the $v' = 6$ resonance taken one after the other under identical conditions (see figure caption) save that in the upper trace the nozzle-skimmer distance was 8 mm and in the lower 15 mm. There is a striking improvement in resolution in the lower trace. It is known (private communication, Professor R. Campargue) that a nozzle-skimmer distance of 8 mm yields a very poor beam of high kinetic temperature and increased angular divergence, the

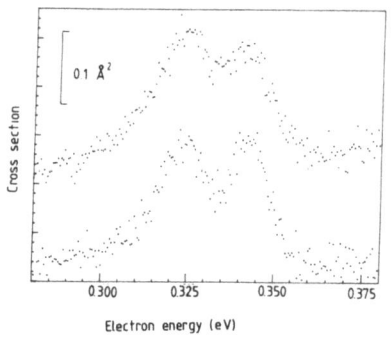

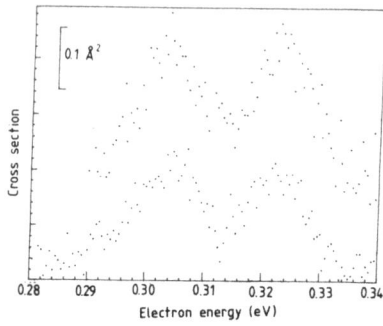

Fig.3. Two traces of the v' = 6 resonance with non-resonant background removed. Upper trace nozzle-skimmer distance (X_s) = 8mm, in lower = 15 mm. $\Delta(h\nu)$ = 0.2 Å, E = 0.4 V/cm, P_o = 1300 mbar

Fig.4. Two traces of the v' = 6 resonance: upper P_o = 1000 mbar, lower P_o = 800 mbar. X_S = 15 mm, $\Delta(h\nu)$ = 0.1 Å, E = 0.4 V/cm.

latter creating little spectral broadening (1.6 meV: Stephen and Burrow). The change in appearance of the spectra therefore arises from the very much lower rotational temperature of the beam with a 15 mm nozzle-skimmer distance. In fig.4 the spectra are increasingly sharpened by reducing the stagnation pressure to 1000 and 800 mbar. A further sharpening of the spectrum may be obtained by seeding oxygen in helium to 30% with a total stagnation pressure of 1500 mbar. The resulting spectrum (for conditions specified in the caption) is shown in fig.5 and shows the second peak in the doublet with a width of ∿5.5 meV (FWHM), which we believe to be the narrowest peak ever observed in gas phase electron scattering. The doublet splittings allow us to derive a value for the spin-orbit coupling constant for O_2^- of 18.75±1.0 meV in agreement with values given in Stephen and Burrow and Land and Raith. Our absolute energy scale has not been calibrated but agrees within 10-20 meV with scales given by these same authors.

IV Brief Discussion of Results

The questions which we pose are (i) what was the instrumental energy resolution in the primary electron beam, (ii) what partial waves contribute to the scattering, (iii) what limit is set by our data on the natural widths of the scattering resonances? We have attempted to answer these

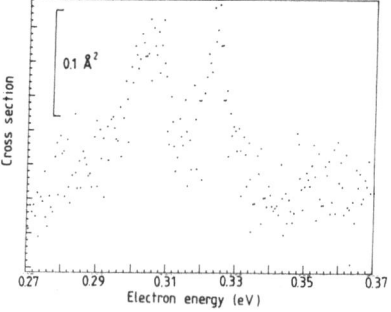

Fig.5. v' = 6 resonance: O_2 seeded in He. $\Delta(h\nu)$ = 0.2 Å, E = 0.2 V/cm, P_o (O_2 + He) = 1500 mbar, X_S = 15 mm.

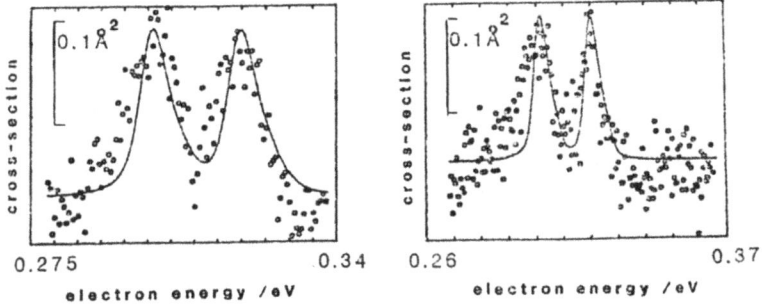

Fig.6. Simulation of lower trace of fig.4 and of fig.5. γ_e = 3.5 meV, γ_r = 0.1 meV, ℓ = 2,4,6 equal in contribution.

questions by modelling the resonance structure using theoretical expressions for relative rotational cross-sections (irrespective of initial v and final v ≡ v') taken from Fiquet-Fayard 1975. Various difficulties hamper us (i) there exists at present no good ab initio theory to describe electron scattering by O_2 though this should soon be set right by work in progress of Noble and Burke, (ii) as is evident from the data of Linder and Schmidt only data for v' = 5 and 6 may be treated as vibrationally elastic and can therefore be represented by a single resonance profile. We have chosen for this reason only to model our spectra for v' = 6 for which we have the most information, (iii) we do not have a precise knowledge of the rotational populations of the molecular beam. We have used a two-temperature model to calculate rotational populations with parameters taken initially from Venkateshan et al. 1982 with a predominant lower temperature component with T = 20K (Gallagher and Fenn 1974; Gaveau 1984) for unseeded beams.

Using a Gaussian distribution of kinetic energies in the electron beam of width γ_e and Lorentzian lineshapes for transitions to O_2^- of width γ_r we may compute the doublet envelope and fit our computations to our observations by choosing suitable values of γ_e and γ_r. Setting the latter to be negligibly small and equal to 0.1 meV we can reproduce our spectra approximately using γ_e = 3.5 meV with equal contributions to the scattering from partial waves with ℓ = 2,4 and 6, making some adjustments to the parameters taken from Venkateshan et al. for the rotational populations. Fits to the unseeded and seeded beam data of fig.4 (lower trace) and fig.5

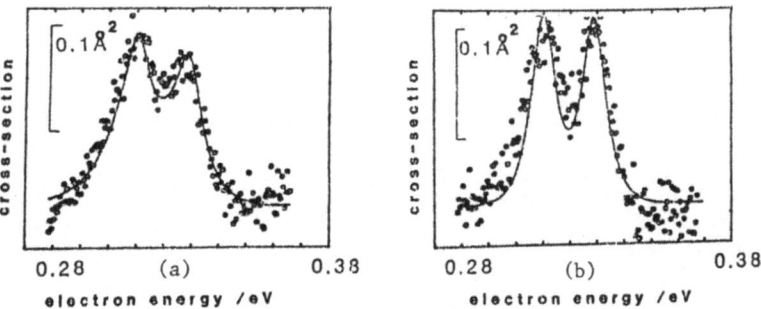

Fig.7. Simulation of fig.3 upper trace in (a) with T_R = 273 K and lower trace in (b) with T_R = 100 K. γ_e = 3.5 meV, γ_r = 2.0 meV, ℓ = 2,4,6 in ratio 1:0.67:0.33.

are shown in fig.6. The difference in widths of the two halves of the spin doublet, apparent in the seeded beam data of fig.5, cannot be reproduced by theory at present. An alternative parameter set involving γ_e = 3.5 meV, γ_r = 2.0 meV and contributions of ℓ = 2,4 and 6 in the ratio 1.0:0.67:0.33 yields an equally reasonable fit to the experimental data. We are therefore unable to distinguish between these two alternatives, in the first place an intermediate of long lifetime (= 6 ps) with equal contributions from partial waves, ℓ = 2,4 and 6 and secondly a shorter O_2^- lifetime of 0.3 ps with decreasing contributions from higher partial waves. In passing one may note that either alternative reproduces quite accurately the higher temperature data of fig.3 as may be seen in fig.7.

V CONCLUDING REMARKS

From the analysis of our experimental data it would appear that the energy spread in the incident electron beam was $\sim$3.5 meV FWHM. The energy spread in the photon bandwidth (for the data of fig.5) is $\sim$4.0 meV and thus the photoionisation source yielded an electron beam of energy width dictated by the photon bandwidth, the design intention. This has enabled us to observe rotational phenomena in electron-molecule scattering at near-rotational resolution. Data of this kind coupled with theoretical studies can yield valuable information with regard to the dominant partial waves contributing to the scattering. Combining this information with angle-resolved data and including energy analysis of the scattered electrons should provide a deeper understanding of the dynamics of electron-molecule encounters in the low energy regime.

REFERENCES

Campargue, R., 1984 , J.Phys.Chem., 88:4466.

Celotta, R.J., Bennett, R.A ., Hall, J.L. Siegel, M.W. and Levine, J. 1972, Phys.Rev., A6:631.

Field, D., Ziesel, J.P., Guyon, P.M. and Govers, T.R., 1984, J.Phys.B: At.Mol.Phys., 17:4565.

Fiquet-Fayard, F., 1975, J.Phys.B At.Mol.Phys., 8:2880.

Gallagher, R.J. and Fenn, J.B., 1974, J.Chem.Phys., 60:3487.

Gaveau, M.A., Ph.D. Thesis, Paris XI - Orsay, 1984.

Herzenberg, A., 1969, J.Chem.Phys., 51:4920.

Huber, K.P. and Herzberg, G., 1979, Constants of Diatomic Molecules, Van Nostrand Reinhold Co.

Koike, F. and Watanabe, T., 1973, J.Phys.Soc. Japan 34:102.

Land, J.E. and Raith, W., 1974, Phys.Rev. A9:1592.

Linder, F. and Schmidt, H., 1971, Z.Naturf., 26a:1617.

Martinez, G., Sancho, M. and Read, F.H., 1983, J.Phys.E: Sci.Instr. 16:631.

Moore, C.E., 1971, 'Atomic Energy Levels', Natl.Stand.Ref.Data Ser. Natl. Bur.Stand. 35.

Parlant, G. and Fiquet-Fayard, F., 1976, J.Phys.B.: At.Mol.Phys., 9:1617.

Stephen, T.M. and Burrow, P.D., 1986, J.Phys.B.: At.Mol.Phys., 19:3167.

Ventkateshan, S.P., Ryali, S.B. and Fenn, J.B., 1982, J.Chem.Phys., 2599.

Yoshino, K., 1970, J.Opt.Soc.Am., 60:1220.

Zecca, A., Brusa, R.S., Grisenti, R., Oss S. and Szmythowski, C., 1986, J.Phys.B.: At.Mol.Phys., 19:3353.

NEGATIVE ION RESONANCE ELECTRON SCATTERING FROM

ORIENTED, PHYSISORBED O_2

R. E. Palmer, P. J. Rous and R. F. Willis

Cavendish Laboratory
Madingley Road
Cambridge CB3 0HE, UK

ABSTRACT

The physisorption of diatomic molecules on a crystalline solid surface at low temperature provides a means of orientating the molecular axis and thus studying low energy electron scattering from an oriented molecule. The cross-section for vibrational excitation of one monolayer of O_2 on graphite at 25K has a peak near 9eV electron energy. The angular distribution of vibrationally inelastic electrons has been measured; comparison with calculated distributions, which include interference effects arising from multiple elastic electron scattering within the molecular layer, allows us to determine the orientation of the O_2 molecule on the surface and to identify the partial wave content of the resonance. The dominant partial wave ($p\pi$) is consistent with the $^2\Pi_u$ compound state but not the $^4\Sigma_u$.

Molecules can be bound on a surface in two ways:
(i) chemically (chemisorption), involving electronic charge transfer and often molecular dissociation - binding energies are typically a few eV;
(ii) by weak Van der Waals polarisation forces (physisorption), where the molecule adsorbs intact and without charge transfer (binding energies typically ~50meV).
In this paper we address the physisorption regime, where much of the molecular identity is retained, but the surface acts as a template which organises the molecules into two dimensional solid and fluid phases. Here there is a meeting point between the domains of atomic and molecular physics on the one hand and solid state physics on the other.

It has been known for some years now that the negative ion resonances which dominate the cross-sections for vibrational excitation of diatomic molecules in the gas phase by low energy electrons ($\leq$20eV) [1] are also observed when the molecules are physisorbed [2-8]. What effects does the surface have? The attractive interaction between the temporary negative ion and its positive image in the (metallic) surface tends to lower the resonance energy by about 1eV [2,9], and also to decrease the resonance lifetime by lowering the centrifugal barrier [9,10]. Calculations by Gerber and Herzenberg [9] indicate also that the breaking of the inversion symmetry of a homonuclear diatomic molecule by the presence of the surface can allow partial waves of lower l to be mixed into the wave functions of the incoming and outgoing electron waves which couple to the resonant state. The latter phenomenon parallels the difference in partial waves coupling to the N_2 $^2\Pi_g$ and CO $^2\Pi$ gas phase resonances, which is manifest in the different angular distributions of inelastically scattered electrons which are observed [1]. One would expect, then, that the experimental angular distributions obtained in the case of physisorbed molecules would prove a sensitive probe of the symmetry of the resonance on the surface. Here we present the first such measurements, for O_2 physisorbed on graphite. We expect also that the angular profiles measured will reflect the orientation of the molecular axis on the surface, and hope indeed to be able to work backwards and derive the orientation from the angular distributions observed.

An electron can be added in various ways to the $^3\Sigma_g$ ground state of neutral O_2 to form a negative ion. In the 6-11eV energy range where our angular measurements have been made, calculations for the isolated O_2 molecule [11] show two resonances intercepting the Franck-Condon region which can be formed by electron addition to the $^3\Sigma_g$ ground state without violating the selection rules governing the attachment process [12]. These are the $^2\Pi_u$ (Feshbach) resonance and the $^4\Sigma_u$ (shape) resonance. The former has been assigned a dominant contribution to the dissociative attachment cross-section for both gas-phase [1] and condensed O_2 [13] - in each case the cross-section has a peak near 7eV. Wong et al [14], who observed a broad peak in the vibrational excitation cross-sections for gas phase O_2 centred near 9.5eV, reasoned that the dominant contribution to this enhancement was probably due to the $^4\Sigma_u$ resonance. They thought a major contribution to the vibrational enhancement from the $^2\Pi_u$ state was unlikely, given the lower energy of the peak in the dissociative attachment cross-section. For the purposes of our analysis of the vibrationally inelastic angular profiles presented later, we have treated both these resonant states as possible causes of the vibrational excitation.

When physisorbed on graphite at low temperatures, O_2 forms a whole range of different phases, depending on the number of molecules on the surface and on the temperature. The system has been the subject of extensive investigation by a variety of techniques which have provided details about the phase boundaries and the lattice parameters in the various phases. The reader is referred to a paper by Toney and Fain [15] for references to this work, and for a phase diagram. The phase we have studied is the $\zeta 2$ phase, prepared by methods described in detail elsewhere [16], the most dense monolayer phase. Molecular packing considerations, based on the lattice constants of the $\zeta 2$ phase and on the charge density distribution of the unperturbed O_2 molecule [17], suggest that the molecular axis in this phase is vertical or close to vertical [18], in contrast, for example, to the lower coverage δ phase for which a parallel orientation has been proposed [19]. The validity of this type of molecular packing argument can only be checked, however, by an experimental determination of the orientation of the molecular axis, something which has been lacking to date.

The experiments reported used a hemispherical electron monochromator and similar analyser. The monochromator was fixed on a flange but the analyser was free to rotate on a turntable. The graphite sample was mounted on a continuous flow helium cooled cold finger, also rotatable, and was resistively heated to ~1300K to remove contaminants from the surface prior to cooling and adsorption of O_2. Details of the sample stage may be found elsewhere [20]. The type of graphite used was Highly Oriented Pyrolitic Graphite (HOPG), an artificial material formed by cracking hydrocarbons over a heated substrate [21]. Crystals of HOPG have the advantage of being larger than single crystals of natural graphite, but the disadvantage of being polycrystalline, in the sense that they consist of crystallites typically 1μm in diameter, the c-axes of which are parallel to within 1°, but which have complete azimuthal disorder. For a single crystal, the hexagonal symmetry of the surface would lead to contributions to the electron scattering from six adsorbate domains; for HOPG, there will be contributions from a complete azimuthal range of domain directions.

An electron energy loss spectrum obtained from a monolayer of the $\zeta 2$ phase of O_2 on graphite can be seen in Fig. 1(a). The frequencies of the vibrational transitions labelled are in all cases close to the values in the gas phase, reflecting the absence of charge transfer in the physisorption interaction. This spectrum was obtained with an electron beam energy of 8.5eV, where the vibrational modes are especially strong, as Fig. 1(b) demonstrates. Here the intensity of the fundamental mode is plotted against electron beam energy, and shows a broad resonance centred at about 9eV. Previously [22], noting that the energy of this peak is about 0.5eV down from the gas phase peak, consistent with

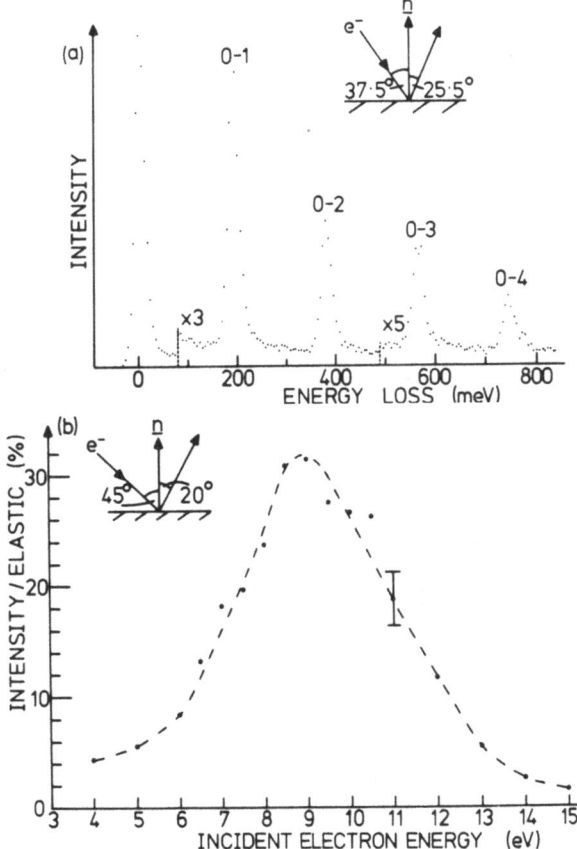

Fig. 1 (a) Electron energy loss spectrum obtained from a monolayer of O_2 on graphite at a sample temperature of 25K. Scattering geometry shown inset. The loss peaks correspond to vibrational excitations of the O_2 molecule and are labelled accordingly. (b) The intensity of the v = 0-1 loss peak (normalised to the elastic peak intensity) plotted as a function of incident electron beam energy for the scattering geometry shown inset.

an image induced shift, and taking on board the gas phase assignment ($^4\Sigma_u$) [14], we attributed the enhancement in Fig. 1(b) to the $^4\Sigma_u$ shape resonance. The analysis of the angular distribution data we have obtained, however, has forced us to question this assignment, as we shall see.

By rotating the electron analyser and measuring the intensity of the v = 0-1 loss peak as a function of angle, we have been able to obtain angular profiles of the electrons ejected out of the resonance centred near 9eV [16]. Figure 2 shows two such profiles, for two different angles of incidence (65° and 50° with respect to the normal to the crystal). In each case there is a concentration of intensity away from the specular direction towards the surface normal, with a peak at 15° from the normal. How do we interpret these profiles? In the case of both the $^4\Sigma_u$ and $^2\Pi_u$ states the partial wave of lowest l consistent with the symmetry of the resonance is a p wave (pσ and pπ, respectively). Emission of an electron from the temporary negative ion into such a partial wave followed

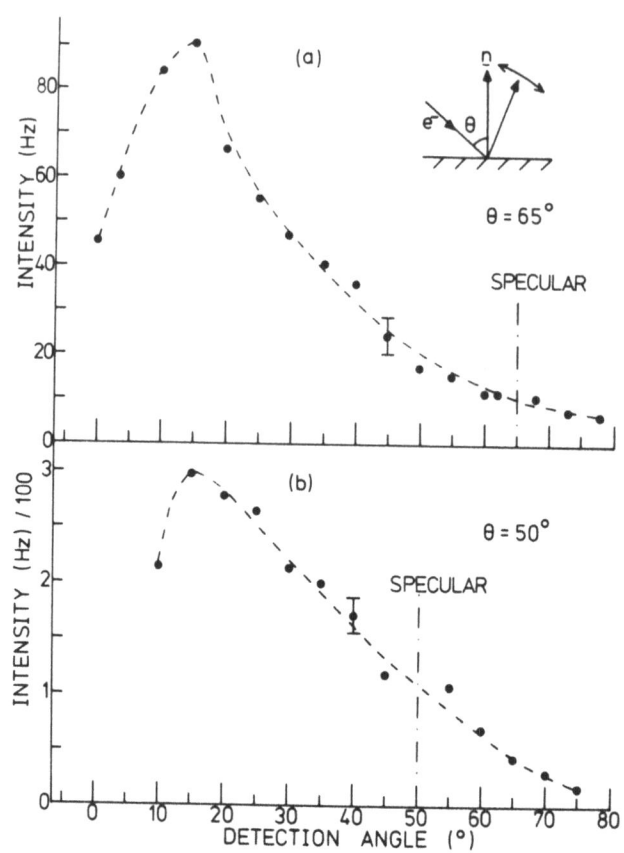

Fig. 2 Angular distributions of electrons detected after exciting the v = 0-1 transition in a monolayer of O_2 on graphite. Angle of incidence (a) 65° to the surface normal, (b) 50°. Incident electron beam energy 8.5eV in each case. The temperature was 23K in (a) and 24K in (b).

by detection would give, respectively, $\cos^2\theta$ and $\sin^2\theta$ angular distributions with respect to the molecular axis [23]. But the profiles in Fig. 2 have much sharper peaks than these functions, and, moreover, they are not symmetric about the peak position - this is especially evident in Fig. 2(b). It turns out that there are two factors which complicate such a picture of the angular distributions: (i) elastic electron scattering in the molecular layer, and (ii) averaging over all the domains on the HOPG surface. We address these issues in turn.

(i) A molecule physisorbed on a solid surface is not an isolated molecule. Consider an electron wave emitted out of a temporary negative ion state. Although initially a single partial wave state may be sufficient to describe the wave, the wave can be elastically scattered by both the atoms in the surface and the other adsorbed molecules, once or many times, before ultimately reaching the electron detector. These waves will interfere and they have to be coherently summed before one can obtain the angular distribution which will be detected.

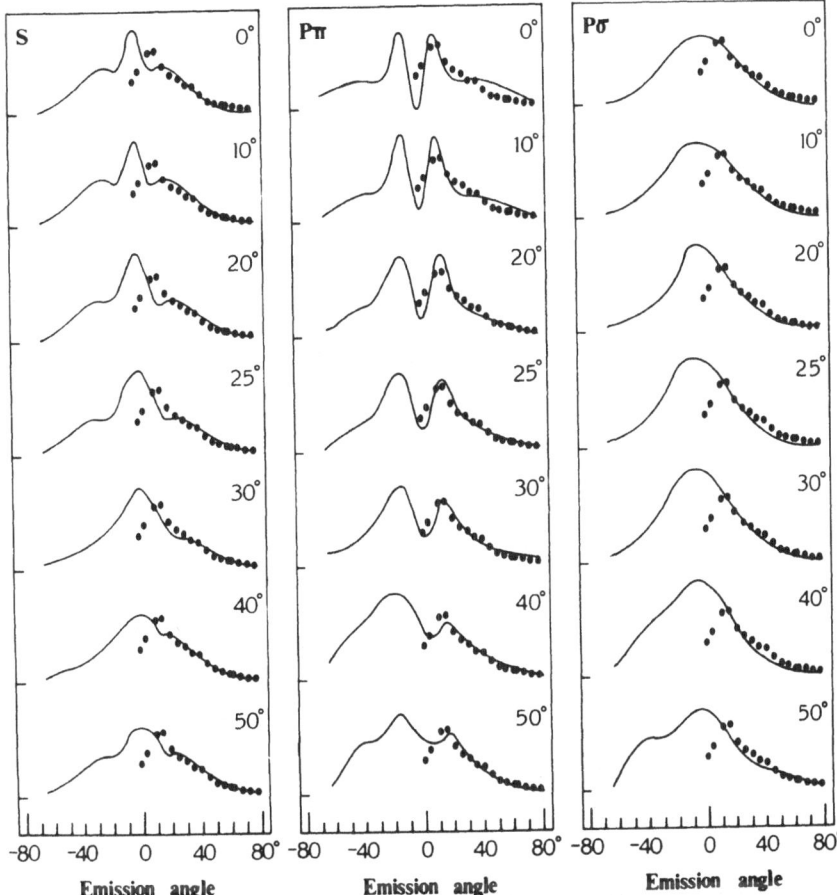

Fig. 3 Calculated angular distributions (solid curves) for the ζ2 phase of O_2 on graphite compared with experimental results (filled circles) for an incident electron angle of 65° to the normal and incident electron energy 8.5eV (i.e. experimental points as in Fig. 2(a)). The three panels show calculated profiles for electron capture/emission by a temporary negative ion via three different partial waves, s, pπ, and pσ, as a function of the tilt of the O_2 molecular axis from the vertical in the range 0-50°.

(ii) Because of the different domains present on the HOPG surface, one needs to account for more than one molecular orientation (except if the molecular axis is exactly vertical), even though within a given domain (for the ζ2 phase of O_2 on graphite) the low energy electron diffraction pattern (LEED) shows each molecule has the same orientation [18]. Thus, although the emitted electron pattern from a molecule of given orientation cannot depend on the incident electron direction, since the captured electron "forgets" where it came from [23], the capture cross-sections for molecules in different domains will not be the same, i.e. each domain contributes a different amount to the detected electron intensity, the relative amounts depending on the incident angle.

We have calculated the angular distributions expected for O_2 molecules in the ζ2 phase on graphite taking account of these two factors. Multiple elastic scattering within the O_2 layer [24] was handled with standard LEED-type wavefunctions [25], using a

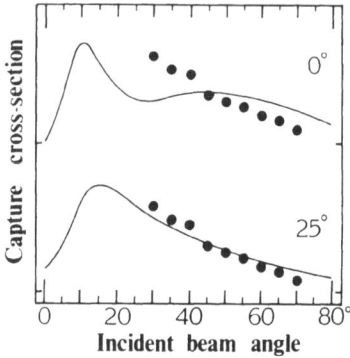

Fig. 4 Experimental and calculated electron capture cross-sections for a monolayer of O_2 on graphite. The experimental points (filled circles) represent the intensity of the v = 0-1 transition observed at a fixed angle (32.5°) to the crystal normal as a function of the angle of incidence of the electron beam. The temperature was 24K. The calculated curves (solid lines) are for electron capture/emission via a $p\pi$ partial wave and for two different molecular orientations, vertical (upper curve) and a tilt of 25° from the vertical (lower curve).

differential cross-section for elastic scattering from the O_2 molecule calculated by the method of Dill and Dehmer [26]. Full details of these calculations will be given elsewhere [27]. The LEED state was matched to a partial wave state describing the electron when initially emitted from the O_2 molecule. Calculations were done for a wide range of molecular tilt angles and for several partial waves. To describe emission from the $^4\Sigma_u$ state we used a $p\sigma$ wave, and also tried an s wave to allow for the breaking of inversion symmetry (i.e. dropping of the "u" label) because of the presence of the surface. For the $^2\Pi_u$ state we used a $p\pi$ wave (which cannot couple to an s-wave because $|m_l| = 1$). Figure 3 shows some of the angular profiles calculated for an incident angle of 65°, compared in each case with the corresponding experimental profile. It is easy to see that none of the profiles for the $p\sigma$ or s match the experimental data for any orientation of the O_2 molecule. On the other hand, the profile calculated for $p\pi$ emission and for a molecular tilt in the region of 25° from the normal to the surface gives what we consider to be very good agreement. Thus we are able to reach two important conclusions: (i) only emission into a $p\pi$ partial wave can account for the experimentally determined angular distributions - this is consistent with the $^2\Pi_u$ negative ion state but not the $^4\Sigma_u$ state; (ii) the molecular axis of the O_2 molecule is oriented at 25° (we estimate to ±5°) from the vertical.

A check on these conclusions can be made by measuring and calculating the angular dependence of the cross-section for electron *capture* into the resonance in addition to the electron *emission* cross-section. This was obtained experimentally by rotating the crystal and analyser in step and measuring the v = 0-1 mode intensity for a fixed detection angle with respect to the crystal, and is presented in Fig. 4. Because of the more restricted angular range available for this measurement than for the emitted electron distributions of Figs. 2 and 3, it is not such a sensitive test of the resonant state symmetry. Nevertheless, the best agreement with the calculations [27] is obtained for the same partial wave, $p\pi$, and the same molecular orientation, 25° from the vertical. The calculated profile for this case is shown in Fig. 4, together with a profile for the same partial wave and an upright orientation for comparison. Further details of the analysis of the capture cross-section is given in ref. [27].

We found the result that the $^2\Pi_u$ negative ion, rather than the $^4\Sigma_u$ state, was consistent with the observed angular distributions something of a surprise, given the apparently reasonable argument for the $^4\Sigma_u$ advanced by Wong et al [14] in their interpretation of the gas phase resonance profile. We suggest a couple of possible explanations of the result. We could be observing the same resonant state as the gas-phase workers, in which case their assignment ($^4\Sigma_u$) would be wrong. If one accepts that the $^2\Pi_u$ state dominates dissociative attachment at 7eV, then the $^2\Pi_u$ state would have to manifest itself at different energies depending on the decay channel. Alternatively, vibrational excitation on the graphite surface could be dominated by a different resonance than in the gas phase; this might be a consequence of the lifetime of a temporary negative ion state depending on its charge distribution in relation to the surface and co-adsorbed molecules, which will vary from state to state. The present results clearly do not distinguish these or other possibilities; they merely identify the partial wave involved in the scattering. We believe that detailed calculations of the identity of the vibrational excitation resonance near 9eV, at least for the isolated molecule, are needed before one can resolve the issue.

Acknowledgements

We are most grateful to John Wilkes for his help with the experimental work. We thank the UK Science and Engineering Research Council for financial support and a Research Fellowship for PJR. REP thanks the Royal Commission for the Exhibition of 1851 for support through a Research Fellowship.

REFERENCES

1.   George J. Schulz, Rev. Mod. Phys. 45:423 (1973).
2.   J. E. Demuth, D. Schmeisser and Ph. Avouris, Phys. Rev. Lett. 47:1166 (1981).
3.   L. Sanche and M. Michaud, Phys. Rev. Lett. 47:1008 (1981).
4.   D. Schmeisser, J. E. Demuth and Ph. Avouris, Phys. Rev. B26:4857 (1982).
5.   L. Sanche and M. Michaud, Phys. Rev. B27:3856 (1983).
6.   J. E. Demuth, Ph. Avouris and D. Schmeisser, J. Electron Spec. 29:163 (1983).
7.   L. Sanche and M. Michaud, Phys. Rev. B30:6078 (1984).
8.   R. E. Palmer, J. F. Annett and R. F. Willis, J. Electron Spec. 38:317 (1986);
     R. E. Palmer, PhD Thesis, University of Cambridge, 1986.
9.   A. Gerber and A. Herzenberg, Phys. Rev. B31:6219 (1985).
10. J. W. Gadzuk, J. Chem. Phys. 79:3982 (1983).
11. M. Krauss, D. Neumann, A. C. Wahl, G. Das and W. Zemke, Phys. Rev. A7:69
     (1973); G. Das, A. C. Wahl, W. T. Zemke and W. C. Stwalley, J. Chem.
     Phys. 68:4252 (1978).
12. G. H. Dunn, Phys. Rev. Lett. 8:62 (1962).
13. L. Sanche, Phys. Rev. Lett.53:1638 (1984); R. Azria, L. Parenteau and L. Sanche,
     Phys. Rev. Lett. 59:638 (1987).
14. S. F. Wong, M. J. W. Boness and G. J. Schulz, Phys. Rev. Lett. 31:969 (1973).
15. Michael F. Toney and Samuel C. Fain, Jr., Phys. Rev. B36:1248 (1987).
16. R. E. Palmer, P. J. Rous, J. L. Wilkes and R. F. Willis, Phys. Rev. Lett., to be
     published in Jan. or Feb. 1988.
17. R. F. W. Bader, W. H. Henneker and P. E. Cade, J. Chem. Phys. 46:3341 (1967).
18. Michael F. Toney and Samuel C. Fain, Jr., Phys. Rev. B30:1115 (1984).
19. P. A. Heiney, P. W. Stephens, S. G. J. Mochrie, J. Akimitsu, R. J. Birgeneau and
     P. M. Horn, Surface Sci. 125:539 (1983).
20. R. E. Palmer, P. V. Head and R. F. Willis, Rev. Sci. Instrum. 58:1118 (1987).
21. M. S. Dresselhaus and G. Dresselhaus, Advanced in Phys. 30:139 (1981).
22. R. E. Palmer, J. L. Wilkes and R. F. Willis, J. Electron Spec. 44:229 (1987).
23. J. W. Davenport, W. Ho and J. R. Schrieffer, Phys. Rev. B17:3115 (1978).
24. Because the O_2 layer is incommensurate with the graphite surface, we are not able to
     calculate multiple scattering in the substrate. However, the effective reflectivity

of graphite (seen by the overlayer) that we calculate at 8.5eV is less than 1%, so this is not an important omission.

25. J. B. Pendry, "Low Energy Electron Diffraction", Academic Press, London , 1974.
26. D. Dill and J. L. Dehmer, J. Chem. Phys. 61:692 (1974).
27. P. J. Rous, R. E. Palmer and R. F. Willis, submitted to Phys. Rev. B.

ROVIBRATIONAL THRESHOLD STRUCTURES IN e-HCl AND e-HF COLLISIONS

G. Knoth, M. Rädle, H. Ehrhardt and K. Jung

Fachbereich Physik der Universität Kaiserslautern

D - 6750 Kaiserslautern, West Germany

ABSTRACT

Rovibrational excitations of HCl and HF by low energy (0.3 eV
to 5.5 eV) electron impact have been measured in a cross beam
apparatus in the angular range from 15° to 135°. Rather strong
rotational excitations are observed in the threshold peak in
the v = 1 channel with j_t = 0,1,2 and minor contributions from
j_t = 3 and 4. Non-isotropic angular distributions have been
obtained right at threshold, also indicating participation of
mostly s-, p,- and d-waves.

In the vibrationally elastic channel below the cusp due to the
v = 1 channel no Feshbach-type resonance structure has been
found.

In 1976, Rohr and Linder [1] discovered in the electron impact spectra
of HCl and HF very large peaks close to the thresholds of vibrational
excitation in the electronic ground state of the molecules. Within the
angular range of their measurements they obtained isotropic angular
distributions of the scattered electrons, if the collision energy was
tuned to the maximum of the threshold peaks.

Also the dissociative attachment cross sections show large steps [2],
which correspond with the threshold peaks of the vibrational excitation
channels of the target molecule.

In the last ten years many theoretical papers have been published, which
deal with these narrow threshold peaks. The first approach [3] came to
the conclusion, that a quasibound state (virtual state) in the electron-
molecule potential was responsible for the threshold strucuture. The
calculation was made for fixed internuclear distance and only s-wave
scattering was taken into account in agreement with the experiments of
that time. Other theories [4,5] were based on the model of Feshbach
resonances induced by the nuclear vibration. Some authors [6,7] made
model calculations with non-local potentials or made ab-initio calculations including long-range dipole and/or polarization forces. Now, most
theoretical approaches take into account non-adiabatic effects [6],
since they seem to play an important role in the formation of these
threshold structures.

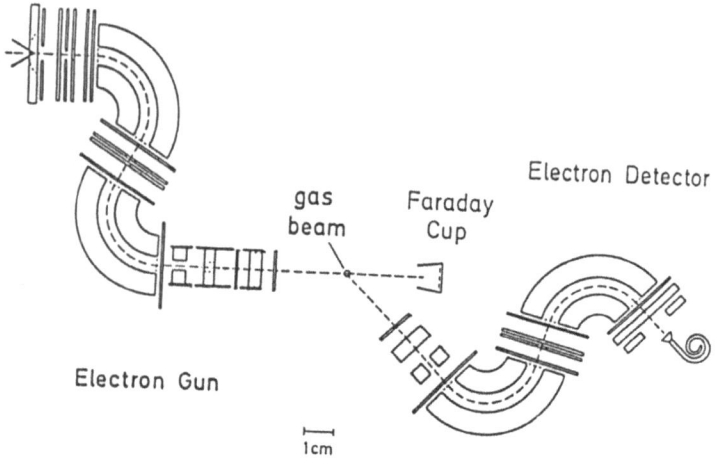

Figure 1: Schematic diagramm of the electron spectrometer.

A quantitative comparison between theory and experiment was hindered by the suspicion that the experimental results of Rohr and Linder [1] might be in error up to a factor two with respect to the absolute cross sections and the width of the structures. Therefore it was required to repeat these measurements with improved techniques and in more detail.

The present measurements cover a wider angular range, they have been performed with tandem systems in the gun and detector in order to reduce the background of stray electrons, which is of great importance especially in forward direction, they have been made with a better energy resolution, so that rotational transitions could be deduced for the first time and a more reliable method was used for the deduction of absolute cross sections.

APPARATUS

A crossed beam electron impact spectrometer has been used (see fig. 1) with $127°$ cylindrical tandem systems for the electron gun and the analyzer of the scattered electrons. The system has been described indetail elsewhere [8]. In short, the electron beam is tunable in energy from 0.08 eV to 5 eV with rather constant transmission, an energy inhomogeneity of ca. 20 meV and a beam current of about 10^{-9} amps.

The beam shape is controlled by a double Faraday cage with 6 and 2 mm diameter. Small angular corrections of the scattered electrons are made by comparison with the well-known differential cross sections of helium [10]. Absolute cross sections are determined by comparison with helium, neon and argon using the flux flow method.

For HCl and HF the rotational constants are 1.31 meV and 2.60 meV respectively for the vibrational ground state. Therefore it is not possible with an electron spectrometer to separate rotational transitions. Consequently the line shape of a rovibrational transition has to be measured very accurately (see fig. 2) and a deconvolution method proposed by Shimamura [10] is used. According to this procedure the cross section for the rotational transition $J_i \to J_f$ with the wave numbers k_i

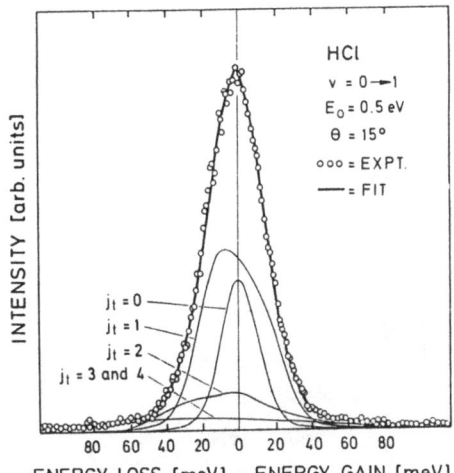

Figure 2: Rotational branch deconvolution of the v = 1 peak in the energy loss spectrum. Circles: Experimental values, full line: Fitted sum of the j_t branches. The j_t = 0 branch is identical with the apparatus profile measured with argon as target gas.

and k_f is related to the cross sections of the transitions $J = 0$ to J_t with $J_t = |J_i - J_f|, \ldots, |J_i + J_f|$ by

$$\frac{d\sigma (J_i, k_i \rightarrow J_f, k_f; \theta)}{d\omega} =$$

$$\sum_{J_t = |J_i - J_f|}^{|J_i + J_f|} C_t \cdot \frac{k_f}{k_t} \cdot \frac{d\sigma (J = 0, k_i \rightarrow J_t, k_t; \theta)}{d\omega},$$

where $C_t = C_t (J_i, J_t, J_f; 0, 0, 0)$ are the squared Clebsch-Gordan coefficients and k_f is the final wave number for the transition $J = 0 \rightarrow J_t$.

In this paper the cross sections for the transitions $J = 0$ to $J_t = 0,1,2,3$ and 4 of the molecular HCl and HF have been determined by a least square fit. Higher J_t are negligible, since the measured line shape does not allow for higher transitions. The initial rotational state distribution of the molecules at the temperature of 400 K shows significant population (larger than 1 %) up to J = 10 in the case of HCl and up to J = 7 for HF.

EXPERIMENTAL RESULTS

Figure 3 represents energy dependences of the (rotationally summed) excitation of the first vibrational level of HCl by electrons as a function of the impact energy in absolute units and at the scattering angles θ = 45 °, 90 ° and 120 °. It shows the resonance like threshold peak with a half-width of about 170 meV. The error bars shown in the graph include all errors which arise from statistics, energy variant trans-

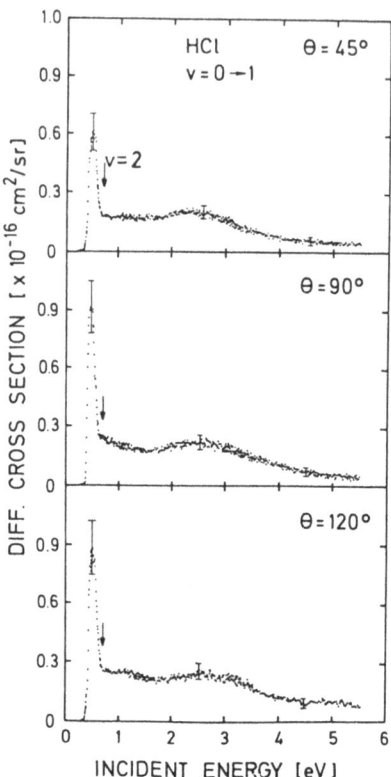

Figure 3: Energy dependences of the $v = 0 \to 1$ excitation of HCl at the scattering angles 45 °, 90 ° and 120 °.

mission, angular errors and normalization to absolute cross section values. The arrows indicate the energy positions of the threshold of the $v = 2$ channel. Beyond the threshold peak to higher impact energies a region (0.7 eV to 1.7 eV) of practically constant cross section is visible. In the region 1.7 eV to 4 eV a cross section is measured which is typical for a broad shape resonance.

The threshold peak has been analyzed with respect to rotational transitions by measuring very accurately the spectral line shape using the deconvolution method of Shimamura [10]. One result is shown in figure 2. The true line shape produced by the apparatus (electron gun plus detector) is shown for the transition $J = 0 \to J_t = 0$ (equivalent to $j_t = 0$, where j_t is the angular momentum transfer of the electrons) and has been measured using argon as target gas. The deviation of the line shape of HCl from the apparatus line shape is due to rotational transitions during the collisions. The rotational transitions with $j_t = 1$ and $j_t = 2$ dominate the cross sections.

Figure 4 shows a plot for HCl of the rotational transition cross sections in absolute units for $j_t = 0,1,2,3$ and 4 as a function of the scattering angle. I is a compilation of peak shape deconvolutions as shown in figure 2 performed at 9 different scattering angles. Total

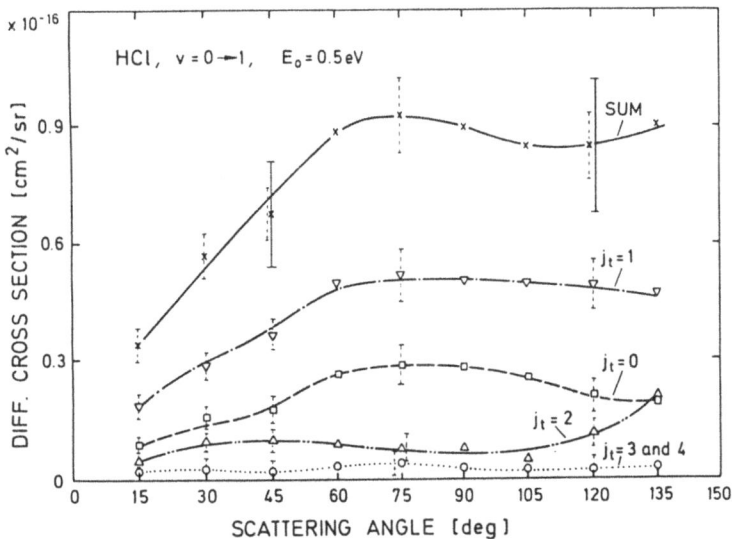

Figure 4: Angular dependences of the absolute differential cross sections for the rovibrational excitation of HCl by 0.5 eV electrons (at the maximum of the threshold peak). The full line error bars refer to the absolute cross sections, i.e. they include the normalization to absolute values. The dotted error bars are relative errors.

errors, including normalization to absolute values, are indicated by full line error bars, partial errors, which include statistics, deconvolution errors and angular errors are shown as dotted error bars.

Striking is the decrease of the partial cross sections to low scattering angles. This deviates markedly from the results of Rohr and Linder [1]. They reported an isotropic angular dependence of the rotationally summed cross section.

Figure 5 exhibits the energy dependences of the vibrationally elastic cross section of HCl close to the opening of the $v = 1$ channel at scattering angles of $60°$, $90°$ and $120°$. Such dependences have been measured in the range from $15°$ to $135°$ with the same results. The cusp structure is well visible, but no resonance structures could be detected in the range 200 meV below the cusp.

In general, the results of HF show the same properties. The spectral line shape for the vibrational excitation at threshold exhibits a distinct broadening, thus giving evidence for simultaneous rotational excitation. The angular dependence of the rotationally summed vibrational excitation cross section at the threshold peak as well as the partial cross sections $j_t = 0,1,2,3$ and 4 are characterized by a decrease to low scattering angles.

DISCUSSION

The isotropic angular dependence of the differential vibrational excita-

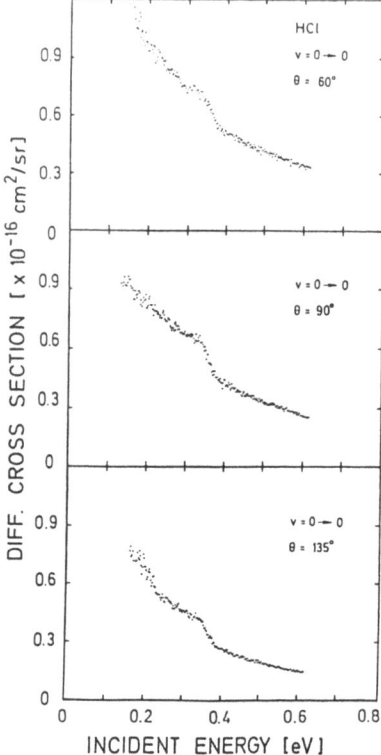

Figure 5: Energy dependences of the vibrationally elastic differential cross section of HCl at the scattering angles 60°, 90°, 120°. The cusp strucuture ($\Delta E_{1/2} \approx 15$ meV) indicates the opening of the $v = 1$ channel at 0.358 eV. No Feshbach resonance structure is detected below the cusp.

tion cross section as measured by Rohr and Linder [1] has led to the conclusion, that the threshold peak originates from pure s-wave scattering. The line shape analysis of the present measurements, yielding partial cross sections for the rotational transitions, demonstrates the existence of rotational excitation with angular momentum transfer $j_t > 0$. These results cleary show that higher partial waves than $1 = 0$ contribute to the scattering process at the vibrational threshold. Most important are momentum transfer values 0, 1 and 2 indicating that at least p-waves must participate in the collisions.

Similarly, the decrease of the differential cross sections in foreward direction of the scatterd electrons can be explained by the interference of s, p and d-waves. Certainly it shows, that outgoing partial waves $1 > 0$ are involved.

The angular anisotropy measured at the threshold peaks of HCl and HF might also be an experimental indication of the importance of the nuclear motion during the passage of the slow scatterd electrons. Such non-adiabatic coupling effects at threshold have been found for the rotational excitation of H_2 [11]. This conclusion (non-adiabaticity) is based only on the similar angular dependencies of H_2 resp. HCl and HF, it has of course to be examined theoretically.

Feshbach resonances can be distinguished experimentally from virtual states, since they are positioned energetically slightly below the threshold of the parent state in the open channel, whereas the virtual state has no such structure. Its influence is only on the cross section right at the threshold of the newly opening channel. In figure 5 the opening of the vibrational channel $v = 1$ is well marked by a cusp structure. Below this cusp we could not detect any rapidly variing structures.

In conclusion, the present experiments show, that the threshold structures of the $v = 1$ channels in HCl and HF seem not to originate from Feshbach type resonances, instead, a non-adiabatic interaction is involved with contributions from several partial waves. These results require further theoretical development.

The authors wish to express their thanks to Drs. R.K. Nesbet, E.S. Chang and F.A. Gianturco for very valuable discussions and for the financial support of the Deutsche Forschungsgemeinschaft (SFB 91).

REFERENCES

[1] Rohr K. and Linder F., J.Phys.B, 9 (1976) 2521
[2] Azria R., Roussier L., Paineau R. and Tronc M., Rev.Phys.Appl., 9 (1974) 469
     Orient O.J. and Srivastava S.K., Phys.Rev.A, 32 (1985) 2678
[3] Dubé L. and Herzenberg A., Phys.Rev.Lett., 38 (1977) 820
     Nesbet R.K., J.Phys.B, 10 (1977) L739
[4] Gauyacq J.P. and Herzenberg A., Phys.Rev.A, 25 (1982) 2959
[5] Kazansky A.K., J.Phys.B, 16 (1983) 2427
[6] Domcke W. and Mündel C., J.Phys.B, 18 (1985) 4491
[7] Rudge M.R.H., J.Phys.B., 13 (1980) 1269
[8] Sohn W., Kochem K.-H., Scheuerlein K.-M., Jung K. and Ehrhardt H., J.Phys.B, (1987) in press
[9] Nesbet R.K., Phys.Rev.A, 20 (1979) 58
[10] Shimamura J., Chem.Phys.Lett., 73 (2), (1980) 328
[11] Jung K., Scheuerlein K.-M., Sohn W., Kochem K.-H. and Ehrhardt H., J.Phys.B, 20 (1987) L 327

PHOTOELECTRON SPECTROSCOPY OF HI AND PHOTOEMISSION FOLLOWING NARROWBAND

VUV EXCITATION

N. Böwering, T. Huth, A. Mank, M. Müller, G. Schönhense,
R. Wallenstein, and U. Heinzmann

Universität Bielefeld, Fakultät für Physik
D-4800 Bielefeld, West-Germany
and Fritz-Haber-Institut der MPG
D-1000 Berlin 33, West-Germany

INTRODUCTION

As a model case for a detailed photoelectron spectroscopy of molecules, the photoelectron emission of hydrogen iodide was studied. The experiments were performed in the autoionization region between the HI^+ $^2\Pi_{3/2}$ and $^2\Pi_{1/2}$ thresholds and between the HI^+ $^2\Pi_{1/2}$ and $^2\Sigma^+$ thresholds. For the region between the HI^+ $^2\Pi$ thresholds where the autoionization features are sharp, a novel scheme for very high resolution photoelectron yield measurements was applied using laser frequency upconversion to the vacuum ultraviolet (VUV).[1] The system is capable to produce rotationally resolved spectra. At photon energies above the $^2\Pi$ limits where the ionization structures are much broader, synchrotron radiation at the Berlin storage ring BESSY was used for angle-resolved photoelectron spectroscopy and measurements of the photoelectron spin polarization[2]. With this apparatus, vibrational excitation in the final states was detected and the energy dependence of the dynamical photoionization parameters was determined.

HIGH RESOLUTION YIELD MEASUREMENTS

For narrow-band excitation, VUV radiation is generated by resonantly enhanced sum-frequency mixing of two pulsed dye lasers in mercury vapor.[3] Both dye lasers are pumped by Nd:YAG lasers. Dye laser 1 is frequency doubled to produce ultraviolet (UV) radiation at λ = 280.3 nm (pulse energy 8 mJ) for excitation of the Hg 6^1S -6^1D two-photon resonance. Dye laser 2 (5 mJ) is tuned in the wavelength range of λ = 600 - 730 nm. Thus, coherent VUV radiation is generated in the range of λ = 113.5 - 117.5 nm in a mixing process which is very efficient due to its two-photon resonant character.

The experimental set-up is shown schematically in Fig. 1. The two synchronized dye-laser beams are combined with a dichroic mirror, and are focussed with matched foci into a heat pipe containing mercury at p = 1.5 mbar and neon at p = 4 mbar. The sum-frequency radiation generated enters the vacuum system through a lithium fluoride (LiF) window and is separated from other UV and VUV radiation by means of a LiF-prism

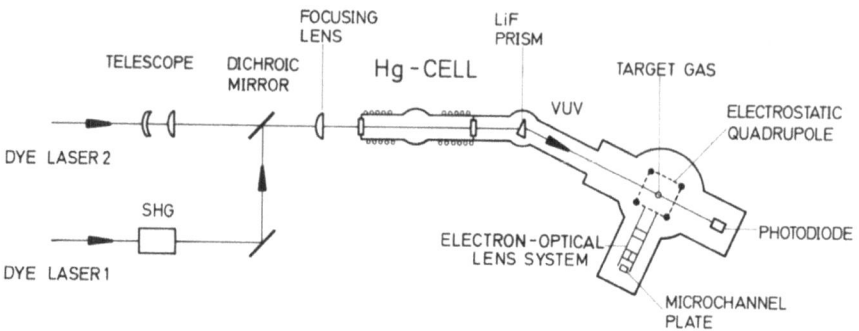

FOCUSING LENS
LiF PRISM
TELESCOPE   DICHROIC   Hg-CELL   TARGET GAS
             MIRROR
                              VUV   ELECTROSTATIC
DYE LASER 2   QUADRUPOLE
     SHG
                        ELECTRON-OPTICAL   PHOTODIODE
DYE LASER 1   LENS SYSTEM
                                    MICROCHANNEL
                                    PLATE

Fig. 1: Experimental apparatus for resonant sum frequency mixing and
         photoionization yield spectroscopy.

and beam dumps. Typically, 10^{10} photons per pulse (about 3 ns FWHM) are
produced at a repetition rate of 10 Hz. The spectral resolution as determined by the bandwidth of the visible laser radiation is calculated to
be 4.5×10^{-4} nm.

In the excitation region the VUV light is crossed at right angles
by a focussed beam of HI molecules generated by a bent capillary plate
used as nozzle. The photoelectrons produced are collected by a quadrupole

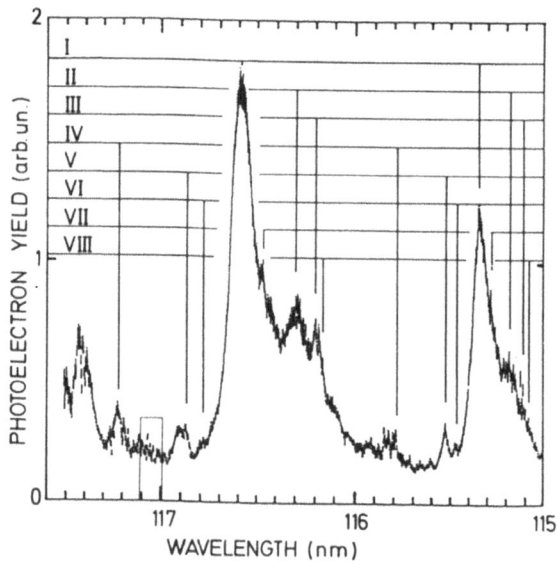

Fig. 2: Coarse scan through n = 6, 7 members of the HI^+ $^2\Pi$ autoionization
         region. Data points are drawn as vertical lines to denote the
         experimental uncertainty. Extending the assignments of Ref. 4,
         several Rydberg series are identified and labelled in the figure.

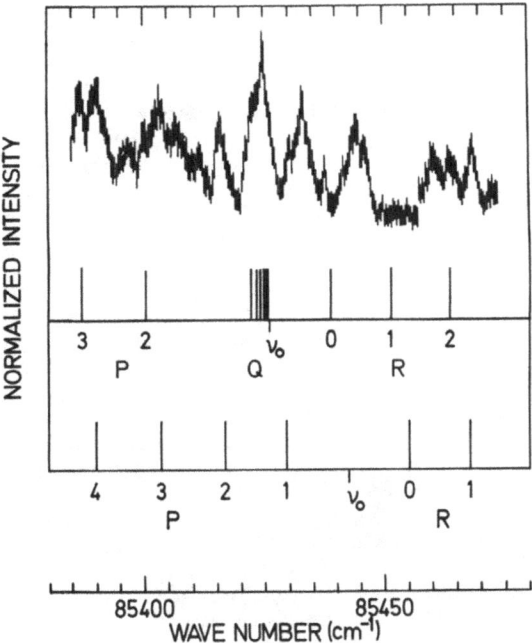

Fig. 3: High resolution scan through the structures around 117 nm (indicated by box in Fig. 2). Step size is 4.5×10^{-4} nm. A tentative assignment of possible rotational fine structure for two transitions is indicated in the lower part of the figure.

field and an electron-optical lens system and detected with a microchannel plate. The photoelectron signal is normalized to the VUV intensity as measured with a vacuum photodiode; each spectral sample is averaged over 20 to 100 laser shots and each scan is repeated several times.

Fig. 2 shows a broad scan with coarse step size of 3×10^{-3} nm over the range of λ = 115 to 117.5 nm. The envelope of the data compares well with the results from photoionization mass spectrometry of Eland and Berkowitz[4] taken at a resolution of 7×10^{-3} nm. The assignment of the Rydberg series by these authors was guided by the analogy to xenon which is the united atom limit of HI. Classification of the main resonances can be accomplished by comparison with the recent theoretical MQDT calculation for n = 6 of Lefebvre-Brion et al.[5]. However, the highly resolved spectra reveal a rich underlying structure which can only be attributed to a small extend to series converging to vibrationally excited levels and is probably largely due to the influence of rotational structure. Additional features may result from L- and S-uncoupling effects[5].

For broad resonances like series I, broadening of the line profiles due to the short lifetime of the autoionizing state dominates and should preclude any identification of rotational bands. Therefore, regions of the spectrum which contain weak but sharp resonances and low background levels were scanned with small step size in order to attempt an analysis of rotational substructure. Fig. 3 shows such a section obtained with

high resolution on an expanded scale. The peaks shown were reproducible
in shape and intensity at equivalent conditions of the molecular beam.
(The rotational temperature is estimated to lie in the range of
100 - 200 K.) Also drawn are rotational branches for two electronic transitions neglecting all broadening and assuming that the rotational constant of the upper state is 1.5% smaller than for the ground state of
HI. The spectra are matched at a possible position of a Q-branch. It
is clear from the spacing of the peaks observed that several spectral
features overlap in this region and the assignment is not unique. Tenta-
tively, a resonance with $\Omega = 1$ at 85426 cm^{-1} and a state with $\Omega = 0$ at
85442.5 cm^{-1} are identified.

VIBRATIONALLY RESOLVED PHOTOELECTRON SPECTROSCOPY

For energy-, angle- and spin-resolved photoelectron spectroscopy
circularly polarized synchrotron radiation emitted out of plane at BESSY
was used. At the exit focus of a normal incidence monochromator
($\Delta\lambda = 0.5$ nm) the photoemission of HI from an effusive molecular beam
is recorded. The photoionization cell is mounted on a liquid nitrogen
cold trap and the background pressure inside the vacuum chamber is less
than 9×10^{-5} mbar. The photoelectrons emitted in the reaction plane
at a definite angle θ are energy-analyzed by a rotatable hemispherical
spectrometer. Intensity spectra are recorded by a channeltron. For spin
analysis, the electrons are accelerated to 100 keV and the two transverse
electron spin-polarization components are measured with a Mott detector.
A detailed description of the apparatus is given in Ref. 6.

Photoelectron spectra were taken at the magic angle ($\theta_m = 54.5°$)
above the HI^+ $^2\Pi$ thresholds with a resolution of $\Delta E \sim 120$ meV. Previously,
no vibrational excitation higher than v = 1 was observed in the final
ionic $^2\Pi$ states (which correspond to the removal of an electron from
a nonbonding orbital) in photoelectron spectra taken with the He I resonance line[7,8], and in the range between the HI^+ $^2\Pi_{3/2}$ and $^2\Pi_{1/2}$ limits[9].
Apparently, levels with v > 1 lie outside the Franck-Condon region. At
photon energies above 13.5 eV levels with v > 1 were not found in this
work, also, and the relative intensity of the (v = 1) levels was very
low, in agreement with the He I spectrum of Ref. 8 (where I(v = 1) /
I(v = 0) = 1/30). However, when tuned to the range above the HI^+ $^2\Pi_{1/2}$
threshold and below the dissociation limit of HI^+ clear evidence of strong
contributions of higher vibrationally excited states of both $^2\Pi_{1/2}$ and
$^2\Pi_{3/2}$ is found as shown in Fig. 4. In this region, the vibrational
branching changes drastically and vibrational levels of HI^+ $^2\Pi_{1/2}$ up
to v = 8 were detected at $\lambda = 92.5$ nm. The enhancement of higher vibrational excitation in the final ionic state is attributed to autoionizing
Rydberg states converging to HI^+ $^2\Sigma^+$. These states are excited in high
vibrational levels and can decay with a large overlap to vibrationally
excited levels of HI^+ $^2\Pi$.

The relevant potential curves are shown in Fig. 5. The Rydberg states
(labelled Ry in the figure) have a larger equilibrium internuclear separation since an electron of the more tightly bound 5pσ-orbital is excited. Lempka et al.[8] concluded from their He I photoelectron spectra
and by comparison with the lighter hydrogen halides that due to the position of the curve crossing of the HI^+ $^2\Sigma^+$ potential curve with the $^2\Pi$
repulsive curves all vibrational levels of HI^+ $^2\Sigma^+$ are predissociating.
Thus, the corresponding Rydberg states can be expected to be strongly
predissociating, as well. Consistent with this picture, data from photoabsorption spectroscopy[10] and photoionization mass spectrometry[11] in
this spectral region do not show distinct vibrational progressions of
Rydberg series. Nevertheless, as demonstrated here, the vibrational

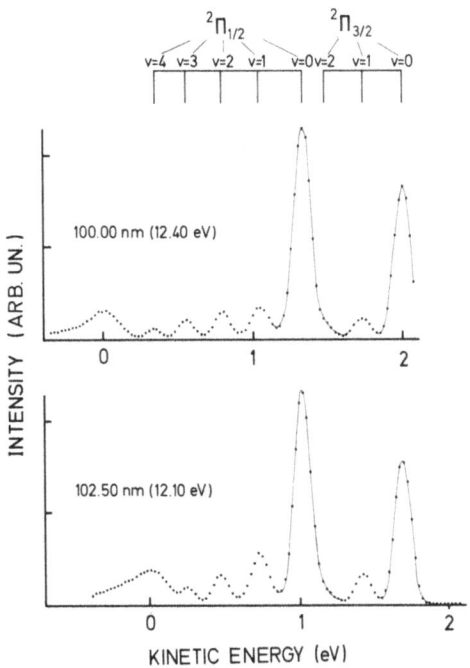

Fig. 4: Photoelectron spectra resolving vibrational excitation of the final $^2\Pi$-ionic states.
(The peak at 0 eV corresponds to a background of slow electrons.)

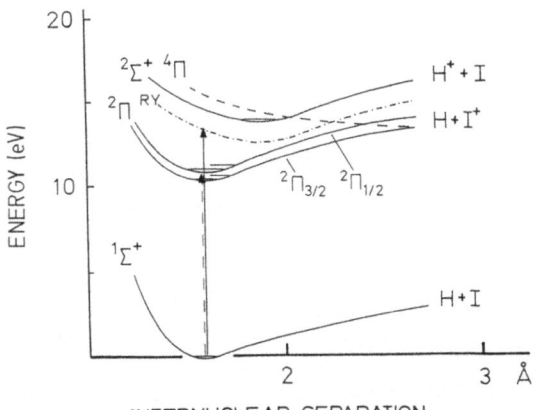

Fig. 5: Potential energy curves of HI. (Adapted from Ref. 8.) Indicated is excitation in the autoionization regions between the $^2\Pi$ states and between $^2\Pi$ and $^2\Sigma^+$.

excitation can be detected in the photoelectron spectra which reflect the vibrational excitation of the intermediate Rydberg state in the distribution for the final ionic states.

There is considerably less information on the ionic ground states for HI than for the lighter hydrogen halides. Spectra of the type of Fig. 4 were analysed to obtain information about vibrational frequencies. The centroids of the vibrational peaks were determined by assuming symmetric Gaussian distributions and the first differences of successive levels, $\Delta G_{v+1/2} = \omega_e - 2\omega_e x_e(v+1)$, were plotted against $(v+1)$ to obtain the vibrational frequency $\omega_e = (2203\pm18)\,\text{cm}^{-1}$ and the anharmonicity constant $\omega_e x_e = (41.5\pm4.8)\,\text{cm}^{-1}$ for the $HI^+\,{}^2\Pi_{1/2}$ state. Since the series of $HI^+\,{}^2\Pi_{3/2}$ is "buried" under the peaks of $HI^+\,{}^2\Pi_{3/2}$, only the first differences $\Delta G_{1/2} = (2157\pm10)\,\text{cm}^{-1}$ and $\Delta G_{3/2} = (1920\pm26)\,\text{cm}^{-1}$ could be determined.

PHOTOELECTRON-SPIN POLARIZATION

The spin-polarization parameters A (characterizing the component $A(\theta)$ of the spin-polarization vector parallel to the photon spin) and ξ (characterizing the component $P_\perp$ perpendicular to the reaction plane) of the photoelectrons for the two ionic fine-structure components $HI^+\,{}^2\Pi_{3/2}$ $(v = 0)$ and ${}^2\Pi_{1/2}(v = 0)$ were determined in the wavelength range of 120 nm to 60 nm using the Mott detector. In addition, the spin polarization asymmetry parameter α was obtained from a measurement of the angular dependence of $A(\theta)$ at $\lambda = 100$ nm. The measurement procedure utilizing the reaction geometry of the experiment is described in Ref. 6.

In comparison to its united atom, xenon, the symmetry is reduced for hydrogen iodide due to the nonspherical molecular potential and the highest degree of photoelectron-spin polarization attainable is 50% since all possible orientations of the molecular axis with respect to the photon momentum have to be averaged[12]. In the nonrelativistic theory the influence of the spin-orbit interaction on the continuous spectrum is neglected and the relationships

(I) $A_{3/2} = -A_{1/2}$ (2) $\xi_{3/2} = -\xi_{1/2}$ (3) $\alpha_{3/2} = -\alpha_{1/2}$

at equal photoelectron energy hold.[12,13]

The experimental results for A and ξ are shown in Figs. 6 and 7, respectively. Also drawn (full curves) are recent nonrelativistic ab-initio calculations (frozen-core static exchange approximation) of Raseev et al.[13]. A comparison of theory and experiment is more favorable for the spin-polarization parameter A than for ξ. Although smaller in absolute value, the experimental data for the A parameter have the same shape of the wavelength dependence as the calculated values. Comparing equal photoelectron energy, the relation (1) is fulfilled approximately for A, but equation (2) for ξ does not hold for the experimental values and the disagreement between experiment and theory is larger. Similar deviations were noted before in experiments with molecules containing heavy halogen atoms[14]. Since ξ is determined from off-diagonal interference terms in the theory and its dynamical factor depends on the sine of phase-shift differences of continuum wavefunctions it is more sensitive to the influence of the spin-orbit interaction in the continuum than the other dynamical parameters in molecular photoionization. (The parameter A is independent of phase-shift differences and the asymmetry parameters β and α have a different phase dependence and are dominated by diagonal terms.)

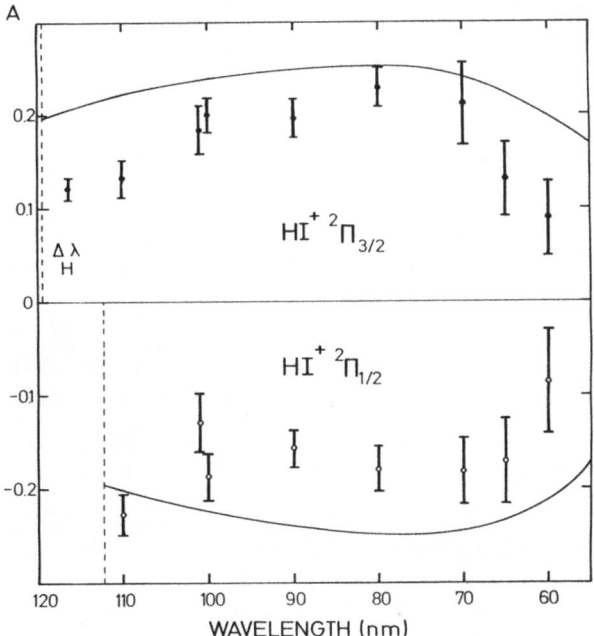

Fig. 6: Spin-polarization parameter A for HI^+ $^2\Pi$ (v = 0).
Data points: this work, full curves: calculation of Ref. 13.

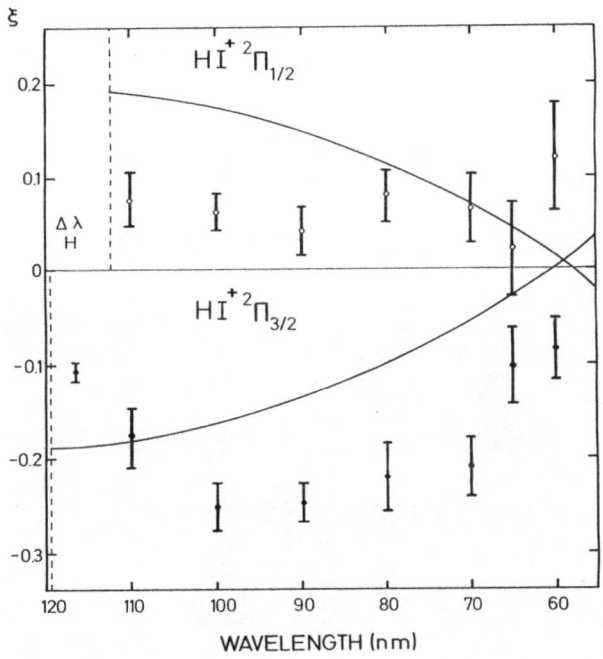

Fig. 7: Spin-polarization parameter ξ for HI^+ $^2\Pi$ (v = 0).
Data points: this work, full curves: calculation of Ref. 13.

From a least-squares fit of the angular distributions measured for $A(\theta)$ at $\lambda = 100$ nm, we have obtained $\alpha_{3/2} = 0.31\pm0.03$ and $\alpha_{1/2} = -0.19\pm0.02$, to be compared with $\alpha_{3/2} = 0.39$, $\alpha_{1/2} = -0.41$ of Raseev et al.[13].

CONCLUSION

Hydrogen iodide has proven to be very suitable to probe the extent and influence of vibrational and rotational structure in the autoionization spectra as well as the spin polarization of the photoelectrons. Furthermore, theoretical calculations are available[5,13] for comparison with the experimental results. Using circularly polarized light, substantial photoelectron-spin polarization was found for the $^2\Pi$ fine-structure components and the energy dependence of the corresponding parameters was examined. The vibrationally resolved measurements have demonstrated that photoelectron spectroscopy with a tunable light source can reveal vibrational excitation which, due to predissociation, is not spectroscopically accessible in photoabsorption experiments. Finally, using laser-frequency upconversion, resolving powers exceeding 2×10^5 can be achieved in the VUV with sufficient intensities for photoelectron emission experiments probing rotational structures in photoionization.

ACKNOWLEDGEMENT

Financial support by the Bundesministerium für Forschung und Technologie (05331AXIO) and the Deutsche Forschungsgemeinschaft (Sfb 216) is gratefully acknowledged.

REFERENCES

1. T. Huth, A. Mank, N. Böwering, G. Schönhense, R. Wallenstein and U. Heinzmann, to be published.
2. N. Böwering, M. Müller and U. Heinzmann, to be published.
3. R. Hilbig and R. Wallenstein, IEEE J. Quantum Electron. 19:1759 (1983).
4. J. H. D. Eland and J. Berkowitz, J. Chem. Phys. 67:5034 (1977).
5. H. Lefebvre-Brion, A. Giusti-Suzor and G. Raseev, J. Chem. Phys. 83:1557 (1985).
6. Ch. Heckenkamp, F. Schäfers, G. Schönhense and U. Heinzmann, Z. Phys. D 2:257 (1986).
7. D. W. Turner, C. Baker, A. D. Baker and C. R. Brundle, in: "Molecular Photoelectron Spectroscopy", p. 59, Wiley, New York (1972).
8. H. J. Lempka, T. R. Passmore and W. C. Price, Proc. Roy. Soc. A 304:53 (1968).
9. T. A. Carlson, P. Gerard, M. O. Krause, G. Von Wald, J. W. Taylor and F. A. Grimm, J. Chem. Phys. 84:4755 (1986).
10. D. T. Terwilliger and A. L. Smith, J. Chem. Phys. 63:1008 (1975).
11. P. M. Dehmer and W. A. Chupka, Argonne National Laboratory Report ANL-78-65, Part I:p. 13 (1978).
12. N. A. Cherepkov, J. Phys. B 14:2165 (1981).
13. G. Raseev, F. Keller and H. Lefebvre-Brion, Phys. Rev. A, to be published, and private communication.
14. G. Schönhense, V. Dzidzonou, S. Kaesdorf and U. Heinzmann, Phys. Rev. Lett. 52:811 (1984).

AB INITIO METHODS FOR ELECTRON-MOLECULE COLLISIONS

L. A. Collins and B. I. Schneider

T-Division
Los Alamos National Laboratory
Los Alamos, NM 87545, U.S.A.

" For all we know and all we guess, we mutually impart" - Iolanthe

INTRODUCTION

If we take the strict definition of "ab initio"- from first principles, our review is at an end since no methods currently exist for
treating the entire electron-molecule collisional process exactly. However,
if we relax this definition to include those techniques that give the most
accurate treatment of the various interactions, then we can discuss
numerous prescriptions. In this sense, "ab initio" takes on a
time-dependent character. The earliest treatments had to rely upon crude
models for the interactions and scattering process. By the mid-1960's with
the advent of electronic computers, the static or direct electrostatic
interaction could be treated accurately while in the next decade, full
treatments of the nonlocal exchange effects, at least for scattering from
the ground state, began to be successfully applied. The last few years have
witnessed a vast expansion of our ability to treat electron-scattering
processes with correlation and multi-channel coupling effects being
included for a variety of small molecular targets. These new methods allow
us to handle such mechanisms as polarization, channel and resonance
interference, and autoionization in a systematic and consistent fashion.

In this review, we shall concentrate on the recent advances in
treating the electronic aspect of the electron-molecule interaction and
leave to other articles the description of the rotational and vibrational
motions. We shall focus on those methods which give the most complete
treatment of the direct, exchange, and correlation effects. Such full
treatments are generally necessary at energies below a few Rydbergs($\approx$
60eV). This choice unfortunately obliges us to omit those active and vital
areas devoted to the development of model potentials and approximate
scattering formulations. The ab initio and model approaches complement each
other and are both extremely important to the full explication of the
electron-scattering process. Due to the rapid developments of recent years,
we must however concentrate on the approaches that provide the fullest
treatment.

That all three effects mentioned above are important in low-energy
electron-molecule scattering is demonstrated in Fig.1 in which we compare
calculations with the linear algebraic method and experimental results
for the total elastic cross section for electron collisions with molecular hydrogen. Even at an energy of 1 Ry., using only the static(S)
interaction leads to errors of over thirty percent. The situation is
improved by an exact treatment of exchange with respect to the ground-
state target and scattering orbitals. However, this static- exchange(SE)
calculation does not reproduce the maximum in the cross section near 4 eV
nor the very low-energy behavior. These features are obtained only by
introducing polarization-correlation effects through virtual excitations
to the closed electronic channels. The full interaction is needed to
properly represent the weak resonance in the sigma-ungerade channel. As
the figure demonstrates, new theoretical and computational techniques are
now available that can treat the full range of the electron-scattering
process.

The ability to treat all aspects of electron scattering on an equal
footing allows us to probe more deeply the basic collisional processes
and possibly to discover new mechanisms not properly contained in models.
Certainly the development of tunable light sources such as the synchro-
tron(Dehmer et. al.,1986), which produces high-resolution spectra over
wide energy ranges, place a heavy burden on the theoretical formulations
since intricate details must now be reproduced and explained. Before
delving into the arcane lore of the methods, we present an example of the
types of processes that can be explored with these more elaborate
treatments. We consider the seemingly very simple event of electron
scattering from the hydrogen molecular ion below the first excitation
threshold. While only elastic scattering is possible, we encounter series
of Feshbach resonances, which result from the temporary trapping of the
incident electron in doubly-excited states of molecular hydrogen. These
compound or autoionizing states must eventually decay to the continuum.
The resonance is characterized by this decay time,τ, or its related
width($1/\tau$). The series converge on the various excited states of the
target ion. For example, if we consider scattering in total symmetry $^3\Pi_u$,
then we have a set of autoionizing states of the form $1\sigma_u n\pi_g$ converging
on the $1\sigma_u$ threshold and $1\pi_u^+ n\sigma_g$ on $1\pi_u$. For certain conditions, these two
series may overlap. We investigate this effect in Fig.2 in which we
follow the resonant width of the two lowest compound states as a function
of the separation of the two nuclei,R. We note that at values of R around
1. bohr, we encounter a strong coupling between the series leading to
rather dramatic changes in the widths. To properly describe such interference requires at least a three-state close-coupling calculation.
These effects may play important roles in such processes as dissociative
photoionization and dielectronic recombination.

The situation at present for ab initio methods is indeed propitious
with many techniques of equal sophistication available. Therefore,
important cross-comparisons are possible in order to determine the range
and validity of the various approaches. The length of this review
precludes a comprehensive treatment of each technique. Instead, we shall
concentrate on the general procedures for solving the electron scattering
problem, illustrating these points with examples from the various
methods. We begin with a brief description of the basic electron-molecule
scattering formulation. We then present several strategies, common to
most of the methods, for reducing the formulation to a more tractable
form. We follow with a short summary of the general methods of solution

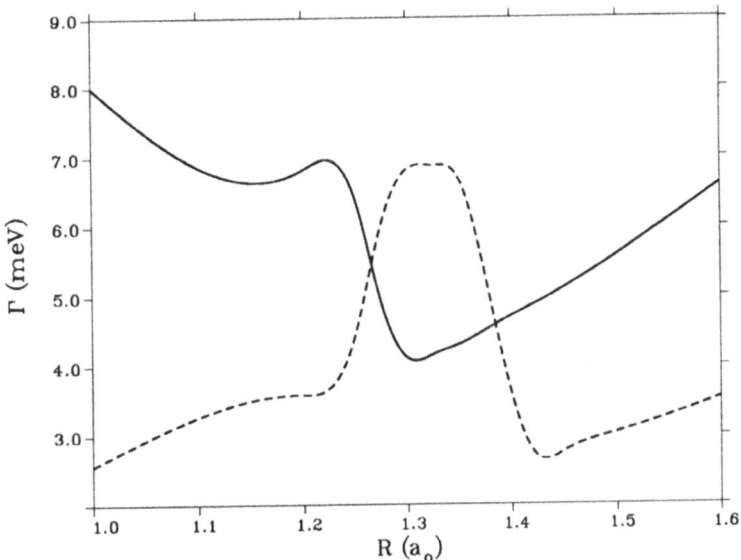

Fig.1. Comaprison of theoretical and experimental total elastic e⁻ + H₂
cross sections as a function of electron energy. Theoretical results(LA
method): chain - static(S); dash - static-exchange(SE); line -
SE+polarization(SEP). Experimental results: triangles - Hoffman et
al.(1982); cross - Dalba et al.(1980)

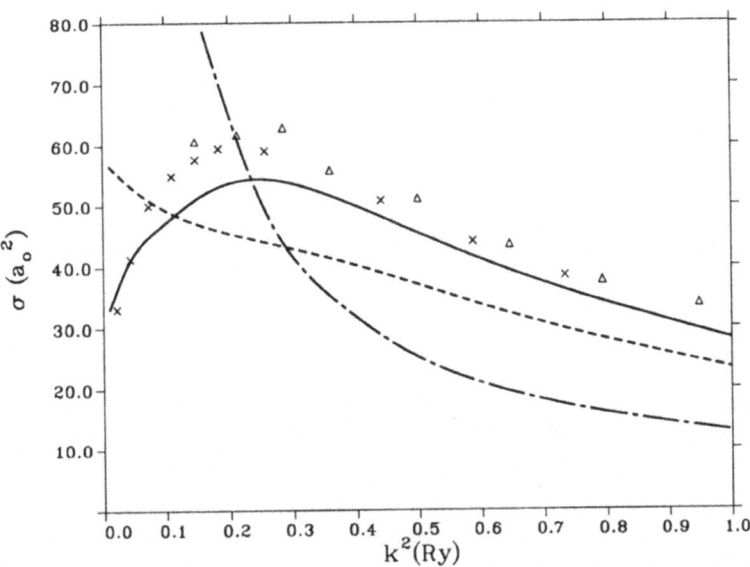

Fig. 2. Resonant widths as a function of internuclear distance,R, for e⁻
+ H₂⁺ scattering below the first excitation threshold($1\sigma_u$) for $^3\Pi_u$
symmetry. Nomenclature: line - lowest resonance; dash - second resonance.

and conclude by comparing the results of several methods and reviewing the general status of the field.

FORMULATION

"And we are right, I think you'll say,
                to argue in this kind of way" - Mikado

Since the details of the general formulation for the scattering of an electron by an n-electron molecule, containing N nuclei, are clearly presented in other reviews(Lane,1980;Norcross and Collins,1982;Morrison, 1983;Buckley et al.,1984), we give only a brief outline of the basic precepts. We sacrifice rigorous formalism for a general, schematic view of the guiding principles behind the various methods. Therefore, at times, we shall mix our "scattering metaphors" in order to evoke a broad view of specific procedures. We describe the collisional process in terms of the solution to a Schrodinger equation(in atomic units) of the form:

$$H \psi(\vec{r},\vec{X}) = E \psi(\vec{r},\vec{X}) ,$$ (1)

where

$$H \equiv T + V + H_{mol} ,$$ (2a)

with T, the kinetic energy of the incident electron; V, the interaction of the free electron with the molecular electrons and nuclei

$$V = - \sum_{\gamma=1}^{N} Z_\gamma (\vec{r} - \vec{X}_\gamma)^{-1} + \sum_{i=1}^{n} (\vec{r} - \vec{x}_i)^{-1} ;$$ (2b)

H_{mol}, the hamiltonian for the molecular target; and $\vec{r}$, the position of the continuum electron. We let $\vec{X}$ represent the coordinates of both the target electrons $\vec{x}_i\{i=1,n\}$ and the nuclei $\vec{X}_\gamma\{\gamma=1,N\}$, and refer all positions to an axis fixed in space. The molecular hamiltonian describes not only the interactions among the electrons and nuclei but also their respective motion. We convert this many-body problem to an effective single-particle one by expanding the total system wavefunction, ψ, in terms of a complete set of target states as

$$\psi(\vec{r},\vec{X}) = \Sigma_\alpha A[F_\alpha(\vec{r}) \phi_\alpha(\vec{X})] ,$$ (3a)

such that

$$H_{mol} \phi_\alpha(\vec{X}) = \epsilon_\alpha \phi_\alpha(\vec{X}) ,$$ (3b)

where A is the antisymmetry operator. Inserting (3) into (1), multiplying through by the conjugate of a representative member, and integrating over all target coordinates, we obtain the following set of coupled integro-differential equations(IDE's):

$$H_\alpha F_\alpha(\vec{r}) = \sum_\beta Z_{\alpha\beta}(\vec{r}) ,$$ (4a)

where

$$H_\alpha \equiv [\vec{\nabla}^2 + k_\alpha^2] \, ,$$

$$Z_{\alpha\beta}(\vec{r}) \equiv \int K_{\alpha\beta}(\vec{r}|\vec{r}_1) \, F_\beta(\vec{r}_1) \, d\vec{r}_1 \, , \tag{4b}$$

$$K_{\alpha\beta}(\vec{r}|\vec{r}_1) = V_{\alpha\beta}(\vec{r}_1)\delta(\vec{r} - \vec{r}_1) + W_{\alpha\beta}(\vec{r}|\vec{r}_1) \, , \tag{4c}$$

with k_α^2 given by $2(E - \epsilon_\alpha)$. The direct electrostatic interaction, $V_{\alpha\beta}$, is local and given by an integral involving the interaction V and two target states, $\langle\alpha|V|\beta\rangle$. The exchange interaction is schematically represented by $W_{\alpha\beta}$, which is a complicated nonlocal term that arises from the constraints imposed on the solution by the Pauli exclusion principle. We extract the collisional information, such as the reactance(K), transition (T), or scattering(S) matrices, from which the cross sections are determined, by matching the solutions F to their proper asymptotic forms.

We also employ an equivalent form of (4) by converting to a set of coupled integral equations(IE's) by introducing a Green's function. In the case of the free-particle Green's function, we have

$$F_\alpha(\vec{r}) = F_\alpha^0(\vec{r}) + \sum_\beta \int G_\alpha^0(\vec{r}|\vec{r}_2) \, Z_{\alpha\beta}(\vec{r}_2) \, d\vec{r}_2 \, , \tag{5a}$$

where

$$H_\alpha \, F_\alpha^0(\vec{r}) = 0 \tag{5b}$$

and

$$H_\alpha \, G_\alpha^0(\vec{r}|\vec{r}_1) = \delta(\vec{r} - \vec{r}_1) \, . \tag{5c}$$

Equation (5) simply gives the coordinate-space representation of the standard Lippmann–Schwinger(LS) equation for the scattering solution F. In many cases, the IE formulation provides a more stable framework from which to instigate a numerical solution. We can derive similar IE's for other scattering quantities such as the T-matrix. In terms of symbolic-operator notation, we write these LS equations in the compact form (Taylor, 1972)–

$$\psi = \psi^0 + G_o^+ V \psi \tag{5d}$$

$$T = V + V G_o^+ T \tag{5e}$$

where G_o^+ represents the out-going free-particle Green's function and T, the transition matrix operator. Despite our efforts, we still have a rather formidable "beastie" to conquer. To aid in this conquest, we devise several general strategies for further reducing the complications with little loss of accuracy.

STRATEGIES

"In short when I've a smattering of elemental strategy"- Pirates of
Penzance

   In (4), we have a complicated set of IDE's, whose coupling term is
both nonlocal and energy dependent. In order to effect a solution, we hatch
a variety of strategies, common to most of the principal methods, to make
the solution more tractable.

S1: Adiabatic-nuclei approximation

   The quantities in (4) all depend on the orientation and separation of
the molecular nuclei, $R \equiv \{\vec{X}_\gamma\}$. Since we are primarily concerned with the
electronic interactions, we shall invoke the adiabatic-nuclei(AN) approx-
imation(Temkin and Vasavada,1967;Hara,1967;Morrison;1987) by which the
motion of the nuclei is assumed to be slow compared to that of the
electrons. In this case, the molecular electrons have time to readjust to
the new nuclear positions, and the electron-molecule collision may be
treated in a sudden approximation with the nuclei frozen at a given value
of R. In this case we can fix R and solve (4) for the scattering solutions
F. By repeating this prescription at different values of R, we obtain a
paramet ic dependence of the collisional information. Quantities such as
rotational- and vibrational-excitation cross sections are then found by
averaging the R-dependent scattering amplitudes over the appropriate
eigenfunctions. For purposes of future discussions, we assume an implicit R
dependence in the scattering wavefunction, and since the molecule is fixed
in space, we refer all quantities to the body frame system centered on the
molecule(z-axis along a major line of symmetry such as the internuclear
axis in a diatomic). The AN approximation has a fairly wide range of
validity but encounters problems when the colliding electron remains in the
vicinity of the molecule for times comparable to the rotational or
vibrational period(Morrison,1987;Jung et al.,1987). This condition arises
for resonances in which a trapping occurs near the target, for energies
close to threshold where the velocity of the incident electron is very low,
and for encounters in very long-ranged potentials. The AN approximation can
be relaxed by directly introducing approximations to the nonadiabatic
effects(Schneider et al.,1979) into (4), by using a vibrational
close-coupling prescription (Morrison and Saha,1986), by utilizing some
form of the frame transformation (Chang and Fano,1972;Norcross and
Collins,1982) or by projection-operator techniques (Mundel et al.,1985). In
addition, we can employ multichannel quantum defect methods(Greene and
Jungen,1985) to obtain detailed rotational, vibrational, and electronic
excitation cross sections from limited theoretical and experimental
information.,

S2: Spatial Dichotomy

   The nonlocal terms in (4) present the greatest challenge to obtaining
a solution. However, these terms are generally quite limited in range.
Therefore, a useful stategy is to divide space into regions, based on the
strength and nature of the interaction. One procedure for effecting this
division is to impose arbitrary boundary conditions on the solution at some
surface,Σ. For example, we might require a particular choice for the

log-derivative -

$$\frac{1}{\psi} \frac{\partial}{\partial n} (\psi) = b , \qquad\qquad\qquad (6a)$$

where $\partial/\partial n$ represents the surface-normal derivative. In the inner region, where all interactions are important, we solve the full Schrodinger equation of the form:

$$(H_b - E) |\psi) = L_b |\psi) , \qquad\qquad\qquad (6b)$$

where

$$H_b = H + L_b . \qquad\qquad\qquad (6c)$$

The standard hamiltonian, H, is given by (2a), and the Bloch(1957) operator L_b guarantees that the solution to (6b) satisfies the proper boundary conditions at Σ. For a spherically symmetric case with b=0, we have

$$L_b = \delta(r - a) \frac{d}{dr} , \qquad\qquad\qquad (6d)$$

where r=a defines the boundary. The rounded brackets signify that the solution and all operations such as integrals are restricted to the volume within the surface.. In the outer region, where the interaction terms become local and usually expressable as inverse powers of r, we employ standard asymptotic expansion(Noble and Nesbet,1984) or propagation schemes(Light and Walker,1976). We obtain a great advantage by this zoning of space. In the restricted inner region($r \leq 10 - 20$ bohr), the full force of the ab initio techniques can be brought to bear while in the outer region, where the interactions are much simpler, less sophisticated techniques can be employed. We could also impose an implicit dichotomy by using basis functions that properly represent the wavefunctions in each region.

S3: P-Q Division

"He minds his P's and Q's, and keeps himself respectable"-Utopia,Ltd.

In many applications, we are only interested in the scattering information for a few of the states in expansion (3). In this case, we employ the standard Feshbach(1958) formalism in order to isolate these states. We use an expansion of the form:

$$\psi(\vec{r},\vec{x}) = (P + Q) \psi(\vec{r},\vec{x}) , \qquad\qquad\qquad (7)$$

where the projection operators are defined as

$$P \psi = \sum_{\alpha=1}^{m} | A(F_\alpha(\vec{r}) \phi_\alpha(\vec{x}))> < A(F_\alpha(\vec{r}) \phi_\alpha(\vec{x}))| \psi > \qquad\qquad (8a)$$

$$Q \psi = \sum_c d_c | \chi_c(\vec{r},\vec{x}) > < \chi_c(\vec{r},\vec{x}) | \psi > . \qquad\qquad (8b)$$

This choice of projection operators is not unique; some prescriptions define P and Q in terms of only the n-electron target functions. In making this separation, we usually enforce the strong orthogonality constraint by which the orbitals that compose χ and ψ are assumed to be orthogonal to all continuum solutions,F. The summation in the first term ranges over a limited number of target states(m), usually those for which we directly seek information. The second term, which consists of n+1 electron functions,is added for completeness. These "correlation" functions,χ, can be used to relax the orthogonality constraints and to represent other correlation effects not included in the first term.

By substituting this expression into (1) and multiplying through in an operator sense by P and Q separately, we derive the following two equations:

$$H_{PP} \, \psi_P \; + \; H_{PQ} \, \psi_Q \; = \; E \, \psi_P \tag{9a}$$

$$H_{QP} \, \psi_P \; + \; H_{QQ} \, \psi_Q \; = \; E \, \psi_Q \tag{9b}$$

where $H_{XY} \equiv \langle X|H|Y \rangle$ and $\psi_X \equiv \langle X|\psi \rangle$. Solving (9b) for ψ_Q and substituting the result into (9a), we obtain the following equation for the P-space function:

$$[\, (H_{PP} - E) \; + \; V_{opt} \,] \, \psi_P \; = \; 0 \quad , \tag{10a}$$

where the optical potential is given by

$$V_{opt} \; = \; H_{PQ} \, (H_{QQ} - E)^{-1} \, H_{QP} \quad . \tag{10b}$$

The derivation so far is merely a formal rearrangement of the original Schrodinger equation. However, this form in terms of the P-space solution has many advantages. For example, we may be able to make approximations to (10b) that allow us to calculate the optical potential directly. In this case, we are left with only having to solve the P-space equations over a small set of functions.

To utilize this division, we must effectively balance the size of the P and Q spaces. In the traditional close-coupling(CC) approach, we set the coefficients of the correlation functions to zero and add target functions to the first term until convergence is reached. This convergence in terms of target functions can be quite slow. In some methods this problem is rectified by adding a few psuedostates directly to the first expansion. These pseudostates are constructed so as to approximate the effects of those states omitted from the original expansion. In other approaches the P-space is kept small and the Q-space expanded, usually with eigenfunctions from a configuration-interaction(CI) calculation on the n+1 system.

S4: Separable expansions

The optical potential, which contains exchange and correlation effects, is a complicated nonlocal, energy-dependent expression. We can greatly simplify its form by making a separable expansion in terms of a discrete set of basis functions$\{x_\gamma\}$:

$$U_{opt}(\vec{r}|\vec{r}_1) \; = \; \sum_\alpha \sum_\beta \chi_\alpha(\vec{r}) \, U_{\alpha\beta} \, \chi_\beta(\vec{r}_1) \quad , \tag{11a}$$

where

$$U_{\alpha\beta} = \langle \alpha \mid U_{opt} \mid \beta \rangle \qquad (11b)$$

and $\langle\alpha|\beta\rangle = \delta_{\alpha\beta}$. Such separable expansions have a wide variety of appli-
cations in electron-molecule scattering. For example, a modified form of
(11) applied to the LS equations can be used to ensure equivalence to
variational procedures such as the Schwinger. In addition, their use in
numerical evaluations of exchange and correlation terms can greatly reduce
computaional times with only minor loss of accuracy (Collins and
Schneider,1986).

METHODS OF SOLUTION: OVERVIEW

"Existence was slow and we wanted variety" - Gondoliers

In this section, we present a brief overview of the plethora of
techniques that have evolved to treat electron-molecule collisions and
illustrate each general area with one of the more heavily utilized methods.
We should probably view it in the spirit of a patter song from a Gilbert
and Sullivan operetta giving a rapid listing with meager explanation in
order to evoke a sense of grandeur for the field. A more detailed
accounting of methods and practitioners is given in the Appendix and in the
article by Buckley,Burke,and Noble(1984). This tact frees us from an
excrescence of citations in the body of this section. We shall only use
references to differentiate methods within a specific area or to introduce
additional material not covered in the Appendix. Comprehensive reviews of
the R-matrix, Schwinger variational, and linear algebraic methods are
scheduled to appear in a future issue of Computer Physics Reports.
Tradition dictates the division of the methods of solution into two
distinct categories: 1) numerical and 2) square integrable. We shall follow
this convention although the boundaries between these approaches has
greatly blurred since most of the current techniques employ both
prescriptions.

Numerical Methods

In most of the numerical methods, we usually convert (4) or (5) to a
set of coupled radial equations by 1)making a single-center(SC) expansion
of the bound and continuum functions of the form

$$F_\alpha(\vec{r}) = \Sigma f_\Gamma(r) Y_{l_\alpha m_\alpha}(\Theta,\phi) , \qquad (12)$$

where Y is a spherical harmonic, and 2)then integrating over all angular
coordinates. The scattering channel is labeled by $\Gamma \equiv (\alpha, l_\alpha, m_\alpha)$ with α
representing the quantum numbers of the target, l_α, the orbital angular
momentum of the continuum electron, and m_α, its projection on the internuclear axis. Much aspersion has been heaped upon SC methods for bound-
state problems and with justification since the form of the electronic
wavefunction near the nuclei is critical to determining the energy.
However, for scattering, the nuclear singularity is less important, and
with proper care SC expansions can yield highly reliable results.

At this juncture, we can work with either the differential or integral equations. The differential equations are usually solved by propagation schemes based on finite difference approximations. The exchange terms,as always, present an additional complication due to their nonlocal nature. We overcome this problem either non-iteratively by introducing additional DE's for the nonlocal terms or iteratively by starting with an approximation for f and the making successive substitutions until convergence is reached. The integral equations may also be solved by direct propagation schemes with the nonlocal terms treated in a similar fashion. Variational prescriptions based on the Kohn procedure and using numerical trial functions have also been developed. Finally, we should note that methods have been devised for diatomic systems by which a quadrature is introduced for both the radial and angular variables. This prescription leads to a set of coupled partial DE's, which can in turn be converted to banded matrix equations. Correlation effects have been introduced through polarized orbitals.

We better illustrate the numerical forms of solution by considering the linear algebraic(LA) method. We work in the IE formulation and make a SC expansion to obtain a set of coupled radial integral equations,which are labeled by the channel index Γ. We employ strategy S2 by imposing log-derivative boundary conditions at a particular radial point,a,which is selected to enclose the region of stong,nonlocal interactions. Inside r=a, we invoke the full LA formation and match to solutions at the boundary determined by simple propagation in the outer zone. In the inner region, we also follow strategy S3 and solve for the P-space solution through an effective optical potential. We now convert to a set of matrix equations by introducing a discrete quadrature of η radial points for the integrals and functions. The resulting set of LA equations is of the form:

$$\underline{M}\,\underline{f} \;=\; \underline{g} \; , \tag{13}$$

where the matrix M is of order $0 = \eta \times m$, the vectors f and g are of length 0, and m is the number of channels in P-space. Each element of the solution vector f is designated by a channel Γ and a discrete radial point. Such linear systems can be efficiently solved by iteration-variation techniques(Schneider and Collins,1986a), which do not require the storage of the full matrix. This approach has two clear advantages: 1) the order of the matrix is not increased by the nonlocal terms and 2) the solution is determined at all points simultaneously. In addition, this procedure takes full advantage of the new vector and multi-tasking computers. We generally construct the optical potential from an entirely square-integrable basis. This approximation to the continuum function allows us to determine this term directly from standard CI molecular-structure programs, especially when we combine the procedure with a separable representation(S4). In addition, the method has been used in conjunction with variationally derived polarization- correlation potentials(Gibson and Morrison,1984).

L^2 Methods

In the L^2 approach, we obtained a solution to (4) by expanding in terms of a set of known functions as

$$\psi(\vec{r}) \;=\; \sum_i c_i\,\phi_i(\vec{r}) \tag{14}$$

Many different forms of these functions have been explored. In analogy with bound-state problems, Slater- and Gaussian-type orbitals(STO and GTO) were first used. This approach has the distinct advantage of utilizing the full panoply of molecular structure programs, which have a long history of development. On the other hand, such basis functions, which efficiently represent the multi-center nature of the wavefunction near the nuclei, have difficulty in tracing the oscillatory form of the scattering solution in more extended regions. For this reason, most square-integrable methods now employ in addition to a GTO basis one that better represents this oscillatory behavior. Such augmented bases are constructed from either numerical functions, which satisfy some model potential, or analytical functions, such as bessels, to better represent the continuum solution. For a simple potential, we have a detailed comparison of thirty-five square-integrable methods(Staszewska and Truhlar,1987). Unfortunately for more complicated interactions, such tests are not available.

Although methods, such as the T-matrix, can arise from a direct expansion of the potential in the LS equations in terms of the finite basis, we derive most of the L^2 methods discussed below from variational prescriptions. For the variational methods, we consider a functional $I[\psi_t]$ of a trial function ψ_t. One particular example is

$$I[\psi_t] \;=\; \langle\psi_t|\;(H - E)|\;\psi_t\rangle \; , \tag{15}$$

where the trial function satisfies a prescribed asymptotic boundary condition. The functional has the properties that 1) for the exact solution $I[\psi] = 0$ and 2) for small variations($\delta\psi$) in the trial function, $\delta I = 0$ (stationary). Many variational principles have been devised ; however, we shall confine our attention to two major approaches in which the functional is defined: 1) over a limited spatial volume or 2) over all space.

The R-matrix, eigenchannel, and finite-volume methods, which form the principal approaches of the first class, differ in details but basically follow a common philosophy. We illustrate these techniques by concentrating on the R-matrix(RM) prescription. The R-matrix simply relates the wavefunction to its derivative

$$\psi(r) \;=\; - \, R \, \frac{d}{dr}[\psi(r)] \tag{16}$$

We begin with strategy S2 and apply arbitrary boundary conditions at a finite surface,Σ. Within this surface, we represent the solution by (7) and thus follow strategy S3. The boundary condition effectively converts the scattering problem in the inner region into a bound-state one of the form of (10). We then expand in terms of a set of discrete functions(14),which satisfy the eigenvalue equations

$$H_b \, |\phi_i) \;=\; \epsilon_i \, |\phi_i) \; . \tag{17}$$

We write a formal soultion to (6b) as

$$|\psi) \;=\; (H_b - E)^{-1} \, L_b \, |\psi) \; , \tag{18a}$$

and by inserting a complete set of eigenfuctions defined by (17), we have

$$|\psi\rangle = -\sum_i \frac{|\phi_i\,)\,(\phi_i\,|\,L_b\,|\,\psi)}{(E\,-\,\epsilon_i)} \qquad (18b)$$

Evaluating this at the surface, using the properties of the Bloch operator, and noting the defintion of the R-matrix, we find

$$R = \sum_i \frac{\phi_i(\Sigma)\,\phi_i^*(\Sigma)}{(E - \epsilon_i)}\quad , \qquad (18c)$$

where $\phi(\Sigma)$ schematically represents evaluation of the function at the surface. Since the eigenfunction solutions (17) are independent of the scattering energy, E, we need only solve the eigenvalue equations once in the inner region. The R-matrix can then be determined over a extensive set of energies from (18c) very economically . By matching to known solutions from the outer region, we derive the scattering information.

Many variational prescriptions have been evolved for treating the scattering problem over all space. Some are based on the solutions to the DE's such as the Kohn and Hulthen methods while others like the Schwinger arise from the IE's. Several forms of the Kohn, based on real-value trial functions(K-matrix boundary conditions) are in use although they all suffer from spurious singularities. This problem appears to be rectified by using out-going(complex)solutions and moving these poles off of the real axis. In addition, complex variational prescriptions have been developed in which the standard trial function expansion is augmented by complex basis functions(CBF). These functions, which have components of the form exp(iαr), contain the flexibility to represent the oscillating continuum part. This choice of basis functions allows scattering quantities to be calculated from standard bound-state programs with some modifications. In addition, projection-operator formalisms have been successfully applied, especially for resonances.

At present the Schwinger(SV) approach has received the most attention. We seek an approximate solution to the LS equations in terms of a set of discrete basis functions{|α>}. To this end, we employ a special form of a separable potential(S4)-

$$V(\vec{r}|\vec{r}') = \sum_{\alpha,\beta} \langle\vec{r}|V|\alpha\rangle\,[V^{-1}]_{\alpha\beta}\,\langle\beta|V|\vec{r}'\rangle, \qquad (19)$$

where the V^{-1} term signifies the inverse of the matrix element $\langle\alpha|V|\beta\rangle$. We then substitute this potential into the LS equation (5e) and obtain the following expression for the T-matrix:

$$\langle\vec{k}|\,T\,|\vec{k}'\rangle = \sum_{\alpha,\beta} \langle\vec{k}\,|\,V\,|\,\alpha\rangle\,[D^{-1}]_{\alpha\beta}\,\langle\beta\,|\,V\,|\,\vec{k}'\rangle\quad, \qquad (20a)$$

where

$$D_{\alpha\beta} \equiv \langle\alpha\,|\,V\,-\,V\,G_o^+\,V\,|\,\beta\rangle\,, \qquad (20b)$$

and k(k') is the initial(final) wavenumber of the continuum electron.

158

This form of the transition matrix is equivalent to the Schwinger variational expression for a trial wavefunction expanded in the same basis. Since the trial function is always found in combination with the potential, the basis need only have the range of V. Therefore, unlike the Kohn-type methods, the Schwinger functions are not required to satisfy the asymptotic boundary conditions. The method can be refined through an iterative prescription by which the numerical solutions are used as a new trial wavefunction. The method has been extended to multi-channel problems by working in terms of a projected LS equation(S3) and modified variational principles. Other related prescriptions, such as the C-functional, have also been developed.

Before leaving this section, we should mention that other L^2 methods, such as moment-theory and linear response, have been successfully applied to photoionization processes. The above exposition serves only to give a flavor of this robust and exciting field with a variety of very different approaches available to tackle the many difficult problems encountered.

"This particularly rapid, unintelligible patter
 isn't generally heard, and if it is, it doesn't matter" - Ruddigore

COMPARISON OF METHODS

In this section, we give a brief comparison of the various methods discussed in the previous section. We shall concentrate on those systems that have been treated by several methods so that we may draw some conclusions on the efficacy of the various approaches.

Hydrogen Molecular Ion H_2^+

This forms the simplest molecular target from which we can scatter electrons. Since we only have a two-electron system, we expect that all of the methods can be carried to a high level of convergence. We consider an electron incident on the ion with an energy below the first excitation threshold($1\sigma_u$). Only elastic scattering is possible; however, we do encounter Feshbach resonances arising from the temporary trapping in the doubly-excited states of the compound system H_2. For scattering in the $^1\Pi_u$ symmetry, the lowest such resonance series is of the form $1\sigma_u n\pi_g$ and converges on the $1\sigma_u$ state of the ion. Other such series, associated with higher excited states, may interfere with this one giving rise to spectacular changes in the resonance width, which is related to the trapping time. In Table 1, we compare the resonance widths as a function of internuclear separation,R, in the linear algebraic (LA), R-matrix(RM), and complex basis function(CBF) approaches(Collins et al.,1986). We employ two forms of the LA method: 1) a n-state CC in which correlation functions are used only to relax the strong orthogonality constraint (LACC) and 2) a single P-space state augmented by an effective optical potential(LAOP). We note that the agreement is excellent for this sensitive quantity. We find comparable agreement among the methods for the positions. The situation is somewhat different for the $^1\Sigma_g$ symmetry. We still have very good agreement on the widths and positions; however, the LACC approach has a much slower convergence.

Table 1. Comparison of Resonance Widths Γ(meV) for the lowest $^1\Pi_u$ Resonances Below the First Excited State for $e^- + H_2^+$ Collisions

method	R(a_o)		
	0.80	1.20	2.00
LACC			
n=4	100.45	14.03	11.88
6	85.66	13.65	12.06
8		15.31	12.34
LAOP	100.8	14.35	11.35
RM	90.52	12.96	11.02
CBF		10.96	13.67

Target nCC bases: 4CC,$[1\sigma_g,1\sigma_u.1\pi_u^+]$; 6CC,$[4CC,2\sigma_{g,u}]$; 8CC,$[6CC.1\pi_g^+]$
References: (Collins et al.,1986; Schneider et al.,1987)

This reinforces our initial dictum to maintain the greatest flexibility in the P and Q spaces. These resonances can have profound effects on the photoionization cross sections (Raseev, 1984; Noble and Bruke, 1986b).

Hydrogen

The next simplest system to consider is molecular hydrogen. In Fig. 1., we have already shown that the LA method is capable of reproducing the most salient features of the low-energy elastic scattering. The agreement with other ab initio methods, such as the Schwinger and R-matrix, is excellent(see Gibson et al,1984). In addition, model potential representations of the polarization-correlation effects give reasonably good results(Gibson and Morrison,1984). Calculations have also been performed for inelastic scattering. The most strigent test for the methods arises for singlet-triplet transitions, since the coupling terms depend only on the exchange interaction. In Fig. 3, we present the total inelastic scattering cross section as a function of electron energy for the excitation from the ground to the first excited triplet state. We compare the SV(Lima et al.,1985), RM(Baluja et al.,1985), and LA (Schneider and Collins,1985) methods and find excellent agreement. The LA and SV calculations were performed at a two- state level while the RM results included some extra correlation functions. The lower curve represents an LA calculation in which the strong orthogonality constraint is invoked but no terms are introduced into Q-space to relax this condition. This case corresponds to several older calculations(Chung and Lin,1978;Weatherford,1980) and demonstrates the need to include all allowed relaxation effects in such calculations. The agreement with experiment (Nishimura and Danjo,1986;Khakoo et al.,1987) is also quite good.

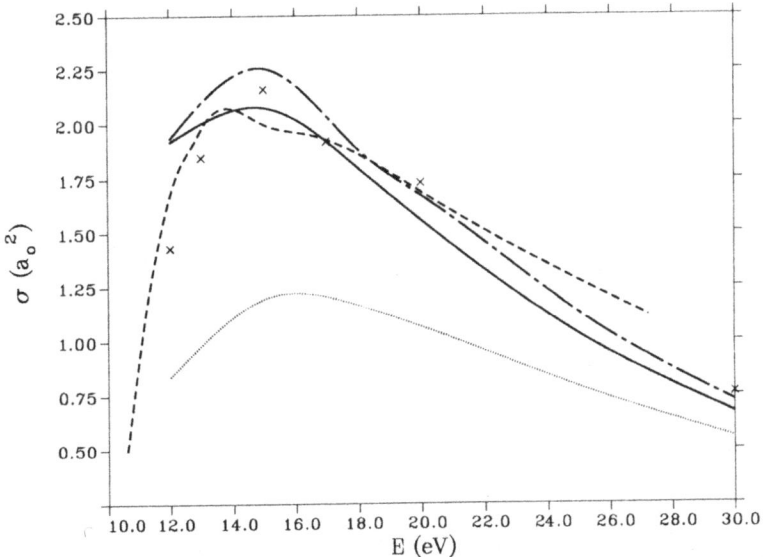

Fig. 4. Comparison of elastic integrated cross sections(a_o^2) for electron scattering from molecular nitrogen in the $^2\Sigma_g$ symmetry. Nomenclature: line - LA(Schneider and Collins,1984); dash - RM(Burke et al.,1983); chain - SV(Huo et al.,1987). Experiment: cross - Kennerly(1980), total cross section.

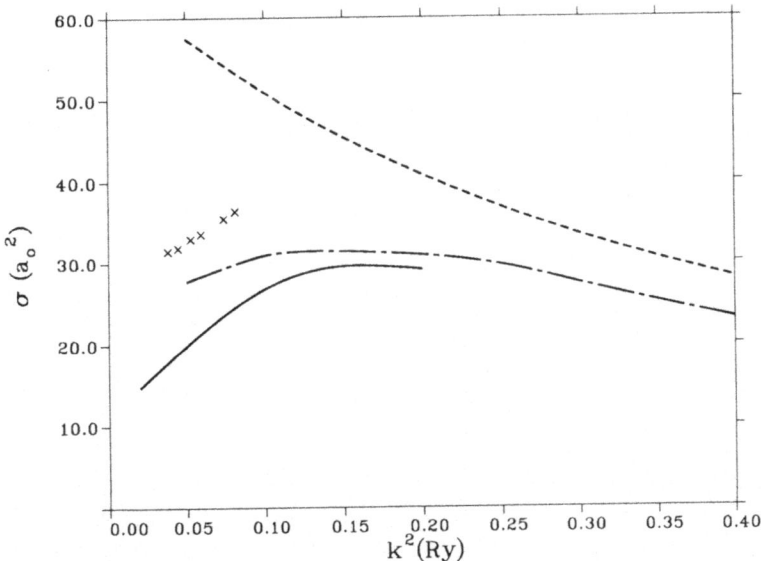

Fig.3. Comparison of theoretical methods and experimental results for electron impact excitation of molecular hydrogen for the transition $X^1\Sigma_g^+$ $\rightarrow b^3\Sigma_u^+$. Theoretical results: line - LA(Schneider and Collins,1985); dash - RM(Baluja et al.,1985); chain - SV(Lima et al.,1985); cross - LA,no relaxation. Experimental: cross - Nishimura and Danjo(1986).

161

Nitrogen

The situation in nitrogen is not as clear. For the $^2\Pi_g$ symmetry, which has a well-known shape resonance at about 2 eV, several methods have successfully reproduced the vibrational structure observed in the elastic scattering cross section. Results with the R-matrix method (Schneider et al.,1979;Morgan,1986) in which nonadiabatic terms were introduced directly into the formal equations yielded quite good agreement with experiment and with other prescriptions (Berman et al.,1983). A recent Schwinger multi-channel calculation(Huo et al.,1987a) using projection-operator techniques also produced the structure in the vicinity of the resonance. Therefore, for the resonance symmetry, the ab initio methods are in reasonably good agreement and are capable of being extended beyond the AN level to produce vibrational structure.

For the $^2\Sigma_g$ scattering symmetry the situation is less propitious. For energies above a few eV, we have fairly good agreement among the various methods. However, at lower energies, we encounter significant differences between the RM(Burke et al.,1983) and the LA(Schneider and Collins,1984) and SV(Huo et al.,1987b) methods. This disagreement can most readily be observed in Fig. 3. We also plot the experimental results of Kennerly (1980) for the total integrated cross section. The comparison at very low energies is valid since the dominant component is the sigma gerade. The source of the differences is not precisely certain. The SV and LA results are within about fifteen percent for the range of energies presented. However, they exhibit different shapes at the very low energies. This behavior may arise from the treatment of the long-range polarization terms. These terms are explicitly included in the LA calculations but are difficult to obtain in the SV since basis functions must span a reasonably large space. The results are very sensitive to the types of correlation used in (8) and to the form of the expansion basis (Schneider and Collins,1984). None of the methods includes correlation of the target function explicitly. The case may be similar to one encounter in atomic scattering in which a delicate balance between bound and continuum correlation had to be struck in order to produce accurate cross sections. This situation will only be rectified by more elaborate calculations, which can systematically vary the amount of correlation in each part and the size of the basis. The capability exists in all of these methods to perform such tests. We should also point out that two model potentials (Padial and Norcross,1984;Morrison et al.,1987) are in better agreement with the experiment than any of the "ab initio" methods.

While disagreements still exist in some cases amongst the methods, the situation is encouraging for treating the direct,exchange,and correlation effects in a consistent fashion for moderately-sized molecular systems.

STATE OF THE FIELD

In this section, we attempt to present the current capabilities in accurately solving the Schrodinger equation for electron-molecule collisions. This presentation is to some extent subjective and will certainly be out-of-date by publication time. This is an unfortunate situation for the authors but a very good one for the field since it demonstrates a vitality and sense of profound progress. Many of these methods are in an inchoate phase, and the next few years should witness a vast extention of our ability to accurately handle electron collisions.

We make a somewhat artificial division based the form of the
interaction.

Static and Local Potentials

At present, for electronically elastic scattering, we are able to
treat static and static+local model potentials for polyatomic systems
having up to third-row atoms. In many cases, these local potentials
effectively model both exchange and correlation effects(Morrison and
Collins,1981;Onda and Truhlar,1980) and have been extended to treat
rovibrational excitations. Some local formulations for electronic excitation have been devised, but their development lags considerably behind
the elastic methods.

Static-exchange

In the SE approximation, we take only a single state in expansion(3)
and treat only the direct and exchange terms. The correlation effects must
come from the interaction with other states and are therefore omitted. This
is a fairly restrictive definition since some authors use SE
interchangeably with CC. Exact SE calculations have been performed for a
wide variety of first- and second-row diatomic and polyatomic molecules.
Such systems as H_2, N_2, CO, NO, CN, Li_2, LiH, LiF, HCl, HF, F_2, HCN, H_2O,
and CH_4 have received routine treatment. While correlation effects play an
important role for neutrals, the SE approximation seems much better for
electron-ion collisions in which the long-range Coulomb term plays an
important role. Thus, the approximation gives substantially better results
for photoionization.

Static-exchange + polarization

For elastic collisions, optical potentials and closed-channel CC
schemes have been devised to account for the correlation effects. These
methods have been particularly effective at low energies in which all but
one channel is closed. Such calculations have been applied to H_2^+, H_2, NO,
NO^+, and N_2.

Electronic excitation

Only in the past few years have the _ab initio_ techniques been honed to
effectively treat electronic excitations although some pioneering work
extends back over a decade. The approaches have been applied to several
systems including H_2^+, H_2, N_2^+, and O_2. Except for the smallest systems, the
calculations have been limited to a very few states. For hydrogen, upto
ten-state CC calculations have been performed.

Remaining Problems

Certain problems cited above such as the disagreement among methods
for a particular system can probably be resolved within the present form of
these procedures, requiring careful and systematic convergence tests.

However, one situation remains, which continues to plague atomic collisions also, namely the introduction of pseudoresonances through limitations on the basis expansions. If we could include all target states in (3) necessary for convergence, then we would never encounter these specious resonances. For low-energy collisions, a few isolated states may be sufficient to describe the scattering event. However, as the energy of the electron increases, more and more states become accessible. For energies of a ten eV or more, hundreds of channels may be open. Clearly, even with the great advances in formulations and computer technology, such large expansions are not feasible. The standard remedy is to represent the effects of the higher channels by a few well-chosen pseudostates. While reducing the size of the calculation to a tractable form, these states give rise to spurious resonances. Several prescriptions have been developed to remove the effects of these resonances by analytically continuing into the complex plane and by performing certain restricted averaging(Burke et al.,1981;Slim and Stelbovics,1987). While these procedures provide a practical solution, they are generally applied after the basic collisional calculations have been performed. The search continues for methods that can be directly inserted into the basic formulation.

In this brief review, we have attempted to present an overview of the basic "ab initio" methods that are currently being used to describe electron- molecule collisions. We have tried to motivate our approach by citing specific examples. We have concluded with a status report on the field and some comments on the remaining problems. We end as we began with a quote from Gilbert and Sullivan, which we hope the sympathetic reader will take to heart –

" If this is not exactly right, we hope you won't upbraid.
  You can't get high esthetic tastes, like trousers, ready made.
  Insight into 'scattering', time alone will bring,
  But as far as we can judge, it's something like this sort of thing."
                          Patience(with liberty)

Acknowledgments

We wish to thank Drs. D. Lynch and N. T. Padial for useful discussions and valuable suggestions concerning the manuscript. Work supported under the auspices of the U.S. Department of Energy through the Theoretical Division of the Los Alamos National Laboratory.

Note added in proof: Recent R-matrix calculations on e^- + N_2 at low energies using pseudostates give results in good agreement with the linear algebraic and Schwinger variational methods. Therefore, for the sigma symmetry the accord among the various ab initio methods is now quite good.

APPENDIX

   In this appendix, we list the major _ab initio_ methods for treating
electron- molecule collisions. The selection is by no means exhaustive
but is representative of the field. In citing sources for each technique,
we have tried to follow a pattern. We have attempted to include an early
paper that outlines the general procedure and a later one that gives a
recent application. If we authors have been diligent in referencing their
earlier works(which they usually are), the latter citation should allow a
trace of the intervening material. We place special emphasis, of course,
on those methods in current use. We generally include calculations at the
static-exchange level or higher.

Method	Authors
Schwinger Variational(SV)	Watson,Lucchese,McKoy,Rescigno(1980)
	Lynch,Lee,Lucchese.McKoy(1984)
	Lucchese,Takatsuka,McKoy(1986)
	Huo,Gibson,Lima,McKoy(1987)
	Berman/Kaldor(1981)
	Berman,Walter,Cederbaum(1983)
R-matrix(RM)	Schneider(1975)
	Schneider,LeDourneuf,VoKyLan(1979)
	Burke,Mackey,Shimamura(1977)
	Noble/Burke(1986a)
	Schneider,LeDourneuf,Burke(1979)
	Nesbet,Noble,Morgan(1986)
	Holley,Chung,Lin(1982)
Linear Algebraic(LA)	Collins/Schneider(1981,1984)
	Schneider/Collins(1986a,b)
	Morrison,Gibson,Saha(1987)
Complex Basis Function(CBF)	McCurdy,Rescigno,,Davidson,
	Lauderdale(1980)
	Yabushita/McCurdy(1986)
C-functional	Lee,Takatsuka,McKoy(1981)
	Basden/Lucchese(1986)
T-matrix	Rescigno,McCurdy,McKoy(1974)
	Fliflet,Levin,Ma,McKoy(1978)
	Klonover/Kaldor(1978)
Finite Volume(FV)	LeRouzo/Raseev(1984)
	Raseev(1985)

<u>Kohn variational</u>

    Real   Takagi/Nakamura(1978)
                                      Collins/Robb(1980)

    Complex   Rescigno,McCurdy,Schneider(1987)

<u>Integral equations</u>(IE)

    Iterative   Collins,Robb,Morrison(1980)

    Non-iterative   Rescigno/Orel(1981)
                                        Padial/Norcross(1984)

<u>Differential equations</u>(DE)   Henry/Lane(1969)
                                      Chung/Lin(1978)
                                      Raseev(1980)

<u>Partial DE's</u>(PDE)   Onda/Temkin(1983)
                                      Weatherford,Onda,Temkin(1985)

<u>Projection operator</u>   Hazi(1978)
                                      Berman,Estrada,Cederbaum,
                                            Domke(1983)
                                      Hara/Sato(1984)

<u>Polarized Orbital</u>   Temkin/Vasavada(1967)
                                      Chandra(1986)

<u>Moment-theory</u>   Langhoff(1980)
                                    Herman/Langhoff(1983)

<u>Linear response</u>   Zangwill/Soven(1980)

REFERENCES

Baluja,K.L.,Noble,C.J.,and Burke,P.G.,1985,J.Phys.B,18:851.

Basden,B. and Lucchese,R.R.,1986,Phys.Rev.A,34:5158.

Berman,M. and Kaldor,U.,1981,J.Phys.B,14:3993.

Berman,M.,Walter,O. and Cederbaum,L.S.,1983,Phys.Rev.Lett.,50:1979.

Berman,M.,Estrada,H.,Cederbaum,L.S.,and Domke,W.,1983,Phys.Rev.A,28:1363

Berman,M.,Mundel,C.,and Domke,W.,1985,Phys.Rev.A,31:641.

Bloch,C.,1957,Nucl.Phys.,4:503.

Buckley,B.D.,Burke,P.G.,and Noble,C.J.,1984,in: "Electron-Moelcule
    Collisions," I.Shimamura and K.Takayanagi,eds.,Plenum,New York.

Burke,P.G.,Mackey,I.,and Shimamura,I.,1977,J.Phys.B,10:2497.

Burke,P.G.,Berrington,K.A.,and Sukumar,C.V.,1981,J.Phys.B,14:289.

Burke,P.G.,Noble,C.J.,and Salavini,S.,1983,J.Phys.B,16:L113.

Chandra,N.,1986,J.Phys.B,19:1959.

Chang,E.S. and Fano,U.,1972,Phys.Rev.A,6:173.

Chung,S. and Lin,C.C.,1978,Phys.Rev.A,17:1874.

Collins,L.A.,Robb,W.D.,and Morrison,M.A.,1980,Phys.Rev.A,21:488.

Collins,L.A. and Robb,W.D.,1980,J.Phys.B,13,1637.

Collins,L.A. and Schneider,B.I.,1981,Phys.Rev.A,24,2387.
_____,1984,Phys.Rev.A,.29:1695.
_____,1986,Phys.Rev.A,34:1564.

Collins,L.A.,Schneider,B.I.,Noble,C.J.,McCurdy,C.W.,and Yabushita,S.,
    1986,Phys.Rev.Lett.,57:980

Dalba,G.,Fornasini,P.,Lazzizzera,I.,Ranier,G.,and Zecca,A.,1980,J.Phys.
    B,13:2839.

Dehmer,J.L.,Parr,A.C.,and Southworth,S.H.,1986, in: "Handbook of
    Synchrotron Radiation,Vol.II, G.V.Marr,ed.,North-Holland,Amsterdam.

Feshbach,H.,1958,Ann.Phys.(N.Y.),5,357.

Fliflet,A.W.,Levin,D.A.,Ma,M.,and McKoy,V.,1978,Phys.Rev.A,17:160.

Gibson,T.L. and Morrison,M.A.,1984,Phys.Rev.A,29:2497.

Gibson,T.L.,Lima,M.A.P.,Takatsuka,K.,and McKoy,V.,1984,
    Phys.Rev.A,30:3005.

Greene,C.H. and Jungen,Ch.,1985,Adv.Atom.Molec.Phys.,21:51.

Hara,S.,1967,J.Phys.Soc.Jpn.,27:1593.

Hara,S. and Sato,H.,1984,J.Phys.B,17:1301.

Hazi,A.,1979,in: "Electron and Photon Molecule Collisions,"T.N.Rescigno,
    V.McKoy,and B.I.Schneider,eds.,Plenum,New York.

Henry,R.J.W. and Lane,N.F.,1969,Phys.Rev.,183:221.

Herman,M.R. and Langhoff,P.W.,1983,Phys.Rev.A, 28:1957.

Hoffman,K.R.,Dababneh,M.S.,Hsieh,Y.F.,Kauppila,W.E.,Poi,V.,Smart,J.H.,
    and Stein,T.S.,1982,Phys.Rev.A,25:1393.

Holley,T.K.,Chung,S.,and Lin,C.C.,1982,Phys.Rev.A ,26:1852.

Huo,W.M.,Gibson,T.L.,Lima,M.A.P.,and McKoy,V.,1987a,Phys.Rev.A(in press).

Huo,W.M.,Lima,M.A.P.,Gibson,T.L.,and McKoy,V.,1987b,Phys.Rev.A(in press).

Jung,K.,Scheuerlein,K.M.,Sohn,W.,Kochem,K.H.,and Ehrhardt,H.,1987,J.Phys.
    B,20:L327.

Kennerly,R.E.,1980,Phys.Rev.A,21:1876.

Khakoo,M.A.,Trajmar,S.,McAdams,R.,and Shyn,T.N.,1987,Phys.Rev.A,35:2832.

Klonover,A. and Kaldor,U.,1978,J.Phys.B,11:1623.

Lane,N.F.,1980,Rev.Mod.Phys.,52:29.

Langhoff,P.W.,in: "Theory and Applications of Moment Methods, B.J.Dalton
    et al. eds.,Plenum,New York.

Lee,M.T.,Takatsuka,K.,and McKoy,V.,1981,J.Phys.B,14:4115.

LeRouzo,H. and Raseev,G.,1984,Phys.Rev.A,29:L214.

Light,J.C. and Walker,R.B.,1976,J.Chem.Phys.,65:4272.

Lima,M.A.P.,Gibson,T.L.,Huo,W.M.,and McKoy,V.,1985,J.Phys.B,18:865.

Lucchese,R.R.,Takatsuka,K.,and McKoy,V.,1986,Phys.Rep.,131:147.

Lynch,D.,Lee,M.T.,Lucchese,R.R.,and McKoy,V.,1984,J.Chem.Phys.,80:1907.

McCurdy,C.W.,Rescigno,T.N.,Davidson,E.R.,and Lauderdale,J.G.,1980,
    J.Chem.Phys.,73:3268.
Morgan,L.,1986,J.Phys.B,19:L439.
Morrison,M.A.,1983,Austr.J.Phys.,36:239.
               ,1987,Adv.Atom.Molec.Phys.(in press).
Morrison,M.A. and Collins.L.A.,1981.Phys.Rev.A,23:127.
Morrison,M.A. and Saha,B.C.,1986,Phys.Rev.A,34:2786.
Morrison,M.A.,Gibson,T.L.,and Saha,B.C.,1987,Phys.Rev.A(in press).
Mundel,C.,Berman,M.,and Domke,W.,1985,Phys.Rev.A,32:181.
Nesbet,R.K.,Noble,C.J.,and Morgan,L.,1986,Phys.Rev.A,34:2798.
Nishimura,H. and Danjo,A.,1986,J.Phys.Soc.Jpn.,55:3031.
Noble,C.J. and Nesbet,R.K.,1984,Comp.Phys.Comm.,33:399.
Noble,C.J. and Burke,P.G.,1986a,J.Phys.B.,19:L35.
                     ,1986b,Int.J.Quant.Chem.,29:1033.
Norcross,D.W. and Collins,L.A.,1982,Adv.Atom.Molec.Phys.,18:341.
Onda,K. and Truhlar,D.G.,1980,J.Chem.Phys.,22:86.
Onda,K. and Temkin,A.,1983,Phys.Rev.A,28:621.
Padial,N.T. and Norcross,D.W.,1984,Phys.Rev.A,29:1742.
Raseev,G.,1980,Comp.Phys.Comm.,20:275.
        ,1985,J.Phys.B,18:423.
Rescigno,T.N.,McCurdy,C.W.,and McKoy,V.,1974,Phys.Rev.A,10:2240.
Rescigno,T.N. and Orel,A.E.,1981,Phys.Rev.A,24:1267.
Rescigno,T.N.,McCurdy,C.W.,and Schneider,B.I.,1987,Phys.Rev.A(in press).
Schneider,B.I.,1979,Phys.Rev.A,11:1957.
Schneider,B.I.,LeDourneuf,M.,and VoKyLan,1979,Phys.Rev.Lett.,43:1926.
                       ,and Burke,P.G.,1979,J.Phys.B,12:L365.
Schneider,B.I. and Collins,L.A.,1984,Phys.Rev.A,30:95.
                       ,1985,J.Phys.B,18:L857.
                       ,1986a,Phys.Rev.A,33:2970.
                       ,1986b,Phys.Rev.A,33:2982.
                       ,and C.J.Noble,1987,Phys.Rev.A(in press).
Slim,H.A. and Stelbovics,A.T.,1987,J.Phys.B,20:L211.
Staszewska,G. and Truhlar,D.G.,1987,J.Chem.Phys.,86:2793.
Takagi,H. and Nakamura,H.,1978,J.Phys.B,11:1675.
Taylor,J.R.,1972,"Scattering Theory," Wiley,New York.
Temkin,A. and Vasavada,K.V.,1967,Phys.Rev.,160:109.
Watson,D.K.,Lucchese,R.R.,McKoy,V.,and Rescigno,T.N.,1980,Phys.Rev.A,
    20:1474.
Weatherford,C.A.,1980,Phys.Rev.A,22:2519.
Weatherford,C.A.,Onda,K.,and Temkin,A.,1985,Phys.Rev.A,31:3620.
Yabushita,S. and McCurdy,C.W.,1986,J.Chem.Phys.,83:3547.
Zangwill,A. and Soven,P.,1980,Phys.Rev.A,21:1561.

ELECTRON SCATTERING BY POLYATOMIC MOLECULES: RECENT ADVANCES IN THEORY

AND CALCULATIONS

F.A. Gianturco and S. Scialla

Department of Chemistry, University of Rome

Città Universitaria, 00185 Rome, Italy

INTRODUCTION

The numerous physical processes which occur when a beam or a swarm of electrons travels through a molecular gas are of considerable import- ance in a wide variety of apparently unrelated areas of molecular physics and physical chemistry. Thus, the importance of electron-impact excita- tion processes involving ground state molecules and molecules in meta- stable excited states has been recognized in the study of ionospheric and auroral processes in planetary atmospheres. Moreover, the discovery and development of discharge-pumped lasers has resulted in a corresponding increase of interest in several energy transfer processes in electron- molecular collisions because of their important role in creating the necessary population inversion[1,2].

The experimental side of the above processes has therefore received a renewed impulse in the last few years and has resulted in the measure- ment of many of the electron impact excitation cross sections involving some of the most common diatomics like H_2, N_2, O_2 and others[3].

The corresponding activity on polyatomic molecules has also been rather intense as far as experiments are concerned and therefore several data on energy transfer probabilities, angular distribution of the scat- tering electrons, and momentum transfer cross sections have been collec- ted for targets like CH_4, H_2O, H_2S, SiH_4 and others[4].

On the theoretical side, on the other hand, the complete development of reliable computational methods based on ab initio procedures has been slow to come and only the last couple of years have seen the emergence of theories that allow us to obtain nearly quantitative agreement with experimental observables like differential and integral cross sections, partial and total, and momentum transfer cross sections for scattering processes which involve molecular targets in their ground states[5].

The corresponding quality of agreement for electronic excitation processes or for collisions with metastable species, however, is still to be found and we will not discuss this particular aspect in the presenta- tion of the computational approach followed in our group. What we will try to show is that theoretical methods which can provide quantitatively reliable cross sections for elastic and inelastic scattering of low-energy electrons by molecules are beginning to appear and that non-empirical models can be effectively used to generate such quantities for several polyatomic targets.

In the following section we describe in some detail the theoretical method that we have used in our most recent calculations which involve polyatomic targets. Specific examples and a discussion of our findings are presented in the results section.

THEORETICAL DEVELOPMENT

As is well known, the non-relativistic Hamiltonian for an electron incident upon an N-electron molecule with M nuclei may be written, in atomic units, as:

$$H_{tot} = - 1/2 \, \nabla^2_{N+1} + V_{N+1} + H_N + H_{rot} + H_{vib} \tag{1}$$

$$= H_{el} + H_{nuclei}$$

where the first two terms represent the kinetic energy of the incident electron and its interaction with the target electrons and nuclei, H_N is the electronic Hamiltonian of the isolated molecule and the last two terms are simply the nuclear kinetic-energy operators in the COM of the system. The total collision wavefunction for the total (N+1)-electron system is obviously the solution of the corresponding time-independent Schrödinger equation subject to the appropriate collision boundary conditions.

The usual procedure for obtaining such a wavefunction is to write an expansion of the form:

$$\Psi = \sum_{\alpha} (\phi_i \, \chi_v \, \rho_j)_{\alpha} \, F_{\alpha}(N+1) + \sum_k c_k \, \Phi_k \, (1\ldots N+1) \tag{2}$$

where ϕ_i, χ_v, ρ_j are the electronic, vibrational and rotational eigen-functions of the asymptotic molecular target, which are in general assumed to be closely coupled by the interaction, whilst the second expansion involves a set of known electron correlation functions which describe the target response to the incoming electron. The k-summation therefore includes functions which would otherwise be omitted due to the usual orthogonality constraint between bound and continuum orbitals and the α-summation.

One could, however, approximate the expansion (2) with only one term in the first summation and without any contribution from the second sum. This level of approximation is clearly the simplest way to describe the N-electron target wavefunction and corresponds to the single-configuration representation of an unperturbed Hartree-Fock (HF) ground state of a closed-shell $^1A_1$ molecular target plus a continuum electron. Moreover, the further approximation of holding the M nuclei fixed in space during the collision process disregards the effect of the H_{nuclei} operator on the total wavefunction and yields a set of exact static-exchange (ESE) equations for the unknown continuum orbital within the fixed nuclei (FN) approximation. The referred frame of reference is the one which is rigid within the molecular system (BF frame)[6].

$$\{-1/2 \, \nabla^2_r + V_{1A_1} \, (\underline{r}) - 1/2 \, k^2\} \, F_{\gamma_i} (\underline{r}) =$$

$$= \sum \, \{(\epsilon_s - 1/2 \, k^2) \langle \phi_s | F_{\gamma_i} \rangle + \langle \phi_s | |r-r'| | F_{\gamma_i} \rangle \} \phi_s \tag{3}$$

where the (N+1) subscript has been dropped from the radial variable and the spin part of all electrons has been integrated out. The s-summation now runs over the occupied, single particle LCAO-MO's which describe the single determinant used to represent the closed-shell electronic ground state of the M bound electrons. The ε_s represents the single particle energies for each of the bound orbitals ϕ_s. The index γ_i labels the symmetry of the continuum electron or, more precisely, the irreducible representation (IR) of the BF molecular point group to which the continuum wavefunction belongs.

The operator $V\,^1A_1(\underline{r})$ is now the static, direct interaction between the incoming electron and the undistorted electronic configuration of the molecular target. It can be written as a multipolar expansion over symmetry adapted components[7]:

$$V\,^1A_1(\underline{r}) = \sum_{h,\ell} V_{h\ell}\,^1A_1(\underline{r})\; X_{h\ell}\,^1A_1(\underline{r}) \tag{4}$$

which is centered around the COM of the molecule in question and where the coefficients can be obtained by numerical quadrature over the molecular charge distribution yielded by an SCF-HF calculation of all the occupied LCAO-MO of the target molecule. The latter orbitals were obtained by us with a single-centre (SC) basis of STO atomic orbitals for the systems discussed below[8].

The index h labels different bases of the same irreducible representation $^1A_1$ that correspond to the same angular momentum ℓ, while the X coefficients are linear combinations of real or imaginary spherical harmonics:

$$X_{h\ell}^{p\mu}(\underline{r}) = \sum_m b_{sh\ell m}^{p\mu}\; Y_{\ell m}(\underline{r}) \tag{5}$$

here p denotes the chosen IR and μ distinguishes each component of the basis if that IR has dimensions greater than one. The index m assumes only positive values[9].

The RHS of Eq. (3) contains the exchange kernel which has been written in terms of two contributions to express more clearly its physical meaning. The first contribution reminds us that the explicit orthogonalisation of the continuum orbital to each bound MO of the same symmetry is an exact requirement for closed-shell targets and has the further advantage of ensuring in the unknown continuum orbital the precise nodal structure of the short range region that contains the bound orbitals within the ESE approximation.

The second contribution to the exchange kernel contains the attractive corrections of the two-electron repulsion term due to the Pauli exclusion principle.

The continuum orbital can also be expanded over symmetry-adapted functions:

$$F_{\gamma_i}(\underline{r}) = \sum_{h\ell} r^{-1}\, g_{h\ell}^{\gamma_i}(r)\; X_{h\ell}^{\gamma_i}(\underline{r}) \tag{6}$$

and therefore, when projecting Eq. (3) into the basis functions X one obtains a set of coupled equations for each channel γ_i, which is here the ground state IR of a closed-shell target:

$$\left\{\frac{d}{dr^2} - \frac{\ell(\ell+1)}{r^2}\right) + k^2\right\} \, g^{1A_1}_{h\ell\ell o} \, (r)$$

$$= 2 \sum_{h',\ell'} \{V^{1A_1}_{h\ell,h'\ell'}(r) + W^{1A_1}_{h\ell,h'\ell'}(r)\} \, g^{1A_1}_{h\ell} \tag{7}$$

where the label ℓ_0 reminds us of the dependence of the g's on the initial, asymptotic angular momentum ℓ_0 of the impinging electron.

The first group of matrix elements on the RHS of Eq. (7) represents the direct interaction via the static potential of Eq. (4):

$$V^{\gamma_i}_{h\ell,h'\ell'} = \langle X^{\gamma_i}_{h\ell} \, |V^{1A_1}_{st}| \, X^{\gamma_i}_{h'\ell'} \rangle \tag{8}$$

and each term of (8) describes one of the static multipole moments caused by the undistorted molecular charge distribution at the value r of the scattering variable.

No short-range terms as those given by the k-summation of Eq. (2) have been introduced at this level of approximation.

The effect of the exchange operator W of Eq. (7) on the continuum radial functions could be schematically written as:

$$W^{1A_1}_{h\ell,h'\ell'} (r) \, g^{1A_1}_{h'\ell'} (r) = \int_0^\infty K^{1A_1}_{h\ell,h'\ell'} (r,r') \, g^{1A_1}_{h'\ell'} r') \, dr' \tag{9}$$

where the inner operator K can be expressed in terms of the radial functions obtained by expanding the bound HF orbitals over the same symmetry-adapted functions used in Eq. (6).

In order to circumvent the extra difficulty introduced by the exchange kernel of Eq. (9), especially for polyatomic targets, we will report below some of the recent approximations that we have employed to solve ESE equations for non-linear molecules.

The Electron Gas Model Exchange

The exchange contribution of Eq. (3) could be rewritten as:

$$\frac{\left\{\Sigma \, \phi_s^*(\underline{r}') \, F^*_{\gamma_i} (\underline{r}) \, |\underline{r} - \underline{r}'|^{-1} \, \phi_s(\underline{r}) \, F_{\gamma_i} (\underline{r}')dr'\right\} \, F_{\gamma_i} (\underline{r})}{F^*_{\gamma_i} (\underline{r}) \, R_{\gamma_i} (\underline{r})} \tag{10}$$

where the term in brackets could be approximately evaluated by replacing both the bound and continuum orbitals with plane waves and by replacing the sum over bound functions by an integral in $\underline{k}$ space up to the Fermi level of the free electron gas[10]. The so-called Hara's modification (Hara Free Electron Gas Exchange)[11] uses for the continuum electron a local moment which is obtained from the conservation of energy based on the asymptotic electron energy and on the potential felt by the outermost bound electrons:

$$1/2 \, K_F^2(\underline{r}) + V'(\underline{r}) = - I \tag{11a}$$

and since:

$$1/2 \, k^2(\underline{r}) + V(\underline{r}) = 1/2 \, k_0 \tag{11b}$$

by putting $V'(\underline{r}) = V(\underline{r})$ one produces an expression for the local moment of the continuum electron:

$$K(\underline{r}) = \{k_0 + 2I + K_F^2(\underline{r})\}^{1/2} \tag{12}$$

The final expression which can then be obtained for an energy dependent, local form of exchange is therefore given by the familiar expression[10]:

$$V_{ex}^{1A_1}(\underline{r}) = \frac{2}{\pi} K_F(\underline{r}) \left\{ 1/2 + \frac{1-\eta^2}{4\pi} \ln \left| \frac{1+\eta}{1-\eta} \right| \right\} \tag{13}$$

where:

$$K_F = \left(3\pi^2 \rho(\underline{r})\right)^{1/2} \tag{14a}$$

$$\eta(\underline{r}) = [k_0 + 2I + K_F^2]^{1/2}/K_F \tag{14b}$$

Here $\rho(\underline{r})$ is the target HF electronic density in the $^1A_1$ IR and I is the first molecular ionisation potential. The model is usually referred to as the HFEGE model.

One immediately notices that in the limit of $r \to \infty$ the asymptotic energy of the electron in Eq. (12) is given by $1/2\ k_0 + I$, which is incorrect. One could therefore modify the above result by introducing an asymptotic adjustment in the quantity of Eq. (14b):

$$\eta(\underline{r})\ 1\ [k_0^2 + K_F^2]^{1/2}/K_F \tag{15}$$

which is called the AAFEGE model.

The Semiclassical Approximations

If one disregards for the moment the orthogonality requirements in Eq. (3), the ESE equations can be written as:

$$\left[-1/2\ \nabla_r^2 + V_{ST}^{1A_1}(\underline{r}) - 1/2\ k\right] F_{\gamma_i}(\underline{r}) = \sum_s \phi_s(\underline{r}) \int d\underline{r}' \frac{\phi_s^*(\underline{r}')\ F_{\gamma_i}(\underline{r})}{|\underline{r} - \underline{r}'|} \tag{16a}$$

or as:

$$\left[-1/2\ \nabla_r^2 + V_{ST}^{1A_1}(\underline{r}) - 1/2\ k^2\right] F_{\gamma_i}(\underline{r}) = -\ \mathscr{L}\ (\underline{r},k^2)\ F_{\gamma_i}(\underline{r}) \tag{16b}$$

which can be written as:

$$(\nabla_r^2 + K_0^2(\underline{r}))\ F_{\gamma_i}(\underline{r}) = 0 \tag{17a}$$

with:

$$K_0^2(\underline{r}) = k^2 - 2\ V_{ST}^{1A_1}(\underline{r}) - 2\quad (\underline{r},\ k^2) \tag{17b}$$

where an energy-dependent, local form of the exchange interaction has been introduced to indicate the quantity we are seeking to calculate.

If one now performs a Taylor expansion around $\underline{r}$, the point in space where each exchange integral is computed[12,13]:

$$\int d\underline{r}_1 \; \phi_s^*(\underline{r}_1) \; |\underline{r} - \underline{r}_1|^{-1} \; F_\gamma(\underline{r}_1)$$

$$= \int d\underline{r}' \; \frac{1}{|\underline{r}'|} \; \exp\left[(\nabla_{\phi_s} + \nabla_{F_{\gamma_i}}) \; \underline{r}'\right] \; \phi_s^*(\underline{r}) \; F_{\gamma_i}(\underline{r}) = I \tag{18}$$

where $\underline{r}' = \underline{r}_1 - \underline{r}$ and the ∇ operators involve, according to their subscripts, either the bound electron wavefunctions ϕ_s or the continuum function $F_{\gamma_i}(\underline{r})$. By making use of spherical polar coordinates and integrating over the volume element one can then solve the integral and obtain[13]:

$$I = - \frac{4\pi}{|\nabla_{\phi_s} + \nabla F_{\gamma_i}|^2} \; \phi_s^*(\underline{r}) \; F_{\gamma_i}(\underline{r}) \tag{19}$$

Since the bound functions are producing slowly varying amplitude factors within the molecular volume, while the continuum functions are faster oscillating functions of the spatial variable[14], one could disregard the ∇_{ϕ_s} operator with respect to the $\nabla_{F_{\gamma_i}}$ operator in Eq. (19) and

therefore obtain, after a few simple steps[13], the following expression for the semiclassical exchange (SCE) interaction:

$$V_{ex}(\underline{r},k^2) = 1/2 \left[E_0 - V_{st}^{1A1}(\underline{r})\right] - 1/2 \left\{\left[E_0 - V_{st}^{1A1}(\underline{r})\right]^2 + 8\pi \sum_s |\phi_s|^2\right\}^{1/2} \tag{20}$$

The main point of the above approximation is that the local momentum of the bound electrons can be disregarded with respect to that of the impinging projectile. This is acceptable at rather large collision energies where it is realistic to assume that electron-molecule collisions modify only slightly the velocity of the continuum electron.

If one, however, wants to deal with lower collision velocities a way of including more correctly the effects of the bound electron momenta on the velocity of the projectile starts from the full form of Eq. (19) and rewrites as follows the gradient operators appearing in it:

$$\left|\nabla_{\phi_s} + \nabla_{F_{\gamma_i}}\right|^2 = \left|\nabla_{\phi_s} + \nabla_{F_{\gamma_i}}\right|^2 + \left|\nabla_{F_{\gamma_i}}\right|^2 + 2 \left|\nabla_{\phi_s}\right| \cdot \left|\nabla_{F_{\gamma_i}}\right| \cos \mathscr{V}_i \tag{21}$$

where the classical meaning of $\mathscr{V}_i$ is given as the angle formed between the direction of the local wavefunction of the impinging electron and that of the bound electron. If one then writes, from Eq. (17b), that:

$$\nabla_{F_{\gamma_i}} \; F_{\gamma_i}(\underline{r}) = - K_0^2(\underline{r}) \; F_{\gamma_i}(\underline{r}) = - 2 (E_0 - V_{st}^{1A_1} - V_{ex}^{1A_1}) \; F_{\gamma_i}(\underline{r}) \tag{22}$$

The effect of ∇_{ϕ_s} could be seen, on the other hand, as equal to the average of the square of the local wavevector for the bound electrons[14]:

$$\nabla_{\phi_s} \; \phi_s^*(\underline{r}) \sim < K^2(\underline{r}) > \phi_s^*(\underline{r}) \tag{23}$$

One can then use the FEG model, as discussed before, to treat the bound electrons' distribution[10] and therefore one obtains for the above average that:

$$< K^2 > = 3/5 \ K_F^2(\underline{r}) \tag{24}$$

where K_F was defined in Eq. (14a).

One can therefore write in a similar way the cross product which appears in Eq. (21) and generate all the average momentum values pertaining to the bound electrons by simply making use of the FEG model discussed above.

Once the integral of Eq. (19) is fully computed one can take advantage of some of the results employed to yield the SCE expression to obtain finally a modified semiclassical exchange (MSCE) interaction in an energy-dependent local form[13,15].

$$V_{ex}^{1A_1}(\underline{r};k^2) = 1/2 \left[E_0 - V_{st}^{1A_1}(\underline{r}) + 3/10 \left[3\pi^2 \ \rho(\underline{r})^{2/3} \right. \right. \tag{25}$$

$$- 1/2 \left\{ \left[E_0 - V_{st}^{1A_1}(\underline{r}) + 3/10 \ (3\pi^2 \ \rho)^{2/3} \right]^2 \ 1 \ 4\pi \ \rho(\underline{r}) \right\}^{1/2}$$

where $E_0 = 1/2 \ k_0^2$ is the scattering electron asymptotic energy.

One sees, by comparing Eqs. (25) and (20) that in the MSCE potential the local velocity of the continuum particle, in a classical sense, is modified by both the existence of the static potential and by the local velocity of the bound electrons. Thus the present modification is expected to be weaker at low collision energies where most of the interesting features which are typical of electron scattering from molecules are experimentally observed.

The Correlation Effects

To improve upon the previous ESE equations one now needs the inclusion of the electronic correlation. The effect of this correlation is to correct the average motion of all the electrons so that, within the molecular volume, each electron surrounds itself with an additional Coulomb hole from which other electrons are excluded.

Thus, one is attempting to introduce an electron-electron interaction not on the average but in every region of configuration space.

When one considers the scattering particle sufficiently outside the space of the target electrons, then the above correlation (which is a short-range effect) takes the form of charge polarisation of the inner bound electrons. Qualitatively, the induced polarisation arises from the distortion of the target charge distribution by the time-varying electronic field of the projectile and gives rise to an additional attractive term in the potential energy of the (N+1) electron system. Its simple asymptotic form for a molecule of T_d symmetry is given by[16]:

$$V_{pol}(\underline{r}) \underset{r \to \infty}{\sim} - \frac{\alpha_0}{r^4} - \frac{\alpha_2}{r^4} \times 1/2 \ (3 \cos^2 \mathscr{V} - 1) - \frac{\alpha_2'}{r^4} \times 1/2 \ \sin^2 \mathscr{V} \cos 2\phi \tag{26}$$

where α_0, α_2 and α'_2 are linear combinations of the molecular dipole polarisabilities defined along the three main axes, i.e. of α_{xx}, α_{yy} and α_{zz}.

Unfortunately, such an equation only applies in the asymptotic limit and its proper form at smaller distances is not known in general nor can one clearly define for all molecules when the asymptotic region is reached by the projectile.

Earlier attempts for polyatomic targets employed a parametric cut-off function to modify Eq. (26) and fixed the parameter by adjusting a computed quantity to a specific scattering observable of the system[16]. Later attempts introduced a polarisation potential based on the methods of Pople and Schofield and the work of Temkin[17] but had to define ad hoc procedures to separate inner and outer regions of polarisation within the total molecular charge density.

A more recent development has been the use of the theory of an inhomogeneous electron gas to obtain correlation-polarisation forces for atomic systems[18], for diatomic molecules[19] and for polyatomic molecules[20].

Briefly, the method considers N interacting electrons in the presence of an external field $V(\underline{r})$ and starts by showing that the electron density $\rho(\underline{r})$ in the ground state determines the potential which acts on the system, as well as all the other properties of the system and that all these quantities could be considered as functions of the total density[14]. One could then try to describe the density in a real situation where interactions are present by defining an effective potential that acts on the equivalent system without interaction and that can be obtained by solving an SCF problem as given by the following equations for a free electron gas (FEG)[21]:

$$\{-1/2 \ \nabla_r^2 + V_{eff}(\underline{r})\} \ u_i(\underline{r}) = \varepsilon_i \ u_i(\underline{r}) \tag{27a}$$

$$\rho(\underline{r}) = \Sigma \ n_i \ |u_i(\underline{r})|^2 \tag{27b}$$

$$V_{eff}(\underline{r}) = V(\underline{r}) + \int \frac{\rho(\underline{r}) \ dr'}{|\underline{r} - \underline{r}'|} + d/d\rho \ E_{ex}(\rho) + d/d\rho \ E_{corr}(\rho) \tag{27c}$$

One can also define the quantity $E_{corr}(\underline{r})$ as being the single-particle correlation energy at the point in space $\underline{r}$. The latter quantity can be related to the N-electron quantity of Eq. (27c) by the following relation:

$$V_{corr} = d/d\rho \ E_{corr}(\rho) = d/d\rho \ \{\int \rho(\underline{r}) \ \varepsilon_{corr}(\underline{r}) \ d\underline{r}\} \tag{28}$$

At this point one can replace $\varepsilon_{corr}(\underline{r})$ with the average correlation energy of an electron in an homogeneous electron gas for which the density can be obtained in a local form from well known equations in which a parameter r_s appears and is defined as follows:

$$4/3\pi \ r_s^3 \ a_0 \ \rho(\underline{r}) = 1 \tag{29}$$

It constitutes the radius of an imaginary sphere containing one electron[22].

Within the high density regions the Hamiltonian is dominated by the particle kinetic energy and therefore the perturbing potential can be treated only by the lower orders of perturbation, thus yielding a specific functional form for $\varepsilon_{corr}(r_s)$[23].

In the low-density regions, on the other hand, the potential interaction dominates and it was found long ago[24] that it reaches a minimum when the particles (electrons) are localized at the nodal positions of a periodic lattice of cfc symmetry and their displacements from equilibrium are treated perturbatively. In the intermediate region one can then resort to an interpolation procedure between the two previous regions[25].

By using such analytic expressions from the FEG model to describe $E_{corr}(\underline{r})$ in Eq. (28) one obtains the following expression for the correlation potential:

$$V_{corr}(\underline{r}) = (1 - \frac{r_s}{3} \frac{d}{dr_s}) \; \varepsilon_{corr}(\underline{r}) \tag{30}$$

which therefore can be analytically obtained and expanded over symmetry-adapted functions:

$$V_{corr}^{1A_1}(\underline{r}) = \sum_{h,\ell} C_{c,h\ell}^{1A_1}(r) \; X_{h\ell}^{1A_1}(r) \tag{31}$$

where the V_c coefficients are obtained by angular integration from the full potential V_{corr}[20].

It is interesting to note, however, that the above development of the short-range correlation forces includes in its form the treatment of exchange terms between interacting particles within the high-density region[23]. In some sense, therefore, the correlation forces are specific for the interaction with an extra electron to the FEG description of the target density. One could therefore obtain a modified expression for correlation forces arising from the perturbing presence of a positron, whenever the positronium channel can be disregarded or considered negligible. The final result is also a functional expression for the high-, low- and intermediate density regions that differs however from the previous form for electrons and can be used to describe more realistically polarisation forces in the case of positron scattering from polyatomic molecules[26].

The final scattering equations which have to be solved for the electron-molecule system therefore contain static, exchange and polarisation contributions (SEP) to the full interaction and include the constraint of continuum orthogonalisation to the bound MO's of the same symmetry via the usual Lagrange multipliers λ_β:

$$\{-1/2 \nabla_r^2 + V_{st}^{1A_1}(\underline{r}) + V_{ex}^{1A_1}(\underline{r};k^2) + V_{pol}^{1A_1}(\underline{r}) - 1/2 \; k^2\} \; F_{\gamma_i}(\underline{r}) = \sum_\beta \lambda_\beta \; \phi_\beta(\underline{r}) \tag{30}$$

RESULTS OF CALCULATIONS

Since the above computational model contains no adjustable parameters and starts from molecular densities obtained from SCF calculations of near-Hartree-Fock quality, it becomes important to verify its reliability in reproducing experimental findings of several scattering observables related to polyatomic molecules.

It is worth noting at this point that many other approaches have been able to treat scattering from polyatomic targets without recourse to parametric potentials or to some empirical formulation of parts of the full interactions (as done, for instance, in Ref.(27)). More recently, the Continuum Multiple Scattering method (CMS) has been freed from some of its earlier parametric choices and is currently using a parameter-free model which has computed differential and total cross sections for elastic scattering in CH_4[28]. It still does not carry out scattering calculations by fully generating the continuum orbitals and therefore it is difficult to see clearly in it the interplay of all the models employed by this procedure. More rigorous computations have also been carried out by the Caltech group which made use of the multichannel formulation of the Schwinger variational method and applied it to solve ESE coupled equations for CH_4[29] and H_2O[30] elastic scattering processes. The latter approach, although computing the correct exchange integrals on an L^2-basis expansion of the wavefunction, suffers from basis set limitations and is not yet able to include polarisation forces to fully perform conclusive comparisons with experimental findings.

The present model on the other hand has been tested on several polyatomic targets and, as will be shown, solves the correct scattering equations that generate continuum functions via the parameter-free model potentials discussed before, and allows one to compare computational results with elastic and inelastic scattering observables, integral and differential. It will be evident later that the many comparisons already carried out for several molecular cases indicate nearly quantitative agreement for most of the observed quantities and underlines the efficiency of the present model for attaining a rather good level of confidence in its predictive behaviour.

One of the most studied systems has been the CH_4 molecule and a rather extensive comparison of the numerous, recent experimental data with available theories has already been presented by us[13,20]. An indication of the relative importance of the various contributions to the scattering equations is shown in Fig. 1, where the spherical components of the static (V_{st}), Hara Free Electron Gas Exchange (V_{ex}) and Correlation-polarisation (V_{pol}) potentials are shown and an earlier form of polarisation potential is also reported ($V_{pol}(AT)$)[17]. One clearly sees the strong dominance of the Coulomb interaction within the short-range region and the decrease of the exchange interaction outside the position

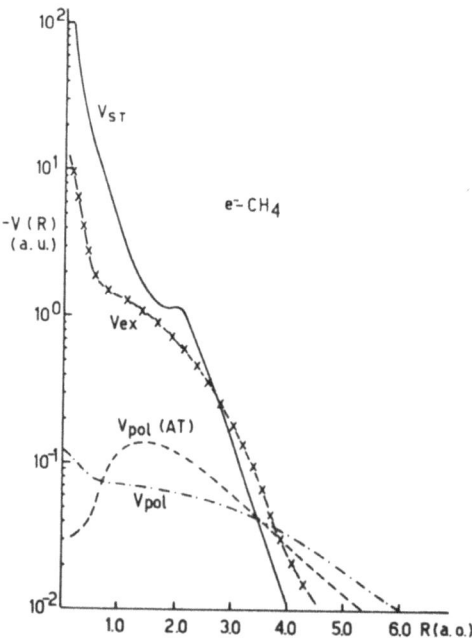

Fig.1. Spherical components of the various contributions to electron-methane interaction. The symbols are explained in the main text.

of the H atom. The present polarisation potential is weaker than the earlier model potential of Jain and Thompson[17] but extends further out than the former, a behaviour which markedly effects, as we shall see later, the low energy scattering cross sections.

In the earlier discussion we showed that local forms of exchange potentials could be obtained within a semiclassical scheme starting from the HF charge density of the molecule. An example of their differences in behaviour is shown, for the CH_4 target, in Fig. 2.

The energy chosen is ~ 0.2 eV, very close to the Ramsauer-Townsend (RT) region of the total cross section. The curve marked with triangles refers to the HFEGE potential adjusted for the correct asymptotic behaviour (AAFEGE potentials), while the simple continuous line refers to the SCE potential discussed before. The open circles are the results from the modified SCE potential (MSCE) recently introduced by us[13]. One clearly sees that to include a local velocity correction makes the exchange interaction follow more realistically the shape of the molecular charge distribution and markedly lowers the exchange forces when a further e^--e^- interaction is included within the MSCE model[13]. A more precise idea of the quality of the present calculations is given by the results of Fig. 3, where the integral elastic computed cross sections around the d-wave methane resonance are shown in comparison with experiments. No adjustments have been made and both computed and measured quantities are on an absolute scale[13].

The full line refers to the SCE model with the orthogonality constraint (OSCE), while the dots and the dashed curves refer to Hara's potential plus orthogonalisation, i.e. to OAAFEGE and to OHFEGE, respectively. The chained curve refers to the OMSCE potential. It is reassuring to see that both semiclassical models are very close to the elastic experimental data ([] and x) from Refs.31 and 32. The experimental points marked with triangles refer to the total cross sections of Jones[33], through which inelastic contributions could be estimated from our calculations.

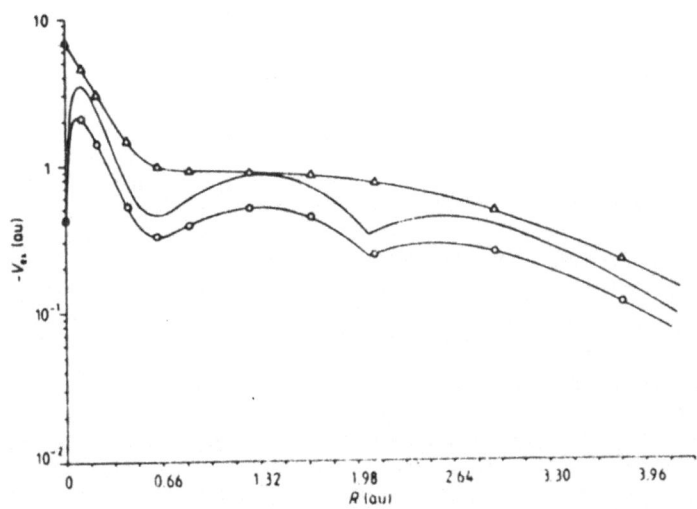

Fig. 2. Radial dependence of the local exchange potentials for e^--CH_4 along the C-H bond direction. The meaning of the symbols is reported in the main text.

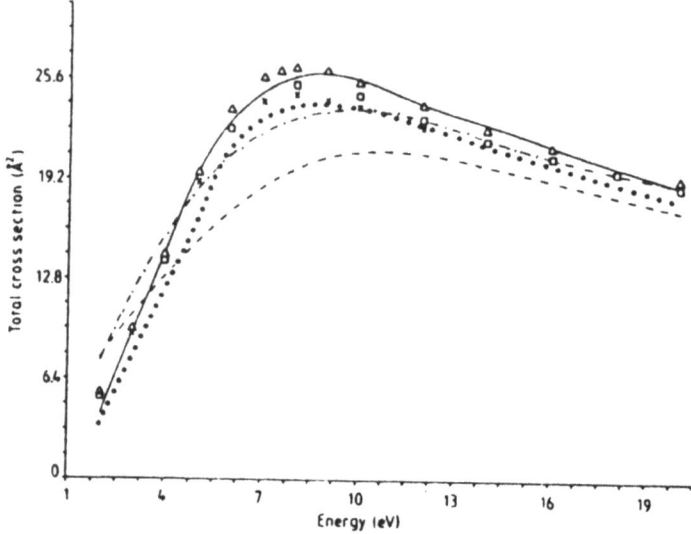

Fig. 3. Computed and measured elastic integral cross sections with different exchange potentials and the V_{corr} discussed in text. See the text for the meaning of the symbols.

It is also interesting to note that the present model fares reasonably well even for a more complicated system like SiH_4, for which we have computed elastic integral cross sections (rotationally summed) in the very structured region below 20 eV[20].

One sees in Fig. 4 that calculations clearly indicate a strong resonant behaviour of the scattering in the t_2 symmetry, as surmised by experiments[20], which primarily attribute to that component the low-energy behaviour of the $e^- - SiH_4$ cross section. The inset also shows that, as the energy decreases, the d-wave component of the t_2 scattering contributes very little to the cross section which is instead dominated by s-wave scattering in the a_1 symmetry[20].

Differential cross sections are also very indicative of the behaviour of the scattering event and tell us a great deal about the dominant forces during the interaction. An example of the quality achieved by computed angular distributions for a polar polyatomic target is shown in Fig. 5 below. The results refer to the H_2O molecule and calculations were carried out, by including static and exchange interactions only at several collision energies and over a wide angular distribution[34]. As one can see, at 20 eV the computed results are very close to the experiments and follow rather well the previous calculations with the correct exchange contributions[30]. The differences between theoretical results in the large angle region are awaiting experimental decision, although momentum transfer cross sections indicate absence of the strong back scattering surmised by earlier calculations and agree more with the present results[36]. In the case of the water molecule, it is also important to compare with experiments the general behaviour of rotationally summed differential cross sections as the energy changes and when polarisation effects are included. The present model also does rather well in this case, as the comparison with the experimental data of ref.36 shows in Fig. 6. As one easily sees, forward scattering dominates the process at all energies and the calculated quantities follow very closely the experimental findings at all the angles observed[36].

180

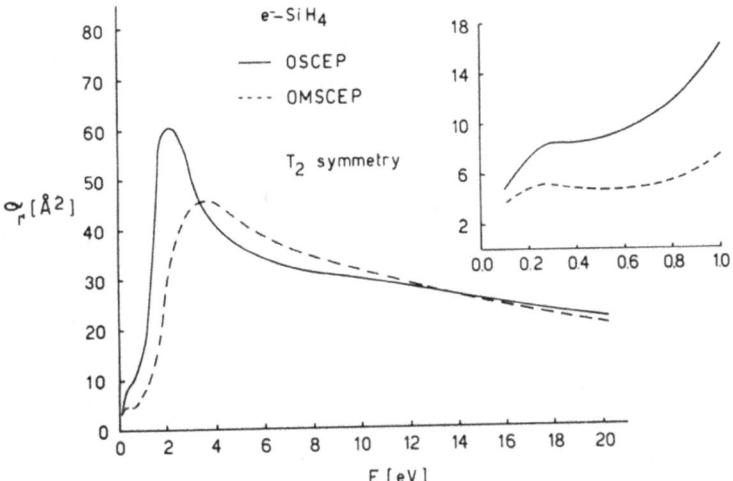

Fig. 4. Computed partial cross sections (t_2 symmetry) with different semiclassical exchange potentials. The inset shows their behaviour below 1.0 eV.

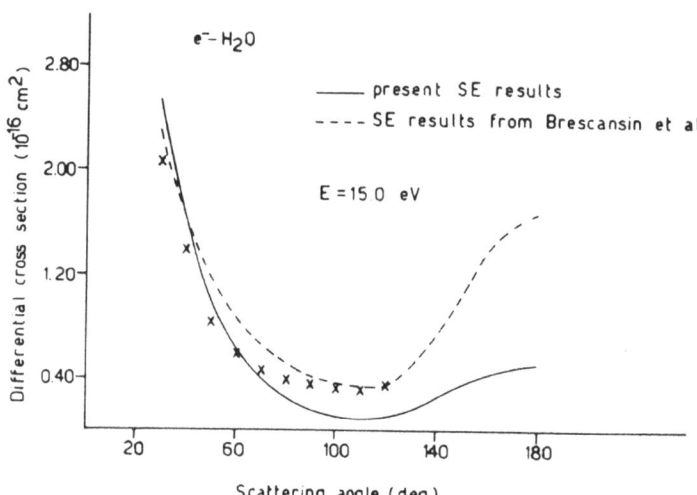

Fig. 5.   Rotationally summed differential cross sections computed with
the present method (solid line) and by the Schwinger variational
approach of ref.30 (dashed line).   The crosses are the experimental values[36].

They are rotationally summed cross sections and indicate three distinct scattering region: (1) a threshold peak dominated by the A_1 scattering state; (ii) an enhancement around σ eγ due to the B_2 scattering state (a rather narrow resonance) and a broad feature between 9-15 eV mainly attributed to the B_1 scattering state.

A general configuration of this behaviour can be obtained from analysis of the eigenphase sums and of their energy dependence within the range of collision energy discussed before. This is shown by the computed results of Fig. 8.

One clearly sees that the A_1 component is varying only slowly with energy but exhibits a sharp increase at threshold. On the other hand, the B_2 component shows a marked shape resonance at low collision energies, dominated by p-wave scattering. Finally, the B_1 component indicates a strong increase with energy but over a broad range of energy values, thus suggesting the broad resonance feature shown by the cross sections of Fig. 7.

In conclusion, the present ab initio method introduces parameter-free models in the treatment of electron-molecule interactions and cor-

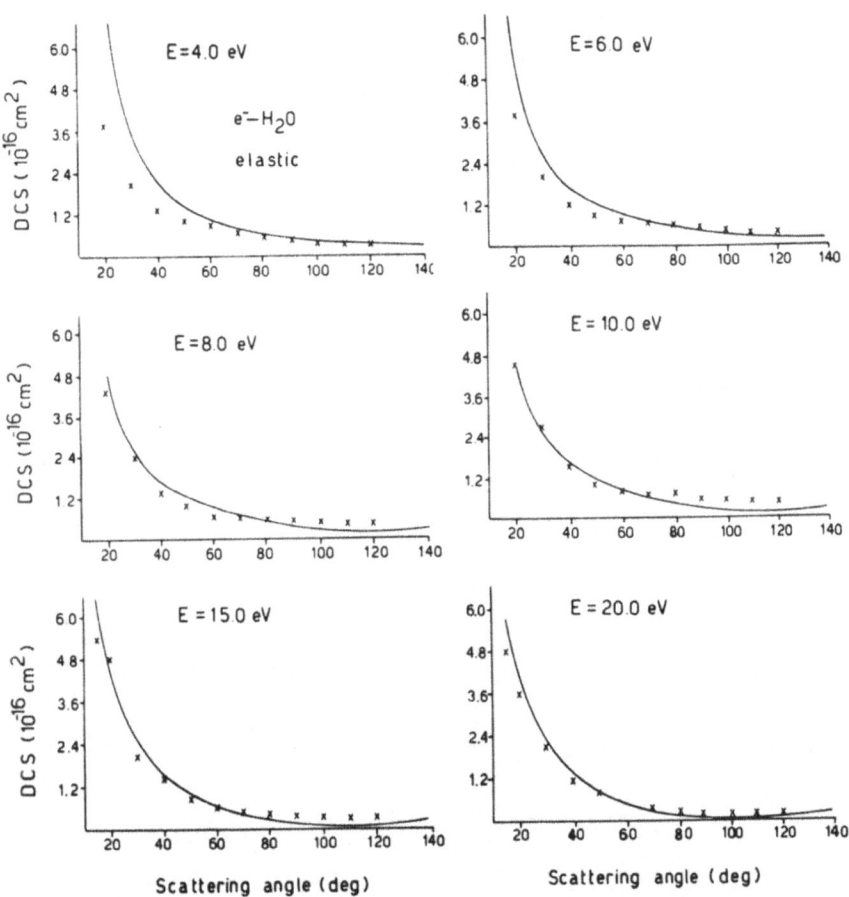

Fig. 6. Computed and measured differential cross sections (rotationally summed) for electron-H_2O scattering. The crosses are the measurements from ref.35.

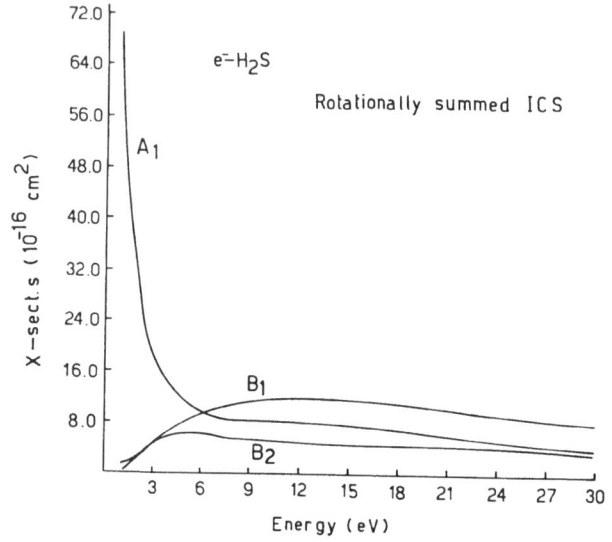

Fig. 7. Computed partial integral cross sections (SEP) for the dominat-
ing symmetries in electron-H_2S scattering.

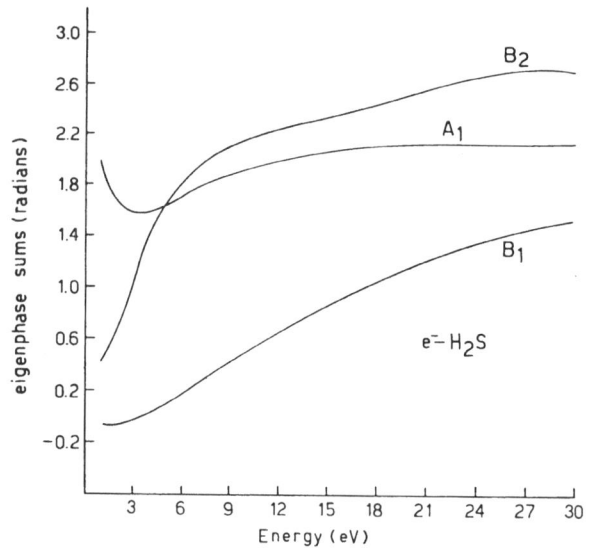

Fig. 8. Computed eigenphase sums as a function of collision energy for
electron-H_2S scattering.

184

rectly solves the coupled channels scattering problems for polyatomic targets. It is the only model calculation which has convincingly explained the cross sections behaviour for CH_4, H_2O, SiH_4 and H_2S targets and the various computational ingredients, albeit limited in scope, are providing us with an efficient and predictive tool to handle low-energy elastic (and rotationally inelastic)[36,37] scattering from more complex targets at a nearly quantitative level of accord with experiments.

ACKNOWLEDGEMENTS

We are grateful to Dr. A.Jain for several useful discussions and comments. We thank Professor R. Moccia for providing us with his recently computed OCE wavefunctions for H_2O, CH_4 and H_2S. The financial support of the Italian Nat. Resarch Council is also acknowledged.

REFERENCES

1. R.C. Whitten and I.G. Poppoff "Fundamentals of Aeronomy", John Wiley, New York (1971).
2. C. Brau, "Excimer Lasers" in Topics in Appl. Phys. 30:87 (1979).
3. L.G. Christophorou, Ed. "Electron-Molecule Interactions and their Applications", Academic Press, New York (1984).
4. F.A. Gianturco and A. Jain, The Theory of electron scattering from polyatomic molecules, Phys. Rep. 143:347 (1986).
5. e.g. see: F.A. Gianturco and D.G. Thompson, Theoretical considerations in the scattering of slow electrons by polyatomioc molecules, Comm. At. Mol. Phys. 16:307 (1987).
6. N.F. Lane, The Theory of electron-molecule collisions, Rev. Mod. Phys. 52:29 (1980).
7. F.A. Gianturco and D.G. Thompson, Computed static potential for AH_n molecules, Chem. Phys. 14:111 (1976).
8. F.A. Gianturco and D.G. Thompson, The scattering of slow electrons by polyatomic molecules, J. Phys. B 13:613 (1980).
9. P.G. Burke, F.A. Gianturco and N. Chandra, Electron-molecule interactions. IV-Scattering by polyatomic molecules, J. Phys. B 5:1696 (1972).
10. J.C. Slater, Quantum Theory of matter, Mc-Graw Hill, New York (1968).
11. S. Hara, A local exchange potential for electron-molecule scattering, J. Phys. Soc. Japan 27:1009 (1969).
12. J.B. Furness and I.E. McCarthy, Semiphenomenological optical model for electron scattering on atoms, J. Phys. B 6:2280 (1973).
13. F.A. Gianturco and S. Scialla, Local approximations of exchange interaction in electron-molecule collisions, J. Phys. B 20:3171 (1987).
14. L.J. Sham and V.Kohn, One-particle properties of an inhomogeneous interacting electron gas, Phys. Rev. 145:561 (1966).
15. F.A. Gianturco, L.C. Pantano and S. Scialla, Low-energy structure in electron-silane scattering, Phys. Rev. A36:557 (1987).
16. F.A. Gianturco and D.G. Thompson, The Ramsauer-Townsend effect in methane, J. Phys. B 9:L383 (1976).
17. A. Jain and D.G. Thompson, Elastic scattering of slow electrons by CH_4 and H_2O using a new polarisation potential, J. Phys. B 15:L631 (1982).
18. J.K. O'Connell and N. Lane, Nonadjustable exchange-correlation model for electron scattering from closed-shell atoms and molecules, Phys. Rev. A 27:1893 (1983).
19. N.T. Padial and D.W. Norcross, Parameter-free model of the correlation-polarisation potential for electron-molecule collisions, Phys. Rev. A 29:1742 (1984).

20. F.A. Gianturco, A. Jain and L.C. Pantano, Electro-methane scattering via a parameter-free model interaction, J. Phys. B 20:571 (1987).
21. W. Kohn and L.H. Sham, Self-consistent equations including exchange and correlation effects, Phys. Rev. 140:A1133 (1965).
22. W.J. Carr and A.A. Maradudin, Ground-state energy of a high-density electron gas, Phys. Rev. 133:A371 (1964).
23. M. Gell-Mann and K.A. Bruckner, Correlation energy of an electron gas at high density, Phys. Rev. 106:364 (1957).
24. E. Wigner, Low-density correlation for an electron gas, Phys. Rev. 46:1002 (1934).
25. W.J. Carr, R.A. Caldwell-Horsfall and A.E. Fein, Anharmonic contribution to the energy of a dilute electron gas, Phys. Rev. 124:747 (1961).
26. F.A. Gianturco and S. Scialla, A modified functional form of positron-molecule correlation interaction, (in preparation).
27. N. Abusalbi, R.A. Eades, T. Nam, D. Thimmalai, D.A. Dixon and D.G. Truhlar, Electron scattering by methane: elastic scattering and rotational excitation cross sections, J. Chem. Phys. 73:1213 (1983).
28. J.E. Bloor and R.E. Sherrod, A CMS method for the treatment of elastic electron-molecule scattering. Total and DCS cross sections for Ar and CH_4, J. Phys. Chem. 90:5508 (1986).
29. M.A.P. Lima, T.L. Gibson, W.H. Huo and V. McKoy, Studies of electron polyatomic molecule collisions: applications to e^--CH_4, Phys. Rev. A32:2696 (1985).
30. L. Brescansin, M.A.P. Lima, T.L. Gibson and V. McKoy, Studies of electron molecule collisions: applications to e^--H_2O, J. Chem. Phys. 85:1854 (1986).
31. B. Lohmann and S.J. Buckman, Low-energy electron scattering from methane, J. Phys. B 19:2565 (1986).
32. J. Ferch, B. Granitz and W. Raith, The Ramsauer minimum of methane, J. Phys. B 18:L445 (1985).
33. R.K. Jones, Absolute total cross section for the scattering of low energy electrons by methane, J. Chem. Phys 82:5424 (1985).
34. F.A. Gianturco and S. Scialla, Low energy electron scattering from water molecules, J. Chem. Phys. in press (1987).
35. A. Danjo and H. Nishimura, Elastic scattering of electrons from H_2O molecules, J. Phys. Soc. Japan 54:1224 (1985).
36. F.A. Gianturco and S. Scialla, Electron scattering by polyatomic polar targets. I. The example of the H_2O molecule, (in preparation).
37. F.A. Gianturco and S.Scialla, Electron scattering by polyatomic polar targets. II. Low-energy resonances in hydrogen disulfide, (in preparation).

RESONANCE COLLISIONS OF ELECTRONS WITH MOLECULES AND IN SOLIDS

Arvid Herzenberg

Applied Physics, Yale University

New Haven, Connecticut 06520, USA

ABSTRACT

The resonance model describes an enhancement of the amplitude of an electron at a collision target, and usually also an increase in the collision time. The importance of the model comes from the reactions made possible by these two effects, reactions which often could not occur in direct collisions.

The resonance model became established in the years 1960-1980, mainly because of its ability to explain vibrational excitation in molecules. In recent years, it has been successfully tested on collisions involving the excitation of very large numbers of vibrational quanta.

The model is still controversial in its application to threshold peaks like those observed in HCl, where the basic mechanism of the amplitude enhancement is not generally agreed. Nevertheless, successful accounts have been given of vibrational excitation and dissociative attachment in HCl near threshold.

Some of the most exciting developments on resonance collisions are happening in solid state physics. Experiments on electron collisions with molecules adsorbed on solid surfaces show vibrational excitation which resembles what we see with the same molecules in the gas phase, yet with intriguing differences which can be traced to the effect of the surface on the trapping barrier.

Resonance collisions of the valence electrons within a metal with the lattice of positive ions are particularly important in the rare earth elements, which have vacant inner atomic f-orbitals very close to the Fermi surface. Like the shape resonances of electrons in small molecules, these resonances are accompanied by a strong coupling between electrons and vibrations. Through this coupling, the resonances in rare-earth atoms seem to lead to an important contribution to acoustic attenuation, whose physical character is quite different from the traditional direct mechanism due to Pippard.

§1. INTRODUCTION

§1.1 Resonance Mechanisms

A resonance occurs whenever a particle is scattered off a target in which the projectile can be trapped temporarily. The simplest example is the scattering of a particle by an attractive potential well surrounded by a repulsive potential barrier[1]. Historically, resonances entered quantum mechanics with Gamow's theory of the decay of α-emitting nuclei, where an α-particle was supposed to be trapped inside a nucleus by the Coulomb barrier[2], but could escape by leakage through the barrier. A different trapping mechanism was suggested by Niels Bohr to explain the vast numbers of resonances observed when slow neutrons (~ 0.1 ev) are scattered by nuclei without a repulsive barrier; Bohr supposed that the incoming neutron becomes accelerated by the attractive potential of the nucleus, and then shares its newly acquired kinetic energy with the nucleons in the target. The incoming neutron becomes trapped because none of the particles has enough energy to hop out of the nuclear potential well until some chance fluctuation concentrates enough energy on one particle[3]. A general theory of Bohr's resonances was developed by Feshbach[4].

There is a time delay which occurs when a particle is scattered at an energy close to a resonance. This delay can be much longer than the time it takes the incident pulse to pass over the target. The time-delay cannot easily be demonstrated for the scattering of electrons with an energy of a few ev by an atom or small molecule, but an analogous experiment may be done with photons whose frequency is close to an excitation frequency of the target. Fig.1 shows the time-dependence of a scattered photon pulse from a molecule which has been illuminated by a short pulse of monochromatic light[5]. The scattered pulse consists of an initial pulse of the same duration as the incident pulse, followed by an exponential tail which comes from the decay of resonant states excited by the incident pulse. In electron scattering at a resonance, the enhanced residence time from the resonant time delay often facilitates processes which could not

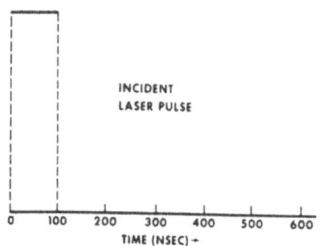
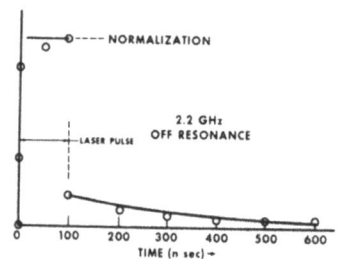

Fig.1 An example of time-delay in the resonance scattering of a pulse of light from an argon-ion laser by an iodine molecule. The left diagram shows the envelope of the incident pulse, and the right picture the envelope of the scattered pulse when the frequency is slightly off-resonance from an absorption line. The scattered pulse consists of a direct part, which lasts as long as the incident pulse, followed by a ringing tail, whose length is determined by the decay rate of the resonance. Open circles: observations; continuous curves: calculated fit. (From ref. 5.)

have occurred within the time the incident pulse takes to pass over the target.

Resonances are characterised by peaks in cross-sections as a function of energy, and by the time-delay. In the late 1930's, there was discovered, in the scattering of neutrons by protons in the spin singlet state, a peak in the elastic cross-section at very low energy ($\lesssim 1$ Mev), which was not associated with trapping or a time delay; this mechanism was called a 'virtual state' by Wigner[6]. Wigner's mechanism occurs when the s-wave dominates the scattering, so that the appropriate part of the incident plane wave in a scattering has the form sin(kr)/(kr), where k is the wavenumber. As kr→0, this wavefunction tends to 1. However, if there is a potential near the origin which shifts the phase of the outgoing wave by π with respect to the incoming wave, the wavefunction outside the potential becomes cos(kr)/(kr); this amplitude becomes 1/(kr) when kr→0, so that there can be a big increase in the amplitude when kr<<1. This enhancement of the amplitude can lead to peaks in excitation cross-sections at very low energies, which have nothing to do with trapping because there is no centrifugal barrier in the s-wave; such peaks arise from the geometrical factor 1/r, together with the interference of the ingoing and outgoing waves near the origin which makes the numerator in the wavefunction non-zero when kr→0.

§1.2 Resonance Poles

A target to which an extra particle has become temporarily attached is said to be in a quasistationary state. As Gamow showed in connection with α-emitting nuclei[2], such a state has a characteristic energy $E_n - i\Gamma_n/2$, where E_n and Γ_n are real and $\Gamma_n > 0$; (the suffix n distinguishes different resonances). $\Gamma_n/\hbar$ gives the decay rate due to the re-emission of the trapped extra particle.

Scattering amplitudes have a pole (or 'resonance') in the complex E plane at $E_n - i\Gamma_n/2$ (E ≡ energy), and $\hbar/\Gamma_n$ is the time delay associated with scattering through the resonance. A resonance has physical significance if the contribution of this pole to the scattering amplitude is dominant when E is real and $E \approx E_n$; a good example is the resonance in the collision $e + N_2$ (e≡'electron') near 2.3 ev. An example where the contribution from the resonance pole is important but not dominant is the collision $e + H_2$ for 2<E<6 ev; in such examples it becomes more difficult to make simple models than when the resonance contribution dominates the scattering amplitude.

Virtual states too are associated with poles in the complex E plane for the scattering amplitude. In the simplest models, these poles fall on the negative imaginary axis in the complex momentum plane; we shall return to them later.

§ 1.3 Vibrational Excitation as a Probe for Resonances

An important difference between electron scattering from atoms and molecules is that molecules can undergo vibrational excitation (and sometimes also dissociative attachment), and in these reactions, the resonant scattering is strongly enhanced with respect to the non-resonant background. (This enhancement occurs only if the short range forces from within the molecule dominate, and where long-range tails, as from a dipole, don't matter.) The enhancement comes from the time-delay $\hbar/\Gamma_n$ in resonant scattering, while there is usually very little time delay in background scattering. The time delay helps the electric field from the projectile electron to get the nuclei moving during the collision in spite of the inertia of their great mass, and this facilitates vibrational excitation; by contrast, in the background scattering, the electron tends to bounce off the heavy nuclei like a fly off an elephant. Therefore vibrational excitation (which is observed through the energy loss of the scattered electron) has traditionally been used as a probe for resonances in scattering. In this lecture, I shall emphasise calculations which treat the vibrational excitation (or dissociative attachment); electron scattering off isolated molecules with fixed nuclei will not be discussed.

§1.4 Resonance Mechanisms, continued

Most resonances in the scattering of electrons by molecules at energies of a few ev depend on the Gamow mechanism: an extra electron becomes trapped in a vacant orbital on the neutral target; the trapping barrier usually comes from the centrifugal potential associated with the orbital angular momentum. Such resonances are usually not called 'Gamow resonances', as they ought to be, but 'shape resonances', because they depend essentially only on the shape of the potential through which the incident electron sees the target. The most thoroughly studied example occurs in the scattering of electrons by N_2 at about 2.3 ev[7,8,9].

Surprisingly, there seem to be no examples in atomic physics of Bohr's mechanism: there an incoming electron would have to be trapped by sharing its initial kinetic energy and some binding energy among several other electrons, so that the ensemble forms a chaotic state. An experiment to look for such resonances was done by Pavlovic et.al.[10] in 1972, by searching for vibrational excitation of N_2 by electrons at energies up to 30 ev; that should have been enough to get a few target electrons excited to form a chaotic state. There was great excitement in the Yale laboratory when the looked-for vibrational excitation was actually found, in a broad peak in the cross-section for vibrational excitation from 20 to 30 ev. However, Dehmer and Dill[11] showed that they could explain the observed peak as a shape resonance due to a <u>single</u> electron trapped in an f-orbital (approximately); they found that there was no need to invoke the idea of the excitation of several electrons. (To this author, it still seems an open question whether it is possible to have <u>any</u> resonance at such high impact energies without mixing with excited background electrons.)

Although there is no evidence in atomic physics for Bohr resonances involving a chaotic state, there are many examples of quasistationary states formed by the excitation of a single target particle, which then forms a highly ordered state with the incoming particle in such a way that both particles feel an attractive potential; (the prototype of this sort of resonance occurs in the scattering of electrons off He atoms at an energy of 19.3 ev.) Such resonances have recently been reviewed by Macek and Watanabe[12].

§1.5 The 2.3 ev Resonance in N_2 as an Example

Electron scattering by N_2 served as a prototype of vibrational excitation at a resonance for twenty years, from about 1960 to 1980. This era began with Schulz' experiments[13] in which individual vibrational excitations were first resolved. A classic paper by Hazi, Rescigno, and Kurilla in 1981 put to rest a large part of the controversy which surrounded the quest for appropriate physical models[14]. In the $^2\Pi_g$ shape resonance which is observed at an impact energy of 2.3 ev, we now have an example where the vibrational excitation is completely dominated by a single isolated electronic resonance; every conceivable theoretical approach has been tried on this example[8,9,15].

For the 2.3 ev – $^2\Pi_g$ resonance, the resonant time-delay is somewhat longer than the time it takes the incoming electron to pass over the target: Consider a wave-packet whose momentum spread Δp is about equal to the characteristic momentum $\sqrt{(2E)}$ (if we take the mass of the electron to be unity). The length of the wavepacket is about $\Delta x \approx \hbar/\Delta p \approx \hbar/\sqrt{(2E)}$, and the passage-time this wavepacket takes to pass over the molecule is $\Delta x/\text{velocity} \approx \hbar/(2E)$. The time-delay is $\hbar/\Gamma$, so that the ratio (time-delay)/(passage-time) is about $2E/\Gamma \approx 10$, for $\Gamma \approx 0.5$ ev from empirical fits of the observed cross-sections[16,17].

The theory of vibrational excitation of N_2 by electrons has been reviewed so often[8,9,15] that another detailed review here would be out of place. The excitation cross-sections for a few quanta of vibration show a peak about 2 ev wide; within the peak, the cross-sections oscillate with a spacing of about 0.2 ev between maxima. This spacing comes from the interference of waves associated with the separation and mutual approach of the nuclei during the residence of the extra electron. Thus, without doing any calculations at all, one can see by looking at the observed cross-section curves as a function of energy that the resonance is associated with vibrations of the nuclei during the residence of the extra electron, and that the extra electron must stay on board for a time at least as long as the vibrational period.

The simplest model of a molecule with a temporarily trapped extra electron rests on the Born-Oppenheimer approximation: because of the large ratio of the masses of the nuclei to the mass of an electron ($\sim 10^4$), and the fact that the forces on both are of the same order, the nuclei move very slowly compared with the electrons when all the electrons are within the trapping region. Therefore within the trapping region, one may assume that the electrons have a wavefunction $\psi(r,R)$, where r and R stand for the co-ordinates of the electrons and the nuclei respectively, and that ψ satisfies a time-independent Schrödinger equation in which R is treated as a constant. Since the electrons are in a quasistationary state in a resonance, ψ has to be a Gamow state[2,18] such that ψ matches on to outgoing waves outside the trapping mechanism which keeps the extra electron attached temporarily.

The energy of such a state is complex, and may be written $E_n(R) - i\Gamma_n(R)/2$, where the dependence on R signifies that the electronic energy will vary with the configuration of the nuclei. The term $i\Gamma_n(R)/2$ comes from the matching to the outgoing waves[18]. The wavefunction for scattering at a resonance may then be approximated by a product $\psi(r,R)\xi(R)$, where $\xi(R)$ is the wavefunction of the nuclei; the function $\psi\xi$ of course describes the wavefunction only in that part of configuration space where all the particles are within the volume prescribed by the trapping mechanism. This approximation contains the assumption that the electrons follow the slowly moving nuclei adiabatically within the trapping region. In addition to $\psi\xi$, the complete wavefunction for scattering must contain terms to describe what happens before the incoming electron enters the trapping region, and after it has leaked out again; in these additional parts of the wavefunction, the extra electron of course does not follow the nuclei adiabatically.

If one inserts the approximation $\psi(\underline{r},\underline{R})\xi(\underline{R})$ into the Schrödinger equation, multiplies by $\psi^*(\underline{r},\underline{R})$, and integrates away the co-ordinates of the electrons, one obtains a wave-equation for $\xi(\underline{R})$, of the form

$$- \frac{\hbar^2}{2M} \nabla^2 \xi + \left(E_n(\underline{R}) - \frac{i}{2}\, \Gamma_n(\underline{R}) \right) \xi - E\,\xi = \rho\, x_0\, \xi \quad,$$

where M is the reduced mass of the nuclei, if we consider a diatomic molecule. ρ contains an electronic matrix element describing the capture into the quasistationary state, and x_0 is the vibrational wavefunction in the initial state of the target. Evidently, the complex electronic energy $E_n(\underline{R}) - i\Gamma_n(\underline{R})/2$ plays the role of a local potential. (See e.g. ref.39) A rule for the conservation of probability can be derived: If one multiplies by ξ^* and subtracts the complex conjugate equation, one obtains

$$\underline{\nabla}\cdot\underline{j} = - \Gamma(\underline{R})\, |\xi(\underline{R})|^2 - 2\, x_0(\underline{R})\, \mathrm{Im}(\xi^*\rho) \,, \quad \underline{j} = \frac{\hbar^2}{2Mi}\left(\xi^*\, \underline{\nabla}\, \xi - \xi\, \underline{\nabla}\, \xi^*\right).$$

$\underline{j}$ is the probability current associated with the transient negative ion. The term $\Gamma|\xi|^2$ describes the loss of probability from the negative ion by the leakage of the extra electron through the trapping mechanism, while the term containing x_0 describes the gain in probability by the capture of electrons from the incident beam.

The cross-section for scattering into the final vibrational state $x_v(R)$ is proportional[39] to the square of an overlap integral of $x_v(R)$ and $\xi(R,E)$. The typical scale of length which enters into the integrand is the wavelength for the nuclear vibrations, which is of order $(m/M)^{1/4}\, a_B \approx 0.1\, a_B$, where m is the mass of an electron, and M the mass of a nucleus. Since both factors in the integrand will usually oscillate, a very high precision in the calculation of $\xi(R,E)$ is needed to get the energy dependence of the vibrational cross-sections right.

As was first pointed out by Bardsley[19] the approximation of an adiabatic theory within the trapping region, with a local complex potential, must break down near a threshold, where the extra electron is very slow at the limit of the trapping region where the matching to the outgoing waves has to be performed. However, the width of the energy band near a threshold where the breakdown occurs does not seem to be well understood. One of the attractions of the 2.3 ev – $^2\Pi_g$ resonance in e+N_2 as a prototype is that the energy is so far from the thresholds of the most important vibrational channels that the breakdown of the adiabatic approximation near thresholds is not very important. Therefore the model with the local complex potential gives a good account of most of the observations[14, 20].

The reader who wants an indication of the state of the theory of resonant vibrational excitation in e + N_2 scattering is recommended to consult references 21 and 22. They report the results of a recent experiment by Allan[21], which observed the excitation of vibrational states up to $v=17$, and an 'ab initio' calculation by Morgan[22] by the R-matrix method[23], which gives a good account of the experiments.

In comparison with the example of the 2.3 ev – $^2\Pi_g$ resonance, cases which we consider below (in §2 and §3) have a time-delay which is no greater than the passage time of an incident wave-packet, and give no evidence of any displacement of the nuclei during the residence of the extra electron. Moreover, the examples we consider below are cases where a local complex potential model is either only of qualitative value, as in §2, or not physically appropriate, as in §3.

§2. THE BROAD $^2\Sigma_u$ RESONANCE IN H_2

§2.1 Properties of the Resonance

The simplest resonance in electron-molecule scattering should occur with H_2 molecules when an extra electron is added in the lowest vacant orbital, $^2\sigma_u$; the resonance would be $^2\Sigma_u$. This resonance has a very large width, of the order of 4 ev (see below) in the vibrational ground-state, which makes the approximation of a local complex potential much more dubious than in the 2.3 ev - $^2\Pi_g$ resonance in N_2. A calculation which avoids the use of a local width function $\Gamma(R)$ has recently been done by Domcke and his collaborators[28].

The $^2\Sigma_u$ resonance in $e+H_2$ is thought to be responsible for a broad peak in the cross-section for vibrational excitation, extending from about 2 to 6 ev. The most important channel is $v=0 \rightarrow v=1$, with a total cross-section of 0.5 A^2 at the maximum; the other channels are weaker by a factor 10 or more[24]. These are very small cross-sections in comparison with the 2.3 ev - $^2\Pi_g$ resonance in N_2, where the vibrational cross-sections are 10 times larger for several channels.

There is good evidence for the $^2\Sigma_u$ character from the differential cross-sections, as a function of scattering angle, for simultaneous vibrational and rotational excitation; they agree reasonably well with a calculation by the rotational impulse approximation[25].

There is no hint of any vibrations of the molecule during the residence of the extra electron: the energy dependence of the cross-section for the excitation $v=0 \rightarrow v=1$ has only a single broad maximum, without minor undulations like those observed in N_2. This is consistent with estimates for $\Gamma(R)$ of the order of 4 ev (see below) in the neighbourhood of the equilibrium separation of 1.4 a_B; this is much larger than the level spacing for vibrations, which should be of the order of a few tenths ev.

Dissociative attachment has a threshold of 3.75 ev when the H_2 target is in its vibrational ground-state. The cross-section is then confined to a narrow peak about 1 ev wide at the threshold, with a maximum of no more than 1.6×10^{-5} A^2. The attachment cross-section increases by about an order of magnitude for each step up the vibrational ladder[26].

The ratio (delay time)/(passage time) at impact energy E is given by $2E/\Gamma$. The maximum in the cross-section for vibrational excitation $v=0 \rightarrow v=1$ occurs at $E = 3$ ev; for $\Gamma = 4$ ev (see below), we get $2E/\Gamma = 1.5$. Therefore there is very little resonant time delay beyond the passage-time of the incident wave-packet. The broad bump observed in the vibrational excitation cross-sections between 2 and 6 ev must therefore be interpreted as due to the weak amplitude enhancement at a broad resonance.

The values of Γ quoted above come from the calculations of the width $\Gamma(R)$ of a local complex potential by many authors; their results agree, at least in order of magnitude[27]. The most recent ab initio calculation[28] gives $\Gamma(R=1.4\ a_B) = 4$ ev at the equilibrium separation of the nuclei, with a monotonic decrease to zero at the crossing ($\equiv R_s$) of the real parts of the energy for H_2 and H_2^- in the respective electronic ground-states. A semi-empirical fit of the dissociative attachment cross-section in the vibrational states $v=0$ to $v=4$ for H_2 and $v=0$ to $v=5$ in D_2 with a local complex potential[29] led to $\Gamma(R=1.4\ a_B) = 7.4$ ev, again with a monotonic decrease to zero at R_s.

§2.2 The Non-Local Calculation

It was first shown by Bardsley[19] that it is possible to go beyond the intuitive argument in §1 which suggested the use of a local complex potential $E_n(R) - i\Gamma_n(R)/2$ to describe a resonance in electron-molecule scattering. (R stands for the co-ordinates of the nuclei.) He expanded the wavefunction in vibrational states of the target molecule, and obtained a complex potential with a non-local operator $\Gamma'(R,R')$, which reduces to a local $\Gamma(R)$ at energies of a few vibrational quanta above threshold. The assumption of a local $\Gamma(R)$ might be expected to break down when Γ is large enough for the details of the matching of the quasistationary resonant state to outgoing waves just outside the trapping mechanism to have a strong effect on the wavefunction.

A treatment of the $^2\Sigma_u$ resonance in H_2 without the assumption of a local complex potential has been given by Domcke and his collaborators[28]. They decompose the wavefunction into a localised part ψ_d corresponding to the trapped electronic state, and a remainder which is orthogonal to ψ_d. This is done with the aid of projection operators Q and P, such that $P + Q = 1$, where Q is defined as $Q \equiv |\psi_d\rangle\langle\psi_d|$. They then derive a wave-equation for the nuclei in which the co-ordinates of the electrons have been eliminated, and calculate all cross-sections from its solution. The contribution of the electrons to the nuclear wave-equation is contained in an optical potential

$$V_{opt} = V_0(R) + \varepsilon_d(R) + \Delta(R, E-\tilde{H}_0) - i\Gamma(R, E-\tilde{H}_0)/2;$$

here V_0 is the electronic potential in the target, $\varepsilon_d = \langle\psi_d|H|\psi_d\rangle$ calculated for fixed nuclei, H is the complete Hamiltonian, and $\tilde{H}_0 = T_N + V_0(R)$, where T_N is the operator for the kinetic energy of the nuclei. E is the total energy. The expression $\Delta - i\Gamma/2$ is defined as

$$\Delta - i\Gamma/2 \equiv \langle\psi_d|H\ \hat{G}_{bg}^{(+)}\ H|\psi_d\rangle$$

where $\hat{G}_{bg}^{(+)}$ is a 'background' Green function constructed from diabatic electronic states orthogonal to ψ_d.

The operators Δ and Γ are non-local because they contain T_N. The local approximation is obtained by replacing $(E-\tilde{H}_0)$ in Δ and Γ by $E_{res}(R)$, the resonance energy for fixed nuclei, so that $\Delta(R) \equiv \Delta(R, E_{res}(R))$, $\Gamma(R) \equiv \Gamma(R, E_{res}(R))$.

§2.3 Results and Comparison with the Local Approximation

For the details of the calculation by Domcke and his collaborators, we must refer the reader to the original paper[28]. They calculated cross-sections for both vibrational excitation and dissociative attachment, for both H_2 and D_2 targets. The molecular axis was regarded as fixed. For comparison with their full non-local calculation, they used the local approximation to calculate the same quantities.

Vibrational excitation, $e+H_2$, $E<6ev$. (From ref.28)

	Non-local:	Local:
$v=0\rightarrow1$:	Within <5% of experiment.	~ 1.5 x experiment.
$v=0\rightarrow2$:	Errors ±25% of experiment.	~ 3 x experiment.
$v=0\rightarrow3$:	~ 2 x experiment.	~ 9 x experiment.

(Note: The maxima of the three experimental cross-sections for vibrational excitation are in the ratio 1 : 0.08 : 0.008 .)

Dissociative attachment, $e+H_2\rightarrow H+H^-$, peak values near threshold. (From ref.28. Cross-sections in A^2. Digits in brackets indicate powers of 10).

Vib.state	Expt.	Non-local theory	Local theory
$v=0$	1.6(-5)	3.0(-5)	28(-5)
$v=1$	5.5(-4)	5.2(-4)	27(-4)
$v=2$	8.0(-3)	4.3(-3)	15(-3)

Obviously, the non-local theory does much better than the local approximation on the whole. The local theory is at best qualitatively right on the largest of the cross-sections quoted: within 50% of experiment on vibrational excitation for $v=0\rightarrow1$, and within a factor 2 on dissociative attachment to a molecule in the state $v=2$.

The non-local and local calculations are about equally good for the largest of the cross-sections for dissociative attachment, for molecules in the state $v=2$; the calculated anwers are respectively too small and too large by a factor 2. Therefore there is an obvious need to extend these calculations to the largest of the cross-sections measured by Allan and Wong[26], in the states $v=3$ and $v=4$. This extension would be particularly interesting because the local approximation should improve as the initial vibrational excitation is raised, so that the journey to the point of stabilisation is confined progressively to separations where Γ becomes smaller.

Is such qualitative success as the local theory achieves worth having? I think that it is, because the theory is so simple that one can make fruitful predictions with it. The cross-section for dissociative attachment in H_2 is now known to increase by about an order of magnitude for each step up the vibrational ladder. My impression is that the experiment in which Allan and Wong[26] discovered this spectacular increase was done only because the model of the local complex potential predicted that the loss of electrons from the negative ion channel should decrease as the vibrational quantum number increased, so that the nuclei wouldn't have so far to go to reach the point of stabilisation.

§2.4 The Effective Range Theory for Broad Resonances

For very broad resonances at low energies, there is an alternative approach to the resonance theory, resting on the following three assumptions:

(i) There is a distance r_c from the centre of the target outside which the wavefunction of the extra electron can be calculated from the long-range potentials (dipole, quadrupole), starting from the radial logarithmic derivative at r_c.

(ii) The potential at distances smaller than r_c from the centre of the target is so strong that the wavefunction of the extra electron does not vary significantly with energy over the limited energy range of interest.

(iii) The electronic logarithmic derivative $f(R) = [(1/r\Psi)(d(r\Psi)/dr)]$ at $r=r_c$ of the complete wavefunction Ψ is practically independent of the energy, because of the strong potential in $r<r_c$, and varies only little with R in the displacements which occur in vibrational excitation or in the movement to the stabilisation point where the negative ion becomes bound.

The approach based on the above assumptions was started by Demkov[30] and Devdariani[31] with the zero-range potential approximation. The method has also been used to fit the threshold peak in the vibrational excitation cross-section of HCl by electrons[32]. Gauyacq and his collaborators have recently applied the method to several problems[33], including dissociative attachment of electrons to H_2 through the broad $^2\Sigma_u$ resonance. Here we shall restrict ourselves to his treatment of dissociative attachment in H_2.

To use the method, one expands the wavefunction Ψ_J, with the extra electron in $r>r_c$, in the form

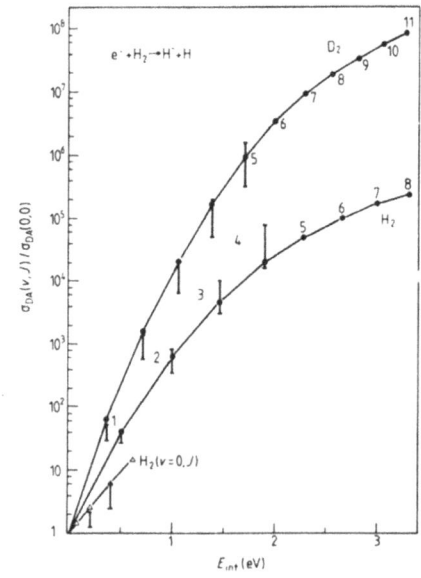

Fig. 2. (From ref.33.) Cross-sections for dissociative attachment at threshold in H_2 and D_2, relative to the cross-section for DA in the level (v=0, J=0). The horizontal scale is the internal energy of the target molecule. The filled circles and open triangles represent the results from ref.33. The error bars are the results of the experiment in ref.26; each bar is labelled with the v-number of the initial vibrational state.

$$\Psi_J = \sum_N A_N^J \phi_{el}(k_{NJ}, r) \, x_{NJ}(R) \, Rot(J)$$

where J is the total orbital angular momentum, x_{NJ} the Nth vibrational state of the molecular target at angular momentum J, k_{NJ} is the wavenumber of an electron leaving the neutral target behind in the state x_{NJ}, $\phi_{el}(k_{NJ}, r)$ is the wavefunction of the extra electron in $r > r_c$ with wavenumber k_{NJ}, Rot(J) is the rotational wavefunction of the neutral with angular momentum J, and A_N^J is a numerical coefficient. One then puts

$$\left[\frac{\partial(r\Psi_J)}{\partial r}\right]_{r=r_c} = f(R)\left[r\Psi_J(r,R)\right]_{r=r_c}$$

and projects on the different vibrational states x_{NJ} in turn to get a set of linear simultaneous equations for the coefficients A_N^J. The numerical solution of these equations is straightforward. The logarithmic derivative f(R) at $r=r_c$ is expanded as $f(R)=f_0+f_1(R-R_e)+f_2(R-R_e)^2+...$, where the expansion is cut off after the first or second power of $(R-R_e)$; R_e corresponds to the nuclear configuration at equilibrium. Several of the coefficients f_i may be obtained from ab initio calculations; for example, f_0 may be obtained from a scattering amplitude calculated with nuclei fixed at their equilibrium positions in the neutral, and f_1 may be adjusted so as to get the curve $E(R)^-$ crossing the potential for the neutral at a stabilisation point obtained in a fixed-nucleus calculation at different R. The remaining parameters have to be regarded as

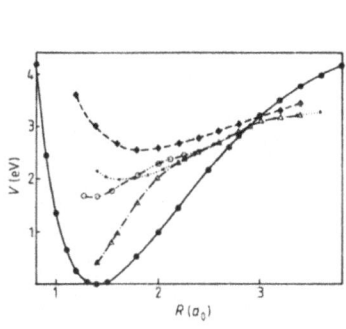 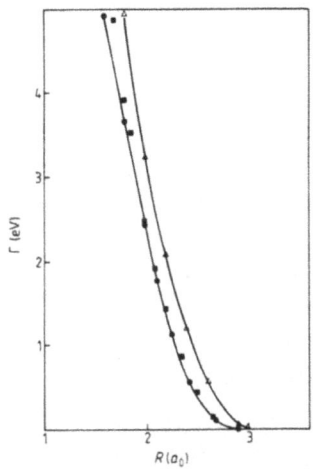

Fig. 3. from ref. 33.
(a) (left): The real parts of the potential energy curves for the H_2-H_2^- system. Filled circles: H_2 potential energy curve (Kolos and Wolniewicz, 1967). Filled diamonds: Bardsley and Wadehra, ref. 29 (semi-empirical local complex potential). Stars: Ab initio calculation (McCurdy and Mowrey, 1982). Open circles and open triangles: ref. 33, from the complex k and complex E planes respectively.

(b) (right): Width of the $^2\Sigma_u$ H_2^- resonance as a function of internuclear distance. Open triangles: ref. 29. Filled squares: McCurdy and Mowrey, 1982 (ab initio). Filled circles: ref. 33.

adjustable; in his calculation of dissociative attachment to H_2, Gauyacq adjusts f_2 to optimise a fit of the experimental cross-sections.

For the method of extracting the cross-section for dissociative attachment from the coefficients $A_N{}^J$, we must refer the reader to Gauyacq's paper. Fig.2 shows the results for the DA cross-section at the threshold for different initial vibrational states. The calculation evidently reproduces very well the spectacular increase of the cross-section with the increase of the initial vibrational quantum number.

The method can be connected to the local complex potential by the equation $f(R) = L^{(+)}(k)$, where $L^{(+)}(k)$ is the logarithmic derivative of the electronic wavefunction just outside r_c when the wavefunction behaves like an outgoing wave with wavenumber k at infinity; if one solves this equation for $k(R)$ (which may be complex), one gets the complex potential curve $E(R) = k(R)^2/2$.

Fig.3a shows the static potential curve for $H_2{}^-$ obtained from $f(R)$ in this way, and compares it with several other calculations and empirical fits of the experiments. It is noteworthy that all the real parts of the complex local potentials in Fig.3a agree very well for separations $R > 2.25$ a_B. (The systematic difference between the curve obtained by Bardsley and Wadehra(ref.29) and the others for $2.25a_B < R < 3a_B$ suggests that the crossing points of the neutral and negative ion potentials used by the authors of the calculations did not all agree.) The agreement between the different authors in Fig. 3a for $R > 2.25$ suggests that different methods for obtaining a static complex potential for the $^2\Sigma_u$ resonance in H_2 are now converging in the range $2.25 < R < 3$ a_B. This represents a considerable improvement over the situation 6 years ago, when Nesbet[27] compared the real parts of the complex energies then available in the literature; the disagreement he found all the way out to the stabilisation point was so chaotic as to lead him to question whether the concept of a static complex potential could mean anything at all.

The present data, Fig.3a, on the real part of the static complex show that the disagreement between different methods is now confined to $R < 2.25$ a_B, where the width Γ becomes much larger; the widths are shown in Fig3b. When $R > 2.25$ a_B, the authors in Fig.3b agree that $\Gamma < 1.5$ev; for $R \approx 1.4$ a_B, they get $\Gamma \approx 5$ev. The region for large R is the one to which dissociative attachment to the higher vibrational levels is most sensitive; according to Fig.2, they have much the largest cross-sections.

§3 VIRTUAL STATES

§3.1 Experimental Facts

In 1976, Rohr and Linder discovered a threshold peak in the vibrational excitation of HCl by electrons[34]. At that time the angular distribution of the scattered electrons seemed to be spherically symmetric, suggesting that the s-wave was predominantly responsible. Therefore the R-L peak looked like a candidate for an analogue in atomic physics of the virtual state in the scattering of neutrons by protons at energies below 1 Mev[17]. Since then, it has been found that threshold peaks occur in the scattering of electrons off many targets. The half-width of the threshold peaks is generally below 0.3 ev[35]. The angular distributions are now known to deviate somewhat from spherical symmetry, so that there must be some admixture of orbital angular momenta above zero[35].

§3.2 General Theoretical Considerations

In §1.1, we discussed the wavefunction at a virtual state with a spherically symmetric static potential; we saw that there is an enhancement over the incident plane wave when $kr \ll 1$, but that there is no barrier behind which the incident particle can be said to be trapped. If the target is a small molecule, the picture is modified by the deviation of the potential from spherical symmetry, in particular by the tail of the dipole potential if there is one, and by the nuclear vibrational and rotational degrees of freedom.

Over the last ten years, there has been much controversy about the relation of the threshold peaks to virtual states. The debate has not been helped by the fact that although it is well understood what constitutes a virtual state in the scattering of a particle off a spherically symmetric potential, there is no general agreement about how this idea should be generalised to more complicated situations. For a spherically symmetric static attractive potential, a virtual state is defined as a pole in the scattering amplitude on the negative imaginary axis in the plane of complex momentum, close to the origin[6]. Can we generalise the idea of the virtual state to targets which are not spherically symmetric and possess internal degrees of freedom?

This author proposes that an appropriate generalisation of Wigner's idea of the virtual state for a spherically symmetric static potential to more complicated situations is the following:

Scattering near a threshold with

(i)   A strong enhancement of the wavefunction at the target relative to an incident plane wave normalised at infinity.

(ii)   There is no trapping mechanism.

(iii)   The scattering must be without time delay.

(iv)   The scattering amplitude must have one (or more) poles close enough to the origin in the complex energy plane to give a threshold spike in the cross-sections.

(v)   The poles must be located so as to give no time delay, unlike resonance poles placed below the real positive physical energy axis.

If this definition does not meet with general approval, a drafting commission should be appointed to produce a better one!

In terms of this definition, we make the following points:

(i)   A virtual state is not a localised quasistationary state, because there is no trapping mechanism.

(ii)   The long-range dipole potential tail splits the single virtual state pole of Wigner's spherically symmetric potential into two poles. If the result still shows threshold spikes and an absence of time-delay, one should still call it a virtual state[36].

(iii)   The poles associated with a virtual state may stray on to unphysical Riemann sheets in the complex energy plane near the

199

origin, so that a search on a restricted part of the many-sheeted plane (the physical sheet for example) may fail to find them[37].

(iv) If there are indeed nuclear-excited Feshbach resonances near the vibrational thresholds in HCl, they should not be regarded as part of the virtual state, because there must be a time delay at a resonance. Instead, one should regard the virtual state as a doorway to the resonances[38].

The tail of the dipole potential has been taken into account in the calculation of the integrated angular cross-sections in the threshold peak in HCl[32]; the wavefunction outside the molecular core is dominated by a single angular mode, which deviates somewhat from spherical symmetry, but still allows a strong enhancement of the wavefunction at the molecule. A calculation of the effects of the variation of the angular mode with angle is still lacking.

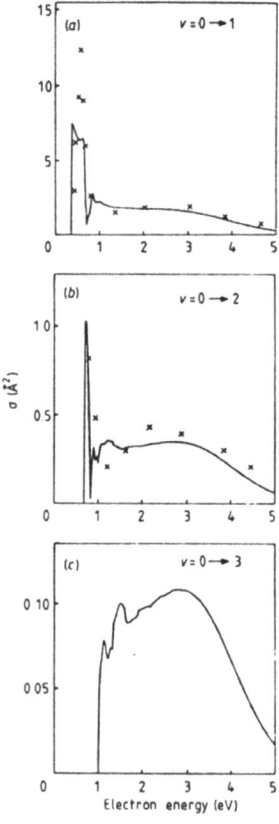

Fig. 4 (From ref. 41) : Integral cross-sections for v=0→1,2,3 vibrational excitation of HCl by electron impact. Full curve: theoretical results; crosses:data of Rohr and Linder (ref.34), rescaled by a factor 0.7 as suggested by Norcross and Padial (ref.42).

§3.3 Why are there so many Threshold Peaks?

Negative ions, with a very small binding energy below 1 ev, are common amongst atoms and small molecules. Therefore almost all atoms and small molecules either have a very weak bound state for an extra electron, or are on the verge of having one. If the s-wave is dominant, as it usually is at low energies, there must then be a large scattering length a, which is related to the s-wave phase shift δ by the effective range formula $k \cot(\delta) = -1/a$, where k is the momentum divided by $\hbar$. If we write $S = e^{2i\delta}$ for the S-matrix in the dominant s-wave, this formula may be rewritten
$S = -[1+(i/ka)]/[1-(i/ka)]$, so that there is a pole at $k = i/a$. This pole corresponds to a weakly bound state if $a>0$, and to a virtual state if $a<0$. If there are some nuclear vibrational degrees of freedom for the electron to couple to, the virtual state will give rise to a threshold peak in vibrational excitation[39]. This argument suggests that most small molecules which do not possess a stable negative ion should show a threshold peak in vibrational excitation.

§3.4 Calculations of Vibrational Excitation and Dissociative Attachment at Threshold Peaks in HCl

Threshold peaks have been observed in HCl and DCl for vibrational excitation and dissociative attachment, for molecules in their first three vibrational states[40]. The dissociative attachment cross-sections at the peaks are very large, of the order of 10 A^2 for vibrational excitation $v=0 \to 1$, and 1 A^2 for dissociative attachment to $v=0$; (the latter is larger by a factor 10^5 than for H_2 in its groundstate!).

Domcke and Mündel[41] have calculated cross-sections by focussing on a broad resonance at ~ 3 ev, which they handled by a technique similar to their treatment of the $^2\Sigma_u$ resonance in H_2. They represented the resonance by a localised wavefunction, and added orthogonal functions to represent the incoming and outgoing electrons. Their formalism contains a large number of parameters, which were all adjusted to reproduce the $^2\Sigma^+$ eigenphase sums obtained by Padial and Norcross[42] in an ab initio calculation for a molecule with fixed nuclei, for energies up to 5 ev. There are no free parameters left after this adjustment. It is remarkable that Domcke et al's results show not only the expected resonance at ~ 3 ev, but also the threshold peaks, without any adjustable parameters or additional components to the wavefunction.

Fig.4 shows the cross-section for vibrational excitation of HCl by electron impact, according to Domcke and Mündel[41]. They also calculated dissociative attachment, and reproduced the dramatic observed increase with the vibrational quantum number, by about a factor ten per step.

Another successful calculation of the dissociative attachment has been done by Gauyacq and Teillet-Billy[43], with the boundary-condition approximation described above for dissociative attachment to H_2.

The threshold peaks obviously signal some singularities in the scattering amplitude at very low scattering energy.

§4. AN ELECTRON RESONANCE IN A MOLECULE ON A SURFACE

Demuth, Schmeisser and Avouris have scattered electrons off N_2 molecules adsorbed on a silver surface, observing vibrational excitation up to the 5th vibrational level.[44] Similar experiments have been done by Sanche and Michaud[45]. The observation of energy losses which are characteristic of the isolated molecule shows that a model with a molecule loosely attached to the surface is appropriate. The results by Demuth et.al. are shown in Fig.5a for $v=0 \rightarrow 1$, together with the corresponding excitation function for isolated N_2 molecules. The measurements in Fig.5a are not absolute, and have therefore been normalised to the gas phase data at the maximum. Evidently, even in the presence of the surface, the excitation function still shows a peak analogous to that in the isolated molecule, but shifted downwards in energy by about 1 ev, and with the vibrational fine structure smeared out.

Andy Gerber at Yale has done calculations[46] to see if the observations can be understood as a modified version of what we know about the isolated molecules. To see whether the absence of vibrational structure on the surface is real or due to limited experimental resolution, one can look at the variation of the vibrational excitation cross-section with the final vibrational quantum number. Fig.5b shows the relative excitation functions for the adsorbed and free molecules; the excitation function on the surface evidently falls off much more rapidly with the degree of vibrational excitation than for the isolated molecule; this would be consistent with an increase of the intrinsic width of the

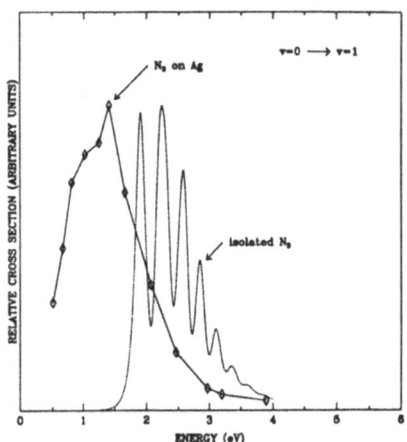
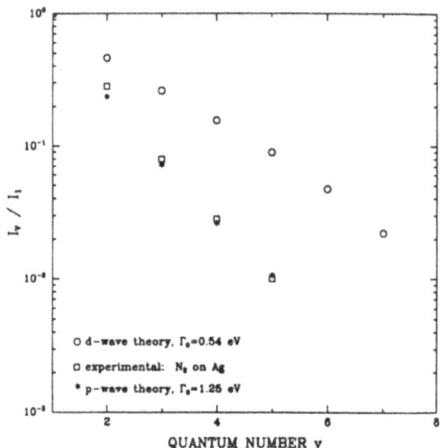

Fig.5 (from ref.46).
(a) (left): The relative vibrational excitation cross-sections for an N_2 molecule on a Silver surface struck by an electron. The full curve with open diamonds represents the experimental cross-section from ref.44. The dotted curve represents the cross-section for gas-phase N_2 (from ref.17). The heights of the two curves have been arbitrarily set equal.

(b) (right): The ratios of intensities I_v/I_0 of scattered electrons which have excited an N_2 molecule into the state $v>0$ and $v=1$. The d-wave theory is for molecules in the gas-phase. (The intensities are integrated over scattering angle.) The experimental data are from ref.44. The stars represent the best fit of the p-wave theory to experiment, with $\Gamma_0=1.25$ ev. (Γ_0 is the width at the equilibrium spacing of the nuclei in N_2.) Comparable fits were obtained in a d-wave theory with $\Gamma_0=1.75$ ev, and a constant Γ theory with $\Gamma=1.0$ ev.

resonance on adsorption[39]. Fig. 5b shows also a satisfactory fit of the surface observations with a local complex potential model; in the calculation it was assumed that the trapping barrier corresponds to a p-wave, rather than to the d-wave as in the isolated molecule, because one expects the distortion of the trapping orbital by the surface must break the inversion symmetry of the isolated molecule, and mix a p-wave into the wavefunction; this would then dominate the decay because a p-wave centrifugal barrier is much lower than a d-wave barrier. The width at the equilibrium separation of the nuclei ($\equiv \Gamma_0$) turns out to be 1.25 ev; if one ignores the variation of the width with the nuclear separation and treats Γ as a constant, one obtains $\Gamma_0 = 1.0$ ev. Although it was not possible either to determine the dependence of Γ on the nuclear separation, or even to decide whether the p-wave Γ or the constant Γ are more appropriate, it appears that a value Γ_0 in the range $1.0 < \Gamma_0 < 1.75$ is needed. This is much bigger than the value $\Gamma_0 \approx 0.5$ ev which fits the isolated molecule[16,17], and quite big enough to account for the absence of vibrational structure without calling on the experimental energy resolution.

To understand the magnitudes of the energy shift and Γ_0, a calculation was done for a molecule with fixed nuclei outside a Sommerfeld potential with a depth of 9.8 ev (the sum of the work-function and the Fermi energy in Silver). The incoming electron was allowed to scatter repeatedly off the surface and the molecule, and the pole in the scattering amplitude corresponding to the multiple scattering was determined. This calculation had to be done for different orientations of the molecule, with the axis either parallel or normal to the surface. The scattering amplitude of the N_2 molecules was taken from the ab initio fixed nucleus calculations of Buckley and Burke[47].

Since the distance of the molecule from the surface is not known, the calculation used the value calculated by Lang[48] for Ar (which is isoelectronic with N_2) physisorbed on a jellium metal with $r_s = 3\ a_B$. Lang gives this distance as $4.25\ a_B$.

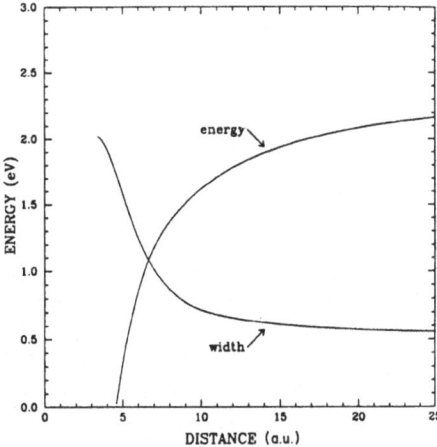

Fig. 6 (from ref. 46): The real($\equiv E$) and imaginary parts($\equiv \Gamma/2$) of the resonance pole $W = E - i\Gamma/2$ of the spherically symmetric 'image potential' model, with the metal surface 'wrapped around' the molecule. The horizontal axis is the distance of the center of the molecule from the surface.

203

The result of this calculation was a disappointment: However the molecule was oriented, the resonance energy did not shift by more than 0.2 ev from the gas-phase value.

One mechanism which had obviously been omitted is the interaction of the extra electron with its image charge in the metal. To repeat the calculation with the inclusion of the image charge would be very difficult, because of the awkward geometry of the problem. However, an estimate of the order of magnitude of the resonance shift can be obtained by 'wrapping' the perturbing potential of the image charge and the Sommerfeld potential of the metal around the molecule in a spherically symmetric fashion. (The radius of the sphere representing the boundary of the metal then becomes equal to the original distance of the molecule from the plane surface.) Moreover, one can replace the molecule by a spherical potential well adjusted to reproduce the 2.3 ev resonance in the isolated molecule. One then has a spherically symmetric problem, for which the complex energy of a quasistationary state can be obtained by matching the wavefunction to outgoing waves at large distances[18]. The results are shown in Fig.6, as a function of the distance of the molecule from the 'wrapped around' surface. There is now a strong shift of the resonance energy, ~ -2 ev for the distance suggested by Lang's calculation. Moreover, the width moves up to ~ 2 ev at the same distance. The shift of the resonance energy is about twice as big as the observation, and the width is about twice the value we need to reproduce the observed decrease of the excitation cross-section with the final vibrational quantum number. However, both these factors two are just the factors by which one would expect the results to be overestimated by 'wrapping' the potentials from the metal around the molecules on all sides, instead of leaving them just on one side of the adsorbed molecule.

The shift of the resonance energy comes from the image potential in the region where the extra electron is localised inside the molecule. The increase of the width comes from the fact that the attractive image potential tears down the barrier from the centrifugal potential; this effect more than compensates for the downward shift of the resonance energy; (taken alone, this would of course reduce the width by making the barrier more difficult to penetrate).

We may conclude that the order of magnitude of the shifts of both the resonance energy and the width may be understood in terms of the image potential of the extra electron in the metal surface.

§5. RESONANCES OF ELECTRONS IN METALS

An atom with a vacancy in an orbital of high angular momentum can show a shape resonance when it is bombarded by valence electrons in a metal. Such resonances are well known for vacancies in the 3d and 4d shells, where the widths are of the order of one or two ev, and in the 4f and 5f shells, where the widths are of the order of a few mev. The resonances in solids were first discussed by Friedel[49]. In metals and alloys containing such atoms, the resonance levels usually lie embedded in conduction bands several ev wide, which provide electrons to collide with the trapping orbits, and into which the electrons can be scattered. There are several reviews of this subject[50,51].

It is a remarkable fact of nature that such resonances frequently exist close to the Fermi level in metals with d and f trapping orbitals. Fig.7 shows[52] experimentally determined densities of states near the Fermi surface in several Ce compounds. These results were obtained by photo-electron spectroscopy below the Fermi energy; above the Fermi energy, they come from observation of the bremsstrahlung photons emitted when electrons from outside the metal are captured into unoccupied states. The peaks within 0.5 ev of the Fermi surface may be interpreted as resonances.

Trapping orbitals near the Fermi edge will fluctuate between being empty and being occupied by a single electron. (Double occupation is usually prohibited by the Coulomb repulsion between a pair of electrons added to the same atom.) Thus the atoms fluctuate between two valence states; materials in which this happens are called 'mixed valence' materials.

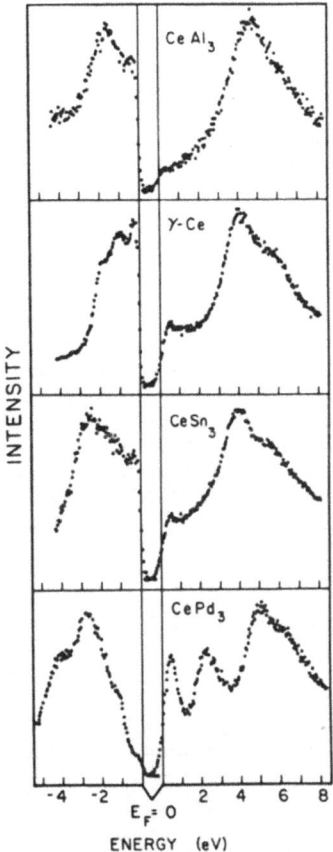

Fig. 7 (from ref.52): Experimental densities of states in a series of Ce compounds, showing a resonance peak due to singly occupied localised 4f orbitals just above (within <<1 ev)   the Fermi edge. The part of the curve to the right of the band marked '$E_f=0$' comes from photons emitted during electron capture into empty orbits; the part of the curve to the left of the band comes from the emission of electrons following photon absorption.

If there are resonances at the Fermi edge, then our experience with the scattering of electrons by small molecules suggests that one should look for strong coupling of electrons to vibrations. A strong electron-phonon coupling is known to arise from the fact that the ionic radius of an ion possessing f trapping orbitals increases by about 20% when an electron is captured.[53] Therefore the capture of an f-electron causes a 'breathing' displacement of the surrounding atoms. This type of coupling has been shown to be important[54] in explaining the phonon softening which occurs in certain mixed-valence compounds at those points in the Brillouin zone which maximise the volume compression about the rare earth ions (the so-called breathing mode deformations).

The transfer of energy from phonons to electrons can be detected by sending a coherent sound wave through a metal, and observing its attenuation by the interaction of the vibration with the conduction electrons. This process is called ultrasound attenuation, because the experiments are usually done at frequencies of the order of 10^8-10^9 sec^{-1}. The inverse of the distance in which the intensity of the sound wave decreases by a factor e is called the 'acoustic attenuation coefficient', and denoted by α.

Pippard[55] has given a well known classical theory of the acoustic attenuation coefficient, based on the idea that the conduction electrons do not follow instantaneously the distortion of the lattice due to the sound wave. Therefore there is a slight imbalance of the charge densities of the oscillating ions and the valence electrons; this gives rise to an electric field, which in turn pumps energy from the phonons into the electrons. In Pippard's theory, the interaction of an electron with the sound wave takes place over a distance of the order of the mean free path.

For comparison with experiment, one needs the dependence of α on the frequency and temperature. If ql<<1, where q^{-1} is the reduced phonon wavelength and l the electron mean free path, this dependence enters Pippard's result in the form $\alpha \propto \omega^2 l$, where ω is the phonon frequency multiplied by 2π. The variation of l with temperature may be obtained from the resistivity ρ, ($l \propto 1/\rho$), so that in conventional metals one expects a monotonic decrease of l as the temperature increases. Mixed valence metals are not conventional, and their resistivities often do not increase with temperature in the conventional way. Nevertheless, for Pippard's mechanism, one should consider only the itinerant electrons, so that the appropriate l should decrease monotonically as the temperature increases. α coefficients which behave in this way are well known. However, in 1986, an example was reported[56] in a mixed valence material where α increases with increasing temperature below 12 ^{0}K, and passes through a maximum at 12 ^{0}K.

It is therefore interesting to ask whether there is another contribution to α from electrons which are scattered through a trapping orbital, and couple to the sound wave by the mechanism pointed out by Sherrington and von Molnar[53]. A calculation has been done at Yale by Andy Gerber, for a single impurity with a trapping level embedded in a sea of itinerant electrons. The trapping level is supposed to be non-degenerate, so that it can be occupied by either a single electron or none. (A more realistic model would have to lift this restriction.) The degree of dilatation of the cage of nearest neighbour atoms around the impurity is described by a 'breathing' co-ordinate Q, and the energy of the trapping level is denoted by $E_f(Q)$. This quantity appears in the Hamiltonian in a term $E_f(Q)f^+f$, where f is the destruction operator for an electron in the trapping level, and f^+ is a creation operator; therefore the Q dependence introduces a coupling between the lattice and an electron in the trapping level. The trapping level is allowed to hybridize with every itinerant electron level with a hybridizing matrix element V, which is taken to be a constant. This model is essentially a simplified version of a well-known model due to Anderson[57]; the simplification consists of dropping the Coulomb interaction term for electrons trapped on the impurity, because we are allowing the trapping level to hold only a single electron. The dependence of $E_f(Q)$ on Q is not usually considered in the Anderson model.

The calculation of the ultrasound attenuation coefficient amounts to a many-body calculation of the self-energy of the phonons, in the approximation based on the diagrams in Fig.8. The double dashed lines in the bubble represent an electron in the trapped orbital, while the full lines represent itinerant states. The transition between the trapped and itinerant states is due to the hybridizing matrix element V. The hybridizing coupling between the trapped and itinerant levels gives the trapped levels a finite width Γ; according to the golden rule, $\Gamma = \pi|V|^2\rho_0$, where ρ_0 is the density of itinerant electron states at energy $E_f(0)$. Because of the dependence of E_f on Q, the trapped level emits and absorbs-phonons, which are represented by the wavy lines. $E_f(Q)$ is approximated by the linear expression $E_f(Q) = E_f(0) - gQ$, where g is the derivative at Q=0; the linear approximation is justified because all the phonons of interest have wavelengths many orders of magnitude larger than the dimensions of the unit cells.

The full derivation will be published elsewhere. The result for the ultrasound attenuation coefficient is

$$\alpha_{MV} = n_{MV} \frac{\hbar\omega^2}{\rho v_s^3} \frac{2}{\pi} \left(\frac{ga}{\Gamma/2}\right)^2 \int_{-\infty}^{\infty} d\epsilon \left(-\frac{\partial f(\epsilon)}{\partial \epsilon}\right) \left[\frac{(\Gamma/2)^2}{(\epsilon - E_f)^2 + (\Gamma/2)^2}\right]^2 .$$

The subscript 'MV' stands for 'mixed valent'; n_{MV} is the number of mixed valent atoms per unit volume; ρ is the density of the material; v_s is the velocity of sound; ω is the frequency of the sound; a is the lattice spacing; $f(\epsilon)$ is the Fermi distribution function at energy ϵ.

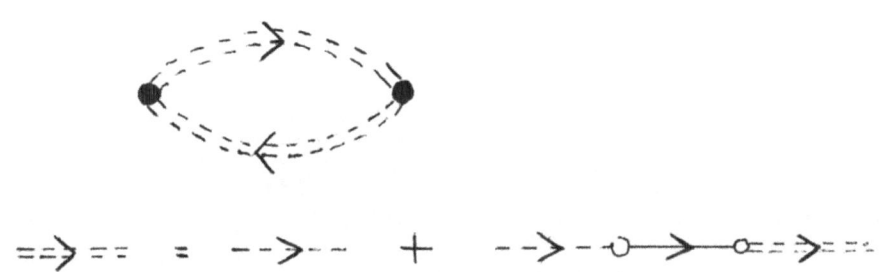

Fig.8. Top: The phonon self-energy in the perturbation theory of ultrasound attenuation due to electrons in resonant levels in rare-earth atoms. The double dashed line denotes a trapped orbital, which is coupled to itinerant states into which it can decay. A filled circle denotes a coupling (g) between a trapped electron and a phonon.
Bottom: The relation between decaying trapped orbitals (double dashed lines), zero-order trapped orbitals which do not decay (single dashed lines), and itinerant orbitals (full line). An open circle denotes a hybridizing matrix element V.

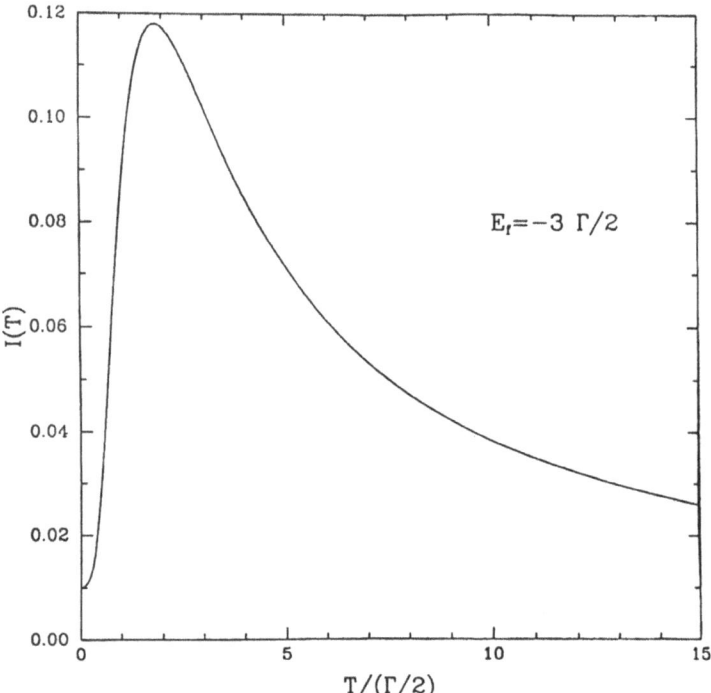

Fig. 9: The calculated temperature dependence of the longitudinal ultrasound attenuation coefficient α in a model mixed-valence compound with an f-level placed three half-widths below the Fermi level. The model predicts a peak in α as a function of T, unlike the conduction band mechanism, which predicts a monotonic decrease as T increases.

One essential ingredient in the formula for α_{MV} is the factor (g/Γ); it is essentially the impulse delivered to the lattice by an electron during its residence in the impurity. This factor is familiar from the vibrational excitation of molecules at resonances in the impulse approximation[39]. The integral $\int d\varepsilon$ ... accounts for the temperature dependence of the result; an example is plotted in Fig.9 (It is worth noting that the result is unchanged if E_f is replaced by $-E_f$, where all energies have their zeros at the Fermi level; the reason is that $\partial f(\varepsilon)/\partial\varepsilon$ is symmetric about $\varepsilon=0$.) Therefore the mechanism is capable of yielding a peak in the temperature dependence.

It is interesting to ask how big the result is in comparison with Pippard's result for attenuation by itinerant electrons. For this purpose, let us take $E_f=0$ (i.e. place the resonance at the Fermi level), and take $k_B T \ll \Gamma/2$, where k_B is Boltzmann's constant. One then obtains

$$\alpha_{MV} = \frac{\hbar\omega^2}{\rho v_s^3} \frac{2}{\pi} n_{MV} \left(\frac{ga}{\Gamma/2}\right)^2$$

To obtain a numerical estimate, one may use data for the mixed-valent compound $Sm_{75} Y_{25} S.$[58] This has $\Gamma\approx100^{\circ}K$. The quantity g ($\equiv\partial E_f/\partial Q$) was calculated to be $\approx 0.4ev/A$,[59] while $a\approx2.9A$ from the sum of the ionic radii of Sm and S.[60] This gives a factor 7.7×10^4 for the quantity $[ga/(\Gamma/2)]^2$. Pippard's result may be rewritten in the form

$$\alpha_C = \frac{\hbar\omega^2}{\rho v_s^3} \frac{2}{15} nk_F l$$

where k_F is the Fermi momentum, l the mean free path, and n the number of itinerant electrons per unit volume. With the numbers quoted above, one gets

$$(\alpha_{MV}/\alpha_C) = \frac{n_{MV}}{n} \frac{1}{k_F l} \frac{15}{\pi} \left(\frac{ga}{\Gamma/2}\right)^2 \approx \frac{3.7\times10^5}{k_F l} \frac{n_{MV}}{n} .$$

If one assumes a typical mean free path such that $k_F l \lesssim 10^3$, the effect from resonant scattering becomes noticeable at a low doping of the order of $n_{MV}/n \sim 10^{-2}$.

NOTES AND REFERENCES

[1] See e.g. Sir H.S.W.Massey, 'Negative Ions', 3rd edn., Cambridge Univ. Press, 1976, p.97-101.
[2] G.Gamow, Z.Phys., 51, 204, 1928.
[3] N.Bohr, Nature, 137, 351, 1936.
[4] H.Feshbach, Ann.Phys.(N.Y.), 5, 357, 1958; 19, 287, 1962.
[5] A.Herzenberg and E.Stryjewski, Phys.Rev.A, 15, 234, 1977.
[6] See e.g. J.R.Taylor, 'Scattering Theory', Wiley, New York, 1972; J.M.Blatt and V.F.Weisskopf, 'Theoretical Nuclear Physics', Wiley, New York, 1952, ch.2, p.68.
[7] G.J.Schulz, Rev.Mod.Phys., 45, 423, 1973.
[8] N.F.Lane, Rev.Mod.Phys., 52, 29, 1980.
[9] D.G.Thompson, Adv.Atomic and Molecular Physics, 19, 309, 1983.
[10] Z.M.Pavlovic, J.W.Boness, A.Herzenberg, and G.J.Schulz, Phys.Rev.A, 6, 676, 1972.
[11] J.L.Dehmer and D.Dill, reviewed in 'Symposium on Electron-Molecule Collisions', edited by I.Shimamura and M.Matsuzawa, University of Tokyo, 1979.
[12] J.Macek and S.Watanabe, Comm.Atomic and Molecular Physics, 19, 313, 1987.

13 G.J.Schulz, Phys.Rev. 125, 229, 1962; 135, A 988, 1964.
14 A.U.Hazi, T.Rescigno, and M.Kurilla, Phys.Rev.A, 23, 1089, 1981.
15 M.A.Morrison, Aust.J.of Physics, 36, 239, 1983.
16 D.T.Birtwistle and A.Herzenberg, J.Phys.B At.and Mol.Phys., 4, 53, 1971.
17 L.Dubé and A.Herzenberg, Phys.Rev. A, 20, 194, 1979.
18 P.L.Kapur and R.E.Peierls, Proc.Roy.Soc., Ser.A 166, 277, 1938.
19 J.N.Bardsley, J.Phys.B, 1, 349, 1968.
20 M.Berman, H.Estrada, L.S.Cederbaum, and W.Domcke, Phys.Rev.A 28, 1363, 1983.
21 M.Allan, J.Phys.B At.Mol.Phys., 18, 4511, 1985.
22 L.A.Morgan, J.Phys.B, 19, L439, 1986.
23 B.I.Schneider, M.LeDourneuf, and P.G.Burke, J.Phys.B At.Mol.Phys., 12, L365, 1979.
24 H.Ehrhardt, L.Langhans, F.Linder, and H.S.Taylor, Phys.Rev. 173, 222, 1968.
25 R.A.Abram and A.Herzenberg, Chem.Phys.Lett., 3, 187, 1969.
26 M.Allan and S.F.Wong, Phys.Rev.Lett., 41, 1791, 1978.
27 R.K.Nesbet, Comments At.Mol.Phys., 11, 25, 1981.
28 C.Mündel, M.Berman, and W.Domcke, Phys.Rev.A, 32, 181, 1985.
29 J.M.Wadehra and J.N.Bardsley, Phys.Rev.Lett., 41, 1795, 1978;
J.N.Bardsley and J.M.Wadehra, Phys.Rev.A, 20, 1398, 1979;
J.M.Wadehra, Phys.Rev.A, 29, 106, 1984.
30 Yu.N.Demkov, Sov.Phys.JETP, 19, 762, 1964.
31 A.Z.Devdariani, Sov.Phys.-Tech.Phys., 18, 255, 1973.
32 L.Dubé and A.Herzenberg, Phys.Rev.Lett., 38, 820, 1977.
33 J.P.Gauyacq, J.Phys.B At.Mol.Phys., 18, 1859, 1985.
34 K.Rohr and F.Linder, J.Phys.B, 9, 2521, 1976.
35 G.Knoth, M.Rädle, H.Ehrhardt, and K.Jung, to be published.
36 A.Herzenberg and B.C.Saha, J.Phys.B, At.and Mol. Phys., 16, 591, 1983.
37 A.Herzenberg, J.Phys.B, At.and Mol. Phys., 17, 4213, 1984.
38 J.P.Gauyacq and A.Herzenberg, Phys.Rev.A, 25, 2959, 1982.
39 A.Herzenberg, in 'Electron-Molecule Scattering', edited by K.Takayanagi and I.Shimamura, Plenum Press, 1964.
40 M.Allan and S.F.Wong, J.Chem.Phys., 74, 1687, 1981.
41 W.Domcke and C.Mündel, J.Phys.B At. and Mol.Phys., 18, 4491, 1985.
42 D.W.Norcross and N.T.Padial, Phys.Rev.A, 25, 226, 1982.
   N.T.Padial and D.W.Norcross, Phys.Rev.A, 29, 1590, 1984; 29, 1742, 1984.
43 D.Teillet-Billy and J.P.Gauyacq, J.Phys.B At. and Mol.Phys., 17, 4041, 1984.
44 J.E.Demuth, D.Schmeisser, and Ph.Avouris, Phys.Rev.Lett. 47, 1166, 1981.
   D.Schmeisser, J.E.Demuth, and Ph.Avouris, Phys.Rev.B, 26, 4857, 1982.
45 L.Sanche and M.Michaud, Phys.Rev.B, 27, 3856, 1983.
46 A.Gerber and A.Herzenberg, Phys.Rev.B, 31, 6219, 1985.
47 B.D.Buckley and P.G.Burke, J.Phys.B, 10, 725, 1977.
48 N.D.Lang, Phys.Rev.Lett., 46, 842, 1981.
   N.D.Lang and W.Kohn, Phys.Rev.B, 7, 3541, 1973.
49 J.Friedel, Can.J.Phys., 34, 1190, 1956;
   Nuovo Cim.(Suppl.), 7, 287, 1958.
50 G.Grüner and A.Zawadowski, Rep.Prog.Phys., 37, 1497, 1974. This review deals with isolated resonant (or 'magnetic') impurities in 'normal' metals.
51 P.A.Lee, P.M.Rice, J.W.Serene, L.J.Sham, and J.W.Wilkins, Comments on Condensed Matter Physics, 12, 99, 1986. This review deals mainly with crystalline materials which have a resonance in each unit cell. There is an introductory section dealing with isolated resonant impurities.

52 Y. Baer, H. R. Ott, J. C. Fuggle, and L. E. de Long, Phys. Rev. B, 24, 5384, 1981.

53 D. Sherrington and S. von Molnar, Solid State Comm., 16, 1347, 1975.

54 P. Entel, N. Grewe, M. Seitz, and K. Kowalski, Phys. Rev. Lett. 43, 2002, 1979.

55 A. B. Pippard, Phil. Mag., 46, 1104, 1955.
See also A. Schmid, Z. Phys. B, 259, 421, 1973.

56 V. Müller, D. Maurer, K. de Groot, E. Bucher, and H. E. Bömmel, Phys. Rev. Lett., 56, 248, 1986.

57 P. W. Anderson, Phys. Rev., 124, 41, 1961.

58 R. Mock, E. Zirngiebl, B. Hillebrands, G. Güntherodt, and F. Holtzberg, Phys. Rev. Lett., 43, 1998, 1979.

59 W. Kohn, T. K. Lee, and Y. R. Lin-Liu, Phys. Rev. B, 25, 3557, 1982.

60 R. M. Martin, J. B. Boyce, J. W. Allen, and F. Holtzberg, Phys. Rev. Lett., 44, 1275, 1980.

ELECTRONIC EXCITATION AS A MULTICHANNEL RESONANT PROCESS

J.P. Gauyacq*, D. Teillet-Billy* and L. Malégat**

*LCAM, Bât. 351, Université Paris-Sud, 91405 Orsay, France
**ER261, Observatoire de Meudon, 92195 Meudon, France

INTRODUCTION

In electron molecule collisions, transient negative ions formed by the capture of the incident electron by the target molecule have been very early recognized to deeply influence the scattering processes. In particular, due to the trapping of the electron, the collision time can be considerably increased thus allowing the target molecule to vibrate during the collision time and thence leading to a vibrational excitation of the target. This process has been observed in quite a few systems[1], and has been extensively studied theoretically[2] both by ab $initio$ approaches and by model calculations. In fact, the above mentionned process can be considered as a two step process : the capture of the incident electron and later the ejection of an electron, the incident and outgoing electron being associated with the same quantum numbers. If the transient negative ion decays by emitting an electron with quantum numbers different from the incident one, the resonant scattering leads to an excitation process, since the target quantum numbers have been modified. The case of a resonant state presenting an incomplete π shell yields a good example of this. Let us look at the $^2\Pi_g$ resonance of O_2^- :

$$e^- + O_2(\ldots\pi_g^2)X^3\Sigma_g^- \to O_2^-(\ldots\pi_g^3)^2\Pi_g \to e^- + O_2(\ldots\pi_g^2) \ X^3\Sigma_g^-, \ a^1\Delta_g, \ b^1\Sigma_g^+ \quad (1)$$

The intermediate resonant state implies three equivalent π_g electrons, it will decay by emitting one of them, thus leaving the O_2 target in any of the three states corresponding to the $(\ldots\pi_g^2)$ configuration. In fact, it is possible to push this idea further, and consider other excitation processes involving electrons from an inner shell. For the O_2^- $^2\Pi_g$ resonance, one can consider the possibility for the resonance to decay by ejecting a π_u electron :

$$e^- + O_2(\ldots\pi_u^4\pi_g^2)X^3\Sigma_g^- \to O_2^-(\ldots\pi_u^4\pi_g^3)^2\Pi_g \to e^- + O_2(\pi_u^3 \ \pi_g^3) \quad (2)$$

The decay by any of the π_u electron will populate any of the six $(\pi_u^3 \ \pi_g^3)$ states of O_2. Indeed the decays by emission of a π_u or a π_g electron should both exist, however with different probabilities. It is possible to go even further and consider the possibility of ejecting more inner electrons. However, one has to consider the energetics of the problem : for a given energy of the O_2^- intermediate, only a fraction of the possible decay channels are energetically accessible. For example, if one considers the center of the O_2^- intermediate, only a fraction of the possible decay channels are energetically accessible. For example, if one considers the center of the $O_2^-(^2\Pi_g)$ resonance located around 0.1 eV above the O_2 ground state at the equilibrium position, all the π_u decay channels are closed and among the π_g decay channels only the ground state channel is open. The excitation process (1) for the (a) and (b) states will then involve scattering in the high energy wing of the $^2\Pi_g$ resonance, where the a and b channels are open (excitation energy of the a and b states : $\sim$1 eV and 1.5 eV resp.). At· this point, it is difficult to guess the importance of this resonant excitation process : it looks efficient from the electronic structure point of view however, implying the far wings of a resonance, it should not be very efficient.

EFFECTIVE RANGE APPROXIMATION FOR THE RESONANT ELECTRONIC EXCITATION

We present below a theoretical study of the electronic excitation in e^--O_2 collisions, via the $^2\Pi_g$ O_2^- resonance : capture of a π_g electron and decay by emission of a π_g or a π_u electron. The method is an extension of the effective range approximation, to deal with electronic channels (the effective range approximation was originally developped to handle vibrational excitation and dissociative attachment processes[3]). The effective range method is in fact very close to the description of the resonant electronic process presented above : a collision involving the formation of an intermediate species. When the collisional electron is far from the molecule, the system is described by channel wavefunctions, product of an O_2 target wavefunctions by a scattered electron wavefunction. In this outer region, the electron interacts with the molecule via a local potential, and it is assumed that angular modes of the electronic motion can be defined. When the collisional electron is at short distance, inside the target electronic cloud, the system is described by a resonant wavefunction $(...\pi_u^4 \, \pi_g^3)^2\Pi_g$ with the $4\pi_u$ and $3\pi_g$ electrons being equivalent. The problem then reduces to matching these two descriptions on a boundary. The wavefunction in the inner region is the resonant $|\pi_u^4 \, \pi_g^3|$. This O_2^- determinant can be expanded with respect to one of his lines to yield an e^--O_2 description :

$$|\pi_u^4 \, \pi_g^3| = \sum_i \alpha_i \, \pi_i \, \Phi_i \qquad (3)$$

where α_i is the fractional parentage coefficient (Slater[4])
π_i is the radial part of the π_u or π_g wavefunction
Φ_i is the O_2 wavefunction in the i state, coupled with the collisional electron spin and angular variables to form a $^2\Pi_g$ sate.
The sum over i implies the sum over all the channels corresponding to the π_g decay ($X^3\Sigma_g^-$, $a^1\Delta_g$ and $b^1\Sigma_g^+$) and to the π_u decay ($c^1\Sigma_u^-$, $A'^3\Delta_u$, $A^3\Sigma_u^+$, $1\Sigma_u^+$, $B^3\Sigma_u^-$, $1\Delta_u$).
In the external region the wavefunction can be written in a similar way, with the electron interacting with the molecule in state i via the local potential V_i :

$$\Phi_{ext} = \sum_i \phi_i \, \Phi_i \qquad (4)$$

where ϕ_i is the wavefunction of an electron scattered by the potential V_i, at the energy corresponding to the channel i.
The wavefunctions (3) and (4) have to be matched at the boundary between the two regions : $r = r_c$. Matching the wavefunctions yields

$$\phi_i(r_c) = \alpha_i \, \pi_i(r_c) \qquad (5)$$

and matching the derivative :

$$\sum_i \alpha_i^2 \, \pi_i^2(r_c) \, \frac{1}{\phi_i} \, \frac{d\phi_i}{dr}\bigg|_{r_c} = \sum_i \alpha_i^2 \, \pi_i^2(r_c) \, \frac{1}{\pi_i} \, \frac{d\pi_i}{dr}\bigg|_{r_c} \qquad (6)$$

One can notice that this matching corresponds to the 9 channel wavefunctions of the outer region condensing into the unique resonant wavefunction in the inner region, because of the three π_g and four π_u electrons becoming equivalent.

Choosing the ϕ_i as the scattering wavefunctions having the asymptotic behaviour corresponding to the process under investigation and bringing them into the equations (5) and (6) then yields the excitation cross section for the various channels.

The above formalism for solving the collision problem requires a few inputs : the radial functions $\pi_g(r)$ and $\pi_u(r)$ on the boundary, the channel potentials $V_i(r)$ and the compound logarithmic derivative appearing in the rhs of (6). For the present study of $e^- - O_2$ collisions, all those quantities have been extracted from *ab initio* calculations.

AB INITIO STUDY OF O_2^- SYSTEM

All the relevant parameters for the ERT treatment were extracted from an SCF calculation of the $O_2^-(\dots\pi_u^4\,\pi_g^3)\,^2\Pi_g$ state. The function $\pi_u(r)$, $\pi_g(r)$ and their derivatives were extracted from the SCF wavefunction. The potential V_i were obtained from the potential (static + local exchange) experienced by a π_g (or a π_u) electron bound in the SCF $^2\Pi_g$ resonance wavefunction. This anisotropic potential, added to the centrifugal potential is then diagonalized at fixed $|\vec{r}|$ to yield a set of adiabatic angular modes and potentials[5] ; the lowest eigen potential for the π_u and the π_g symmetry are then taken as the V_i local potentials.

In the ERT approximation, one assumes that the wavefunction in the inner region is independant of the collision energy ; this requires that the potentials seen by the electrons in the inner region are large compared with the collision energy. As a consequence, the matching radius r_c has to be small enough for the $V_i(r_c)$ potentials to be much larger than the collision energy. In the present O_2 study, we chose the radius $r_c=1.6a_o$. The SCF energy of the O_2^- resonance is around 1.5 eV above the O_2 ground state, rather far from the experimental position around .1 eV[6,8,9].

However, since the ERT formalism only uses the SCF wavefunction in the inner region, the ERT resonance position should be different from the SCF one. It is simply obtained by looking for a complex energy that allows for the solution of equations (5) and (6) with the ϕ_i functions being asymptotically pure outgoing waves. The ERT resonance position is found at 0.25 eV above the O_2 ground state. This improvement is due to the fact that the ERT takes into account the instability of the O_2^- as well as that it introduces an open shell structure into the wavefunction in the outer region.

The complex part of the resonance position (width of the resonance) is presented on figure 1 as a function of the real part (this variation is simply obtained by varying the logarithmic derivative in equation 6). The width very quickly increases with the energy, according to the Wigner threshold law (for a d wave $\Gamma \propto E^{2.5}$).

Also presented are the results of the multichannel R matrix study by Noble and Burke[7] and the results of an adjustment procedure by Parlant and Fiquet-Fayard[8] : these authors determined the $\Gamma(E)$ function that would be consistent with the experimental vibrational excitation cross sections[9]. The good agreement between their determination and our results give confidence in our ERT description of the O_2^- resonance. Indeed this width is the total width of the resonance ; it is possible to analyze it in terms of partial widths corresponding to the various decay channels (see in Teillet-Billy et al[10]).

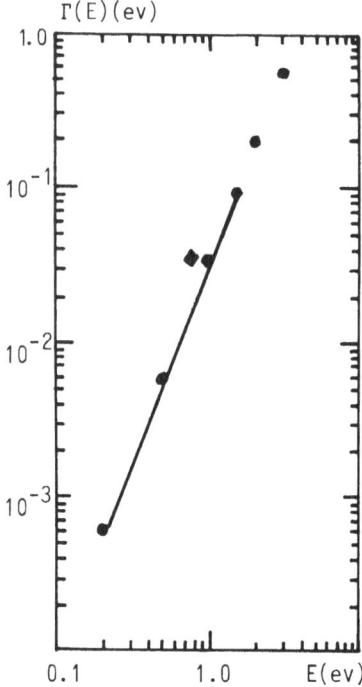

Figure 1 : $O_2^-(^2\Pi_g)$ resonance width function :

● : present ERT results ;

◆ : Noble and Burke[7] ;

——: Parlant and Fiquet-Fayard[8]

EXCITATION CROSS SECTIONS

The figure 2 presents our results for the total cross sections for
excitation to the a and b states together with experimental results and
with the multichannel R matrix results of Noble and Burke[7]. The resonant
contribution is seen to dominate the excitation process, since our results
typically lie at the lower end of the experimental error bars. The differential cross sections for excitation to the a and b states had been
measured by Hall and Trajmar[11] and analyzed by Chang[12] who concluded in a
$^2\Pi_g$ O_2^- state being an intermediate in the process, in accordance with the
present results. It is also worth noting that our results do not present
an energy dependence typical of a resonant process : the center of the
resonance is at 0.25 eV, below the a and b excitation thresholds. The energy
dependence of the theoretical cross section is then the superposition of
the d wave threshold law and of the tail or a resonant energy dependence.
Both cross sections present a drop around 6-7 eV that is due to the
opening of the "6 eV channels" namely the $(\pi_u^3 \, \pi_g^3) c\,^1\Sigma_u^-$, $A^3\Sigma_u^+$ and $A'^3\Delta_u$
states.
Figure 3 presents our results for excitation of the "6 eV group" ($c\,^1\Sigma_u^-$,
$A'^3\Delta_u$ and $A^3\Sigma_u^+$). Very recently the corresponding total cross section has
been measured by Abouaf and Benoit[17]. The two sets of results are consistent : they display the same energy dependence and the theoretical results
typically lie at the top of the experimental error bars. These cross
sections are larger than those for the a and b states, due both to the

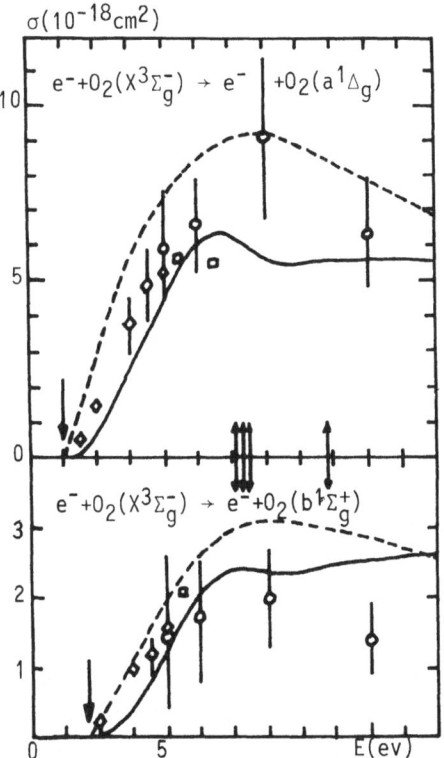

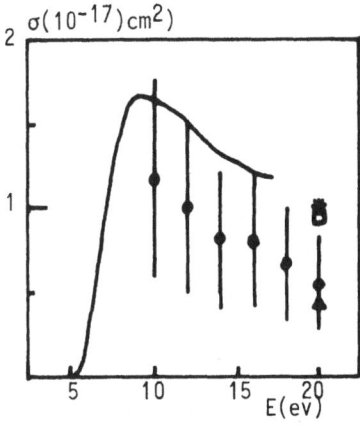

Figure 2 : Total cross sections for excitation of the a and b states. Theoretical results :——: present ; --- : Noble and Burke[7] ; Experimental results : ◇ : Linder and Schmidt[9] ; o : Trajmar et al.[13] ; □ : Hall and Trajmar[11]

Figure 3 : Total cross sections for excitation of the "6 eV group of states" : Present theoretical results :—— ; Experimental results : ● : Abouaf and Benoit[17] ; * : Wakiya[14] ; □ : Trajmar et al[15] ; ▲ : Konishi et al[16].

large fractional paretage coefficient of the 6 eV states and to the relatively easier ejection of a $p\pi_u$ electron than a $d\pi_g$ electron. In the results presented above, the total energy is always rather far away from the center of the resonance, so that no time delay can be expected in the scattering and the nuclear adiabatic approximation should be valid[2]. Indeed, the above calculations were made at fixed internuclear distance R and then averaged over the R distribution in the v=0 $X^3\Sigma_g^-$ state. The same calculation can be used to compute the vibronic excitation i.e. the vibrational excitation, associated with the electronic excitation. Figure 4 presents the ratio of the excitation cross sections to the (v=0) and (v=1) vibrational states of the a and b states. As a main result, the vibrational populations are very different from Franck Condon populations (ratio of .15 and 0.073 for the a and b states). This is due to the thresholds of the "6 eV group" of states which varies rapidly with R in the Franck Condon region and then induces a significant vibronic interaction. Also presented on figure 4 are the recent experimental results of Abouaf and Benoit[17] which confirmed these non Franck Condon populations, in the limited range of collision energies where the experiment was feasible.

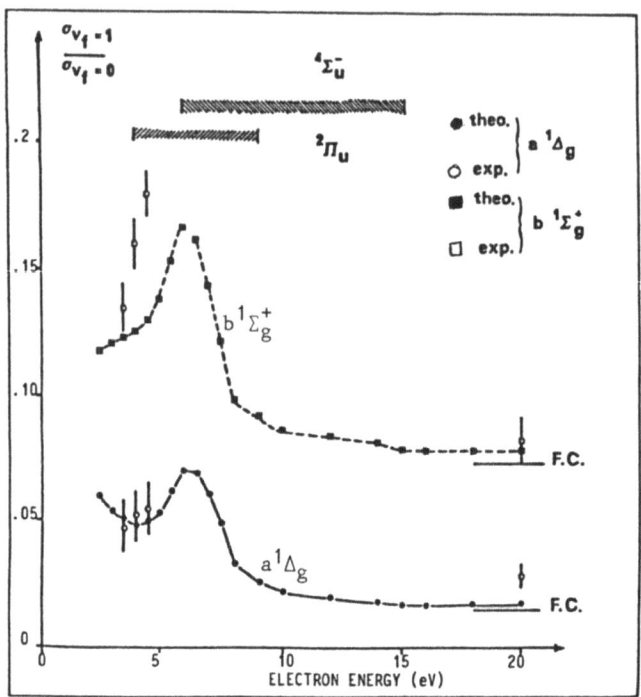

Figure 4 : Ratio of the excitation cross section of the v=0 level and that
of the v=1 level for the a and b states. Full symbols and line : present
theoretical results. Open symbols with error bars : experimental results
of Abouaf and Benoit[17].

CONCLUSION

     Using an extension of the effective range theory, we have presented a
theoretical study of the excitation process to the a,b,c, A and A' states
of O_2, which revealed that they were due to the transient formation of the
O_2^- $^2\Pi_g$ resonance. It has been shown that a very narrow resonance, centered
at low energy can influence the scattering at rather high energies. A
"usual" description of resonant scattering yields a lorentzian shape for
the energy dependence. However, the width Γ involved in this lorentzian
shape is energy dependent; this is very effective, especially if the
resonance is located at low energy, and it results in scattering amplitudes
that can still be sizable far away from the center of the resonance.
     - This resonant process does not correspond to the "traditional" picture
of the incident electron being captured, to form an O_2^- state, that
vibrates along a potential energy curve, the total energy is too far from
the O_2^- energy. The resonant intermediate is rather a way of coupling the
spin and angular momenta during the collision.
     - This excitation process should be rather general. Any shape resonance
of a molecule has a multichannel character and will lead to an excitation
process. This resonant excitation is an exchange process and can be
expected to play a role in the case of "forbidden" transitions. Indeed,
in the case of a dipole allowed transition, direct processes should
dominate.
     - The general situation for an electronic excitation process in electron
molecule collisions consist in a transition from an initial state $\phi_{init} M_{init}$

to a final state $\phi_{fin} M_{fin}$ (ϕ and M are the collisional electron and the target molecule wavefunctions). Both total states have the same symmetry, they are coupled by the electron-electron interaction that leads to an *a priori* small mixture of the two states. In the case of resonant scattering, this mixture is very important, since the initial and final states in fact corresponds to a unique intermediate and the mixing coefficient is simply an angular momentum recoupling term. The resonant process, if it exists, can then be expected to dominate over the other exchange processes.

REFERENCES

1. G.J. Shulz 1973, Rev.Mod.Phys. 45, 378
2. N.F. Lane 1980, Rev.Mod.Phys. 52, 29
3. J.P. Gauyacq 1983, J.Phys.B 16, 4049
   D. Teillet-Billy and J.P. Gauyacq 1984, J.Phys.B 17, 4041
4. J.C. Slater 1960, Quantum Theory of Atomic Structure, McGraw-Hill
5. M. Le Dourneuf, Vo Ky Lan and J.M. Launay 1982, J.Phys.B 15, L685
6. R.H. Celotta, R.A. Bennett, J.L. Hall, M.W. Siegel and J. Levine, 1972 Phys.Rev.A 6, 630
7. C.J. Noble and P.G. Burke 1986, J.Phys.B 19, L35
8. G. Parlant and F. Fiquet-Fayard 1976, J.Phys.B 9, 1617
9. F. Linder and H. Schmidt 1971, Z.Natur.A 26, 1617
10. D. Teillet-Billy, L. Malégat and J.P. Gauyacq 1987, J.Phys.B in press
11. R.I. Hall and S. Trajmar 1975, J.Phys.B 8, L393
12. E.S. Chang 1977, J.Phys.B 10, L677
13. S. Trajmar, D.C. Cartwright and W. Williams 1971, Phys.Rev.A 4, 1482
14. K. Wakiya 1978, J.Phys.B 11, 3931
15. S. Trajmar, W. Williams and A. Kuppermann 1972, J.Chem.Phys. 56, 3759
16. A. Konishi, K. Wakiya, M. Yamamoto and H. Suzuki 1970, J.Phys.Soc. Japan 29, 526
17. R. Abouaf and C. Benoit 1987, XVth ICPEAC (Brighton) Book of Abstracts 312-3 and to be published.

ELECTRON-N$_2$ CROSS SECTIONS AT LOW ENERGIES CALCULATED USING A VIBRATIONAL AVERAGING PROCEDURE

Bidhan C. Saha
University of Oklahoma
Department of Physics and Astronomy
Norman, OK 73019, U.S.A.

ABSTRACT

We calculate e$^-$-N$_2$ elastic and rotational excitation cross sections — within the ground electronic and vibrational state of the target — at low energies and compare our results to available experimental findings. In addition to the static potential we include the effect of polarization via a variationally-determined non-adiabatic polarization potential. We incorporate the effects of exchange exactly employing the linear algebraic method. Using a potential averaging technique we include the effects of the vibrational motion of the target.

1. INTRODUCTION

Low energy e$^-$-N$_2$ scattering cross sections are very important in diverse fields such as discharge physics, laser physics, and the physics of the planetary atmosphere. Over the last two decades the e$^-$-N$_2$ system has enjoyed considerable front-line research both theoretically[1] and experimentally.[2]

The full Lab-frame Close Coupling(LFCC) approximation[3] is computationally intractable, except for simple systems, because a huge number of target states must be included in the target state expansion even at low incident energies. The widely-used approximation for a system like e$^-$-N$_2$ (with 14 bound electrons) is the well-known adiabatic-nuclei (AN) theory,[4] which relies on the separability — the Born-Oppenheimer separation — of the electron molecule wave function. In this method, the nuclear Hamiltonian is absent from the scattering equation and the scattering function depends parametrically on the internuclear separation. These simplifications have made possible the study of low-energy e$^-$- molecule scattering for a variety of target molecules.[1,5]

In this study we start with the rigid rotator approximation, which "freezes" the internuclear separation of the target at equilibrium. It is well-known that this approximation introduces "cusps" at the nucleus in the Legendre projection v_λ of the static component, V_{st}, of the interaction potential. The electron-nucleus interaction produces these cusps. This artificial singularity in the potential makes the convergence of the scattering quantities very time consuming because a huge number of partial waves are needed. There are, however, no contributions from the electron- nucleus term in the exchange and polarization potentials. Instead of using the Legendre projection of the v_λ^{st} (see Eq. 4) in our calculation we use $\left\langle \phi_0|v_\lambda^{st}|\phi_0 \right\rangle$ where ϕ_0 is the ground-state vibrational wave function. To accomplish this goal we solve numerically the nuclear Schrödinger equation to generate the vibrational wave functions and carry out the integration over R. The resulting matrix elements are smooth functions of r. This is the averaging procedure we use in this investigation to calculate $e^- - N_2$ cross sections.

2. THEORY

We begin with the Body-frame (BF) fixed-nuclei (FN) formulation[1,4] in which the scattering function for a particular symmetry at a prescribed energy $E_b = \frac{1}{2}k_b^2$ (in rydberg) can be expressed as

$$\Psi_{E_b,\ell_0}^\Lambda(\vec{r};R) = \frac{1}{r} \sum_{\ell}^{\ell_{max}} u_{\ell\ell_0}^\Lambda(r;R) Y_\ell^\Lambda(\hat{r}) \tag{1}$$

where Λ is the projection of ℓ, along the internuclear axis. The radial-function $u_{\ell\ell_0}^\Lambda(r;R)$ satisfies the coupled integro-differential equations

$$\left(\frac{d^2}{dr^2} + k_b^2 - \frac{\ell(\ell+1)}{r^2}\right) u_{\ell\ell_0}^\Lambda(r;R) = 2 \sum_{\ell'}^{\ell_{max}} V_{\ell\ell'}^\Lambda(r) u_{\ell'\ell_0}^\Lambda(r,R)$$

$$+ 2 \sum_{\ell''}^{\ell_{max}^{ez}} \int K_{\ell\ell''}^\Lambda(r,r') u_{\ell''\ell_0}^\Lambda(r',R)dr' \tag{2}$$

On the right-hand side of Eq. (2), the "direct" matrix elements are

$$V_{\ell\ell'}^\Lambda(r) = \left\langle \ell\Lambda | V_{st}(\vec{r};R) + V_{pol}(\vec{r};R) | \ell'\Lambda \right\rangle$$

$$= \sum_{\lambda=0}^{\lambda_{max}} v_\lambda(r;R) \sqrt{\frac{2\ell'+1}{2\ell+1}} C(\ell'\Lambda\ell;\Lambda 0) C(\ell'\Lambda\ell,00) \tag{3}$$

where $C(\ell'\Lambda\ell,\Lambda 0)$ are Clebsch-Gordon coefficients.

The functions of $v_\lambda(r;R)$ are the coefficients in a single-center expansion of the potentials $V_{st} + V_{pol}$, i.e.,

$$V_{st}(\vec{r};R) + V_{pol}(\vec{r};R) = \sum_{\lambda=0}^{\lambda_{max}} v_\lambda(r;R) P_\lambda(\cos\theta) \tag{4}$$

where $\cos\theta = \hat{r} \cdot \hat{R}$.

The other terms in Eq. (2), the exchange matrix elements, can be expressed in terms of the single-center expansion coefficients of the occupied molecular orbitals, Nocc, of the target. Denoting ϕ_{ℓ,Λ_i}^i as the coefficient for the ith occupied orbital we can write the exchange matrix elements as

$$K_{\ell,\ell''}^\Lambda(r,r;R) = -\sum_{\lambda=0}^{\lambda_{max}} \frac{r_<^\lambda}{r_>^{\lambda+1}} \sum_{\bar{\ell}\bar{\ell}'}^{\ell_{max}^{mo}} \sum_{i=1}^{Nocc} g_\lambda(\ell\ell'\bar{\ell}\bar{\ell}',\Lambda\Lambda_i)$$

$$* \phi_{\bar{\ell}\Lambda_i}^{(i)}(r',R) \phi_{\bar{\ell}'\Lambda_i}^{(i)}(r',R) \tag{5}$$

where $\phi_{\ell\Lambda_i}^{(i)}(r)$ can be obtained from

$$\phi_i^\Lambda(r;R) = \frac{1}{r} \sum_{\bar{\ell}}^{\ell_{max}^{mo}} \phi_{\ell\Lambda_i}^{(i)}(r,R) Y_{\bar{\ell}}^{\Lambda_i}(\hat{r}) \tag{6}$$

In Eq. (5) $r_<^> = \frac{\max}{\min}(r, r')$ and the factor g_λ are the product of four vector coupling coefficients.[6]

To solve the scattering equations (2) numerically we use the linear algebraic method of Collins and Schneider.[7] The coupled equations are first converted into integral equations using the Green's function technique. These integral equations are then transformed into a set of linear algebraic equations by introducing quadrature schemes on all integrals. We adopt a global convergence criteria (always within 1%) throughout our calculations. A detailed list of the computational aspects have been given in a recent publication by Morrison et al.[8]

3. INTERACTION POTENTIALS

For low-energy e^--molecule collisions the construction of an accurate interaction potential is as important as the scattering theory. The present interaction potential consists of three components i) static, ii) polarization, and iii) exchange.

i) The static potential is the average over the ground electronic state of coulomb potentials between the scattering electron and the constituents of the target. It is attractive, non-spherical, short-ranged in nature, and has a weaker long-range tail depending upon the permanent moments of the molecule. We use an augumented GTO [10s6p2d/6s4p2d] basis set[9] to calculate this potential at near Hartree-Fock level of accuracy and obtain the quadrupole moment -0.91 ea_0^2.

ii) The polarization force is long range and has considerable influence on low-energy cross sections. We calculate an ab-initio non-adiabatic polarization potential. This potential is based on a variational calculation of the energy lowering of the target molecule due to the electric field of the projectile. A detailed description of this potential has been shown in a separate publication.[8] This model potential is free from adjustable parameters and requires no scaling. We used the so-called "better than adiabatic dipole" (BTAD) polarization potential

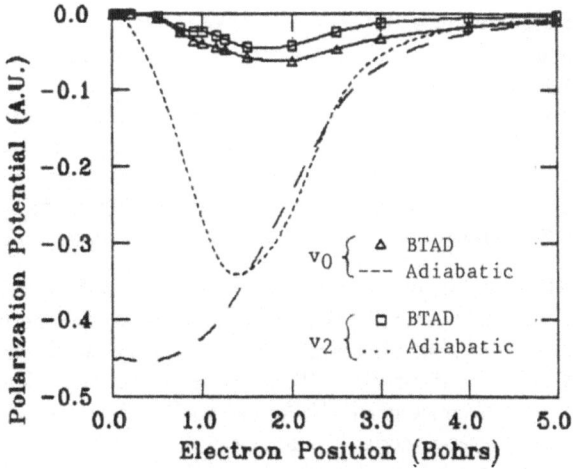

Fig. 1. Spherical and non-spherical projections of e^-- N_2 polarization potential at $R_e = 2.068$ a_0.

in our investigation. In Fig. 1 we plot the spherical ($\lambda = 0$) and non-spherical ($\lambda = 2$) projections of e^--N_2 BTAD potential along with the adiabatic potentials at $R_e = 2.068a_0$. It is apparent from the figure that the BTAD potentials near the target region are noticeably weaker than the purely adiabatic potential. This feature is due to the fact that the BTAD potential incorporates higher order non-adiabatic effects via the non-penetrating[10] approximation. For future studies we present here a convenient analytic form of this potential

$$v_\lambda^{Pol}(r, R) = -D \frac{1 - e^{-(r/r_0)^f}}{(ar^2 + br + c^2)^2} \tag{7}$$

where

For $\lambda = 0$
$$\begin{cases} a = 0.97, \quad b = -1.66, \quad c = 2.83, \\ \qquad\qquad\qquad\qquad\qquad\qquad\qquad 0 \leq r_e \leq 6.0a_0 \\ D = 4.7, \quad r_0 = 1.35, \quad f = 2.62; \\[2mm] a = 0.98, \quad b = 0, \quad c = 0.95, \\ \qquad\qquad\qquad\qquad\qquad\qquad 6.0 \leq r_e \leq 25.0a_0 \\ D = 5.7, \quad r_0 = 1.35, \quad f = 2.62; \end{cases}$$

For $\lambda = 2$
$$\begin{cases} a = 1.6, \quad b = 0.16, \quad c = 1.9, \\ \qquad\qquad\qquad\qquad\qquad\qquad \leq r_e \leq 8.0a_0 \\ D = 4.75, \quad r_0 = 1.59, \quad f = 4.7; \\[2mm] a = 1.0, \quad b = 0, \quad c = 0, \\ \qquad\qquad\qquad\qquad\qquad\qquad 8.0 \leq r_e \leq 25.0a_0 \\ D = 1.68, \quad r_0 = 1.59, \quad f = 4.7; \end{cases}$$

The spherical and non-spherical polarizabilities we obtain at $R_e = 2.068\, a_0$ are $\alpha_0 = 11.42\, ea_0^3$ and $\alpha_2 = 3.37\, ea_0^3$ respectively and compares favorably with experimental values[11] $\alpha_0 = 11.744 \pm 0.004\, ea_0^3$ and $\alpha_2 = 3.08 \pm 0.002\, ea_0^3$.

3. RESULTS AND DISCUSSIONS

We have calculated integral, differential, and rotational-excitation cross sections for low-energy e^--N_2 scattering. We discuss first the cross sections evaluated in the rigid-rotator approximation.

In Fig. 2 we have shown eigenphase sums in the Σ_g and Π_g symmetries, the most important symmetries in this system. The scattering quantity that at low, non-resonant energies is most sensitive to exchange is the eigenphase sum in the Σ_g symmetry. On the other hand, Π_g symmetry dominates in the resonant energy region. Inclusion of the polarization effects alters the eigenphase sums considerably.

Our BTAD total integrated cross sections (the sum of contributions for $\Sigma_g, \Sigma_u, \Pi_g, \Pi_u, \Delta_g$ and Δ_u symmetries) are compared to recent experimental findings[12] in Fig. 3. The most noticeable difference between our calculated cross sections and the experimental data of Kennerly[12] is the absence in the theoretical results of oscillatory structure near the resonance. Using methods such as R-matrix methods,[13] Hybrid theory,[14] Boomerang model,[15] and many-body optical potential theory[16] etc. one can, however, reproduce this structure. Nevertheless, our Π_g eigenphase sums clearly exhibit a shape resonance. By fitting to a Breit-Wigner form we obtain resonance energy $E_r = 2.25\, eV$ and resonance width $T = 0.47\, eV$ which compares favorably to the experimental values $E_r = 2.35\, eV$[17] and $T = 0.41\, eV$.[15]

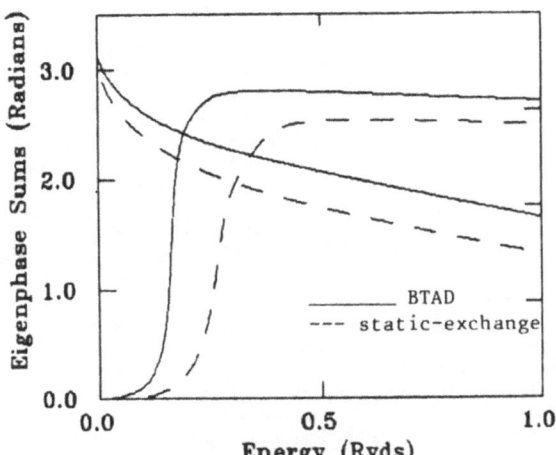

Fig. 2. Eigenphase sums in the Σ_g and Π_g symmetries.

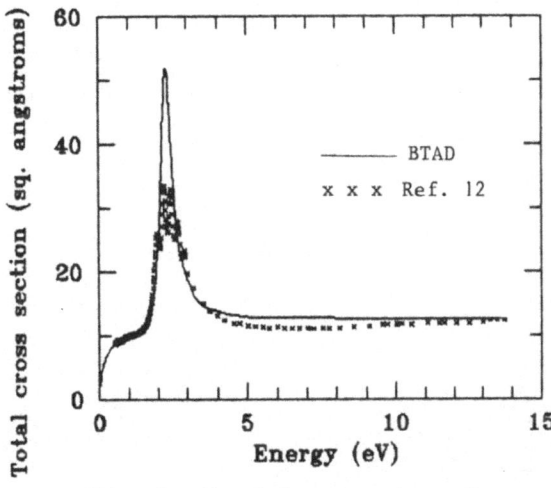

Fig. 3. Total integrated e – N_2 cross sections.

In Fig. 4 our total momentum transfer cross sections are compared to data from analysis of transport properties measured in e^--N_2 swarm experiment.[18]

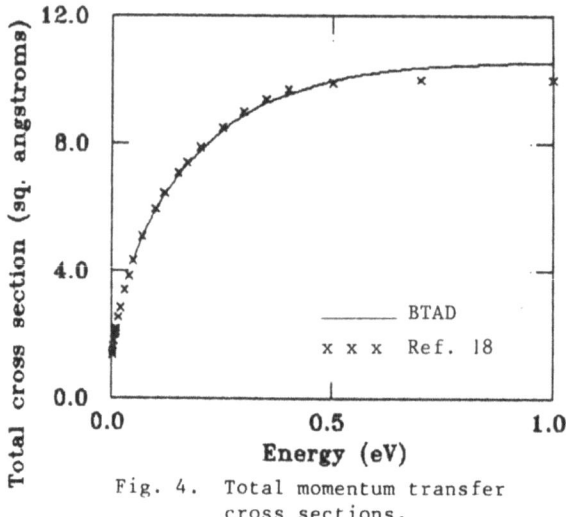

Fig. 4. Total momentum transfer cross sections.

The most stringent test of the theoretical calculation is to compare its angular distributions to experimental results. In Fig. 5 we show our present BTAD results at $E = 0.1\,eV$ along with the measured values of Sohn et al.[19] The theoretical results even at this low energy are in excellent agreement with those of the experiment.

The rotational excitation cross sections play a significant role in modeling the electron velocity distributions in gaseous discharges. In Fig. 6 we plot our BTAD results for $j_0 = 0$ to $j = 2$ excitations along with those recently calculated by Onda[20] using a polarized-orbital potential. In his calculation Onda used free-electron-gas model exchange potential to model the non-local exchange interaction.

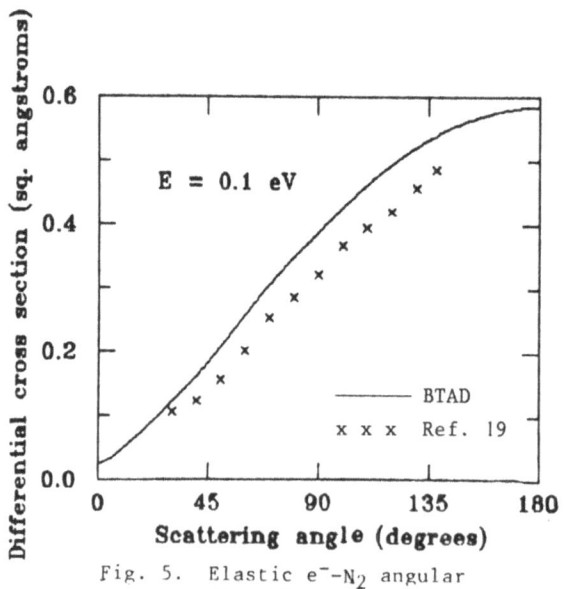

Fig. 5. Elastic $e^- - N_2$ angular distribution at E=0.1 eV.

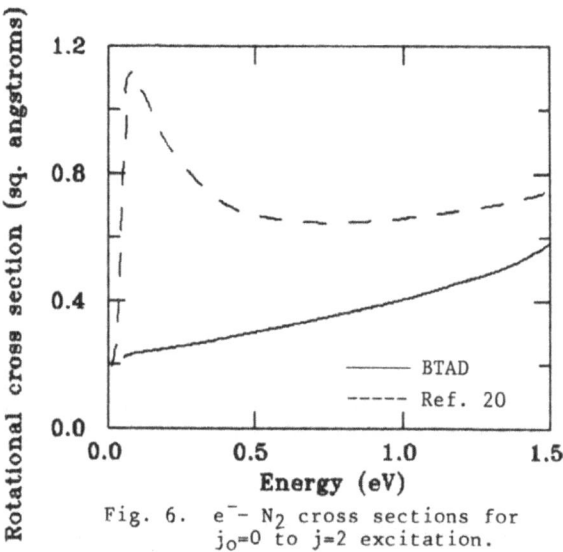

Fig. 6. $e^- - N_2$ cross sections for
$j_o = 0$ to $j = 2$ excitation.

This potential also needs scaling to reproduce the experimental results. It is rather difficult to asses whether the sharp peak around $E = 0.08\,eV$ in Onda's results is due to polarization or the exchange potential or both.

In Fig. 7 we compare the total integrated cross sections obtained by using the ground-vibrationally-averaged-potential, with the rigid-rotator results at non-resonance energies. It is evident from the figure that the approximate inclusion of vibrational effects does not produce much changes on the results obtained using the rigid-rotator approximation. We obtain similar agreement in the resonant energy region too.

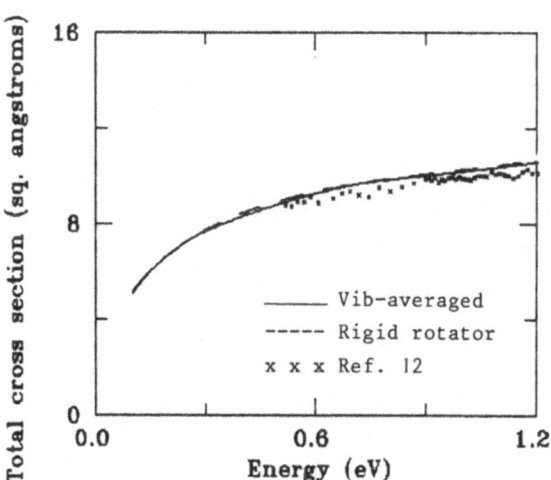

Fig. 7. Low-energy total integrated
cross sections.

4. CONCLUSION

In the present investigation we have shown that the effect of the ground-vibrationally averaged potential on the cross sections is negligible. The averaging procedure is easy to implement and is very economical computationally. Our model potential yields results - total, momentum transfer, angular distributions - in excellent agreement with existing experimental values. The next most important application of this procedure is to evaluate the vibration excitation cross sections.

ACKNOWLEDGEMENTS

I am deeply grateful to Professor Michael A. Morrison for his constructive suggestions and fruitfull discussions throughout the progress of this work and to Wayne Trail, Bill Isaacs, Brian Elza, and Mehran Abdolsalami for their assistance at various stages of this project. The research was supported by Grant No. PHY-805433 from the National Science Foundation.

REFERENCES

1. N.F. Lane, Rev. Mod. Physics, 52:29 (1980); M.A. Morrison, Advan. Atom. and Molec. Phys. in press (1987); and references therein.
2. S. Trajmar, D.F. Register, and A. Chutjian, Phys. Rept., 97:221 (1983); and references therein.
3. M.A. Arthurs and A. Dalgarno, Proc. Roy. Soc. (London) A256:540 (1960); R.J. Henry, Phys. Rev. 2:1349 (1970).
4. A. Temkin, K.V. Vasavada, E.S. Chang, and A. Silver, Phys. Rev. A 186:57 (1969).
5. D.W. Norcross and L.A. Collins, Advan. Atom. and Molec. Phys. 18:341 (1982).
6. M.A. Morrison and L.A. Collins, Phys. Rev. A 17:918 (1978).
7. L.A. Collins and B.I. Schneider, Phys. Rev. A 24:2387 (1981).
8. M.A. Morrison, B.C. Saha, and T.L. Gibson, Phys. Rev. A (in press) (1987).
9. M.A. Morrison and P.J. Hay, Phys. Rev. A 20:740 (1979).
10. A. Temkin, Phys. Rev. A 107:1004 (1957).
11. N.J. Bridge and A. D. Buckingham, Proc. Roy. Soc. (London), A295:334 (1966).
12. R.E. Kennerly, Phys. Rev. A 21:1876 (1980).
13. B.I. Schneider, M. Le Dourneuf and Vo Ky Lan, Phys. Rev. Lett. 43:1926 (1979).
14. N. Chandra and A. Temkin, Phys. Rev. A 14:507 (1976).
15. L. Dubé and A. Herzenberg, Phys. Rev. A 20:194 (1979).
16. M. Berman, H. Estrada, L.S. Cederbaum, and W. Domeke, Phys. Rev. A 28:1361 (1983).
17. G.J. Schulz, Rev. Mod. Phys. 45:423 (1973).
18. A.G. Engelhardt, A.V. Phelps, and C.G. Risk, Phys. Rev. 135:1566 (1964).
19. W. Sohn, K.-H. Kochem, K.-M. Scheuerlein, K. Jung, and H. Erhardt, J. Phys. B 19:4017 (1987).
20. K. Onda, J. Phys. Soc. (Japan) 54:4544 (1985).

VERY LOW ENERGY e-N$_2$ SCATTERING AND INTERIM HYBRID CALCULATIONS

OF THE Π_g RESONANCE

C.A. Weatherford* and A. Temkin**

*Florida A & M University, Tallahassee, FL, USA

**NASA/Goddard Space Flight Center, Greenbelt, MD, USA

INTRODUCTION

The purpose of this article is twofold: based on potentials coming from an MCSCF N$_2$ wavefunction (i) we present 10 state vibrational close coupling results (pending completion of a 15 state calculation) of vibrational excitation cross section over the Π_g resonance range (1 $\leqslant$ E $\leqslant$ 3 eV). The resulting total cross section curve reveals essentially correct peak to valley substructure ratios, in contrast to previous calculations; (ii) we discuss and calculate both fixed-nuclei and $\Sigma_g{}^+$ vibrational close coupling results as a means of assessing two sets of discordant experimental results of very low energy (E < 1 eV).

PRELIMINARIES

The low energy scattering from N$_2$ has provided a very rich testing ground for electron-molecule experiment and theory. In particular many of the underlying theoretical ideas (cf, for example Temkin, 1980) have been developed in response to the challenge presented by many beautiful measurements of the e$^-$-N$_2$ system and its prominent Π_g resonance with all its accompanying substructure.

In this report we present a semi-final summary of our present results which incorporates all of the developments we have made in this field. The basic theoretical approach is the hybrid theory (Chandra and Temkin, 1976), a calculational approach based on the fact that the time scales of the Π_g resonance are such that the rotational motion can be treated adiabatically whereas the vibrational motion has to be treated dynamically. This leads to a method in which vibrational close coupling (i.e. a non-adiabatic effect) and rotational adiabatic-nuclei theory are intimately intermixed.

This basic idea has been supplemented by three further developments: (i) The adaptation and application of the non-iterative partial differential technique (cf. Sullivan and Temkin, 1982) to the electron-molecule scattering problem (Temkin, 1979; Onda and Temkin, 1982). [In the 1982 paper, the non-iterative technique was also applied to calculating an e-N$_2$ polarization potential]. (ii) A more recent generalization of the non-iterative p.d.e. technique to include electron exchange exactly

(Weatherford et al., 1985), and (iii) most recently, generalization of an e-N_2 static potential based on an MCSCF wavefunction for N_2 (Weatherford et al., 1987).

In the present paper we combine all three developments (i) - (iii) in a hybrid calculation consisting mainly of vibrational close coupling for the Π_g resonance region (1.5 < E > 3 eV); however in response to some conflicting experimental results below the resonance region (0 < E < 1 eV), we have also carried out vibrational close coupling calculations there for the dominant E_g^+ partial wave.

Some theoretical aspects and calculational details are given in the next section. In the final section results will be presented and discussed in comparison with some alternative ab initio calculations, and experiments.

SOME THEORETICAL AND CALCULATIONAL DETAILS

The underlying techniques of the present calculations are described in the references mentioned in the above section. The present work generalizes the non-iterative p.d.e. approach to include vibrational states in the close-coupling expansion for any partial waves for which close-coupling is necessary. In particular, close-coupling has been included in the $^1\Pi_g$ symmetry for 1 to 3 eV and in the $^2\Sigma_g^+$ symmetry for 0 to 1 eV.

Electron-molecule vibrational close coupling starts with a generalization of the fixed-nuclei wavefunction (Temkin et al., 1967, 1969) to include vibration:

$$\psi^{(m)} = \sum_{i=1}^{15} (-1)^{P_i} F^{(m)}(\chi_i;R) \Phi_{N_2}(\chi^{(i)};R) \tag{1}$$

x_i are coordinates of the coordinates (space and spin) of the ith electron, and $x^{(i)}$ refers to the collection of the coordinates of the remaining electrons of the target molecule (14 in the case of N_2). The parity of the cyclic permutation which puts the ith electron in the first position is denoted by $(-1)^{P_i}$.

Assuming N_2 is described by a set of $N_{occ}=7$ spatial orbitals, labelled ϕ_α, (two electrons for each spatial orbital), this allows the total Schrödinger equation to be reduced to the form (in a.u.)

$$(\nabla^2 - 2H_{vib} + k^2) F^{(m)}(\vec{r};R) = 2V(\vec{r};R) - 2\sum_{\alpha=1}^{N_{occ}} W_\alpha^{(m)}(\vec{r};R) \phi_\alpha(\vec{r};R) \tag{2}$$

In (2) the spins have been eliminated by pre-multiplication of appropriate spin factors; $W_\alpha^{(m)}$ represents the associated exchange kernel

$$W_\alpha^{(m)}(\vec{r};R) \equiv \int d^3r' \phi_\alpha(\vec{r}';R) \frac{1}{|\vec{r}' - \vec{r}|} F^{(m)}(\vec{r};R) \tag{3}$$

$V(\vec{r};R)$ is the direct static potential associated with Φ_{N_2} [cf. Eqs. (10) and (11)]. The close coupling aspect of the calculation enters by expanding the scattering orbital, $F^{(m)}(r;R)$ in terms of vibrational functions, $\chi_v(R)$, of the N_2 molecule

$$F^{(m)}(\vec{r};R) = \sum_v F_v^m(\vec{r}) \, \chi_v(R). \tag{4}$$

The non-iterative incorporation of exchange, Weatherford et al., (1985), is accomplished by making a similar expansion for $W_\alpha^{(m)}$:

$$W_\alpha^{(m)}(\vec{r};R) = \sum_v W_v^{(m,\alpha)}(\vec{r}) \, \chi_v(R). \tag{5}$$

We also utilize the observation that $\left| \vec{r}' - \vec{r} \right|^{-1}$ is the Green's function for ∇^2:

$$\nabla^2 \left| \vec{r}' - \vec{r} \right| = - 4\pi \, \delta(\vec{r}' - \vec{r}). \tag{6}$$

Operating on (3) by ∇^2 allows (2) to be reduced to the set of Eqs.

$$(\nabla^2 + k_v^2) \, F_v^{(m)}(\vec{r}) = 2 \sum_{v'} V_{vv'}(\vec{r}) \, F_{v'}^{(m)}(\vec{r}) - 2 \sum_{v',\alpha} \phi_{vv'}^{(\alpha)}(\vec{r}) \, W_{v'}^{(m,\alpha)}(\vec{r})$$

$$\nabla^2 \, W_v^{(m,\alpha)}(\vec{r}) = - 4\pi \sum_{v'} \phi_{vv'}^{(\alpha)}(\vec{r}) \, F_{v'}^{(m)}(\vec{r}) \tag{7}$$

The important thing about the coupled Eqs. (7) is that they contain <u>no</u> integral terms, and this is the essence of the non-iterative exchange technique. [$V_{vv'}$ and $\phi_{vv'}^{(\alpha)}$ represent matrix elements of the unsubscripted quantities between vibrational states $\chi_v(r)$ and $\chi_{v'}(R)$]

Equations (7) may be further reduced from 3- to 2-dimensional p.d.e.'s by exploiting the cylindrical symmetry of all orbitals. Without giving any further details (which are given in Weatherford et al., 1985) the equations are reduced to 2-dimensional form

$$\left[\Delta(m) + k_v^2\right] f_v^{(m)}(\underset{\sim}{z}) = 2 \sum_{v'} V_{vv'}(\underset{\sim}{z}) \, f_{v'}^{(m)}(\underset{\sim}{z}) - 2 \sum_{v;\alpha} r^{-1} \, \phi_{vv'}^{(\alpha)}(\underset{\sim}{z}) \, W_{v'}^{(\alpha)}(\underset{\sim}{z})$$

$$\Delta(m) \, W_m^{(\alpha)}(\underset{\sim}{z}) = \left[\frac{(m - m_\alpha)^2 - m^2}{r^2 \sin^2 \theta}\right] W_v^{(\alpha)}(\underset{\sim}{z}) - 2 \sum_{v'} \phi_{vv'}^{(\alpha)}(\underset{\sim}{z}) \, f_{v''}^{(m)}(\underset{\sim}{z}) \, , \tag{8}$$

where $\underset{\sim}{z} = (r,\theta)$ is a 2-dimensional vector, and

$$\Delta(m) \equiv \frac{\partial}{\partial r^2} + \frac{1}{r^2} \left(\frac{\partial^2}{\partial \theta^2} + \cot \theta \, \frac{\partial}{\partial \theta} - \frac{m^2}{\sin^2 \theta}\right) \tag{9}$$

The main new aspect of this calculation is that the static potential is here generated from an MCSCF wavefunction

$$V_{st}(\vec{r};R) = \langle \Phi_{N_2}(\text{MCSCF}) \, V_{e-mol} \, \Phi_{N_2}(\text{MCSCF}) \rangle \, . \tag{10}$$

Repeating, this potential, which was used in the fixed-nuclei calculation of Weatherford et al., (1987), is here used in a vibrational close coupling calculation as described by the above equations. To the direct potential, whose vibrational matrix elements enter Eq. (8), we have also added a polarization term as also described in our previous papers:

$$V(\vec{r};R) = V_{st}(r;R) + V_{pol}(\vec{r};R) \qquad\qquad (11)$$

Note that in Eqs. (8) all exchange terms continue to be described and treated by a target $\Phi(N_2)$ at the SCF level. We consider this to be an eminently justifiable as well as practical concomitant approximation.

For the numerical solution of Eqs. (8) we use nineteen mesh points in θ, and $\Delta r = 0.5$ a_0. The numerical solution was carried in the 0-r rectangle to r = 10 a_0. The K-matrix at that r was propagated as described in Weatherford et al., (1985) to r = 60 a_0 which, as described there, was completely adequate for convergence in r.

RESULTS AND DISCUSSION

Figures 1 and 2 present some of the results obtained in this calculation, which includes 10 vibrational states in the VCC expansion in the Π_g symmetry. For this number of states the present calculation is not converged for processes involving higher (i.e. $v \gtrsim 5$) vibrational states. Thus, we present in the first two figures only the individual vibrational excitation cross sections for $v_i = 0$ to $v_f = 1,2,3,4$. Even for these cross

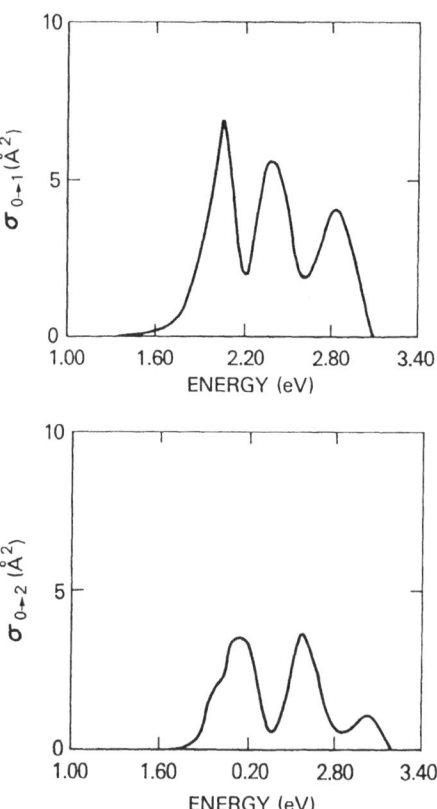

Fig. 1. Vibrational excitation cross section in square Angströms versus energy in eV, using the "Hybrid Theory" with 10 states included in the Π_g symmetry. Upper part, $\sigma_{0\to1}$ lower part, $\sigma_{0\to2}$. Cross sections in all our figures are for the e-N_2 system.

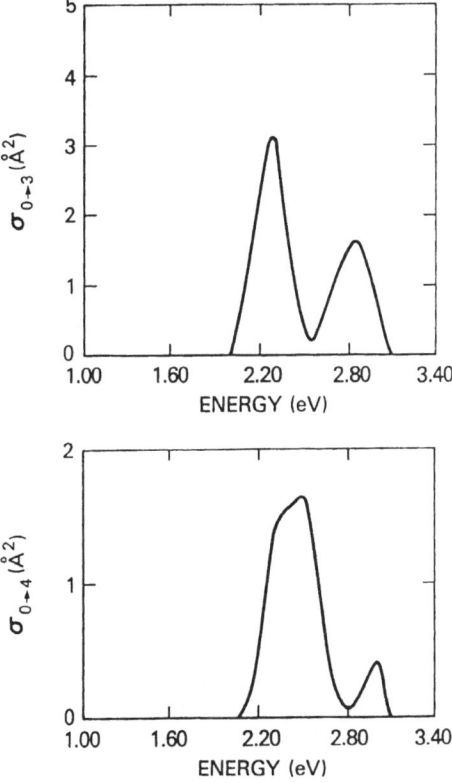

Fig. 2. Same as Fig. 1 for $\sigma_{0\to3}$. Lower part, $\sigma_{0\to4}$.

Table 1. Positions and magnitudes of the first four maxima for the $0 \to v_j$ excitation of the four lowest vibrational states. Present results are compared with the theoretical results of Morgan (1986), Huo et al. (1987a), and the experimental results of Allen (1985)

	Present Calculations		Morgan (1986)		Allan (1985)		Huo et al. (1987a)	
v of final vib. state	peak energy (eV)	absolute cross section (Å^2)	peak energy (eV)	absolute cross section (Å^2)	peak energy (eV)	cross section (Å^2)	peak energy (eV)	normalized cross section (Å^2)
1	2.03	6.94	2.12	6.94	1.95	5.6	1.96	5.75
2	2.12	3.67	2.32	3.39	2.00	3.7	2.00	3.30
3	2.27	3.13	2.40	3.09	2.15	3.1	2.19	2.90
4	2.48	1.65	2.62	1.39	2.22	2.1	2.22	1.79

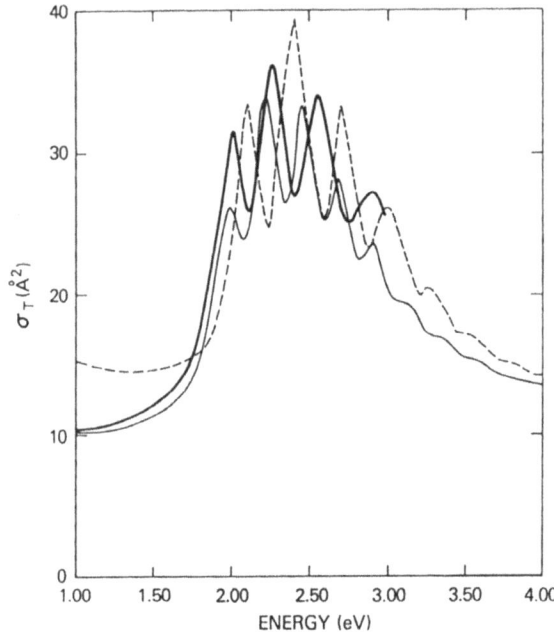

Fig. 3. Total cross section; present "Hybrid Theory" with 10 states in
$^2\Pi_g$ symmetry (thick solid line), experiment of Kennerly (1980)
(thin solid line), R-matrix calculation of Morgan (1986),
(dashed line).

sections our results contain many fewer peaks than are present in experiments (cf. Allan, 1985) or, say, R-matrix type calculations, which theory
is particularly well suited to obtaining essentially all of the vibrational substructure (cf. Morgan, 1986). However within the confines of
the first few peaks a detailed comparison shows that our results are in
even better agreement with experiment than the previous calculations
[except for a most recent Schwinger multichannel calculation of Huo et
al., (1987)]. For example referring to Table I, one sees that both the
position and absolute height (cross section value) of the first peak of
the $\sigma_{0 \to v}(v=1,2,3,4)$ cross sections is in better accord with Allan's
results than are those of Morgan.

In Fig. 3, we present our hybrid calculation using 10 vibration
channels in the Π_g symmetry and fixed-nuclei results for other partial
waves for the total cross σ_T,

$$\sigma_T \equiv \sum_v \sigma_{0 \to v} \, , \qquad\qquad (12)$$

in comparison with the well-known experimental results of Kennerly (1980)
and the most recent R-matrix calculation (Morgan, 1986). Here one sees
even more directly that peak heights, but most importantly the peak
height to valley ratios are much closer to the experimental values than
Morgan, and all previous theoretical calculations. (Again this excludes
the very recent calculation of σ_T by Huo et al., 1987b. There is not a
clear separation between target state and total wavefunction correlation,
but there is no question that some target state correlation has also been
included). Thus we attribute this to improved static potential coming

from the MCSCF N_2 wavefunction. At the same time one sees that the higher energy substructure, which is very suppressed in σ_T, is much better represented in Morgan's (1986) as well as many of the earlier R-matrix and boomerang calculations (Schneider et al., 1979; Dube and Herzenberg, 1979; Hazi et al., 1981).

Finally we turn to the very low energy e-N_2 cross section. In Fig. 4, a number of cross sections, experimental and theoretical, are presented. The solid line is our fixed-nuclei calculation (Weatherford et al., 1987) with an MCSCF target state. The long dashed curve is a fixed nuclei calculation with an SCF target state (Weatherford et al., 1985). The short dashed curve is the calculation of Padial and Norcross (1984). The short dashed-long dash curve is the recent calculation of Morrison et al. (1987). The squares are our present vibrational close coupling results using six vibrational states in the $^2\Sigma_g^+$ symmetry. On the experimental side the dots are the results of Kennerly (1980), the plusses are the results of Jost et al. (1983), and the asterisks are the results of Sohn et al. (1986).

For E > 0.5 eV the agreement of all calculated results tends to support the experimental results of Kennerly (1980) and Jost et al (1983). However below 0.5 eV the results are more ambiguous. Our previous calculations as well as Padial and Norcross support the Sohn et al results at lowest energy but rise to Kennerly's as E → 1 eV. This is also supported by our present Σ_g vibrational close coupling calculation, albeit with a suggestion of a shape more in accord with the Jost experiment, and the

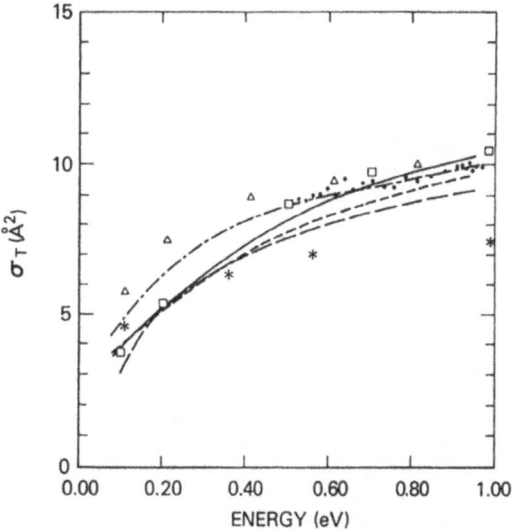

Fig. 4. Total cross section at very low energy; present "Hybrid Theory" with 6 states in $^2\Sigma_g^+$ symmetry using MCSCF target (□); Hybrid Theory with fixed nuclei $^2\Sigma_g^+$ symmetry and MCSCF target (Weatherford et al., 1987) (————); Hybrid Theory with fixed nuclei in $^2\Sigma_g^+$ symmetry and SCF target (Weatherford et al., 1985) (— — — —); Padial and Norcross (1984); Fixed nuclei calculations (—— - ——); A. Kennerly (1980) experiment (...); Sohn et al., (1986) experiment (*); Jost et al., (1983) experiment (+++)

recent calculation of Morrison et al (1987). Because the latter uses a polarization potential which in our opinion is the best thus far derived, we tend to prefer their result.

We therefore intend to use the Morrison et al (1987) potential suitably generalized to other R in further vibrational close coupling and fixed-nuclei calculation both at these very low as well as Π_g resonant energies. One expects, using a simple but cogent time delay criterion $[(\Delta t)_{coll} \ll \tau_{vib}]$, that the fixed-nuclei approximation is well justified for <u>elastic</u> scattering above 0.1 eV (but below the resonance). This is seen to occur in our results and suggests that values achieved with a better V_{pol} will provide a fairly definitive result.

ACKNOWLEDGEMENTS

Charles A. Weatherford's work was supported by U.S. National Aeronautics and Space Administration (NASA) Grant No. NAG5-738 and by U.S. Air Force Grant No. AFOSR-86-0149. Dr. Weatherford would like to acknowledge a grant of computer time from Florida State University for the use of the CYBER 205 at their Supercomputer Computations Research Institute. Aaron Temkin's work is supported under NASA-RTOP 442-36-58-01.

REFERENCES

Allan, M., 1985, <u>J. Phys. B: At. Mol. Phys.</u>, 18:4511.
Chandra, N. and Temkin, A., 1982, <u>Comput. Phys. Commun.</u>, 25:97.
Dube, L. and Herzenberg, A., 1979, <u>Phys. Rev. A</u> 20:194.
Hazi, A.U., Rescigno, T.N., Kurilla, M., 198, <u>Phys. Rev. A</u>, 23:1089.
Huo, W.H., Gibson, J.L., Lima, M.A.P., McKoy, V.M., 1987a, <u>Phys. Rev. A</u>, 36:1632.
Huo, W.H., Lima, M.A.P., Gibson, J.L., McKoy, V.M., 1987b, <u>Phys. Rev. A</u>, 36:1642.
Jost, K., Bisling, P.G.F., Eschen, F., Felsmann, M., and Walther, L., 1983, Abstracts of Contributed Papers, Thirteenth International Conference on the Physics of Electronic and Atomic Collisions, ed., J. Eichler, I.V. Hertel, and N. Stolterfoht, eds., North-Holland, Amsterdam, p.91
Kennerly, R.E., 1980, <u>Phys. Rev. A</u>, 21:1876.
Morgan, L.A., 1986, <u>J. Phys. B: At. Mol. Phys.</u>, 19:L439.
Morrison, M.A., Saha, B.C., and Gibson, T.L., 1987, <u>Phys. Rev. A</u>, 36:3682.
Onda, K. and Temkin, A., 1983, <u>Phys. Rev. A</u>, 28:621.
Padial, N.T., and Norcross, W.D., 1984, <u>Phys. Rev. A</u>, 29:1393.
Schneider, B.I., LeDourneuf, M. and VoKyLan, 1979, <u>Phys. Rev. Lett.</u>, 43:1926.
Sohn, W., Kochem, K.-H., Scheuerlein, K.-M., Jung, K., and Ehrhardt, H., 1986, <u>J. Phys. B: At. Mol. Phys.</u>, 19:4017.
Sullivan, E.C., and Temkin, A., 1982, <u>Comput. Phys. Commun.</u>, 25:97.
Temkin, A., 1979, "Symposium on Electron-Molecule Collisions", Shimamura, I. and Matsuzawa, M., eds., University of Tokyo, Tokyo, p.55.
Temkin, A., 1980 "Electronic and Atomic Collisions", K. Takayanagi and N. Oda, eds., North Holland, Amsterdam, p.95.
Temkin, A., and Vasavada, K.V., 1967, <u>Phys. Rev.</u>, 160:109.
Temkin, A., Vasavada, K.V., Chang, E.S., and Silver, A., 1969, <u>Phys. Rev.</u>, 187:57.
Weatherford, C.A., Onda, K., and Temkin, A., 1985, <u>Phys. Rev. A</u>, 31:3620.
Weatherford, C.A., Brown, F.B., Temkin, A., 1987, <u>Phys. Rev. A</u>, 35:4561.

LOW ENERGY ELECTRON SCATTERING BY DIATOMIC MOLECULES

USING THE R-MATRIX METHOD

C. J. Gillan*, P. G. Burke*, C. J. Noble+
and L. A. Morgan†

* Queen's University of Belfast, Northern Ireland BT7 1NN
+ SERC Daresbury Laboratory, Cheshire, England WA4 4AD
† Royal Holloway & Bedford New College, Surrey, England
  TW20 OEX

INTRODUCTION

The R-matrix method was first introduced by Wigner (1946a,b) and
Wigner and Eisenbud (1947) to study nuclear reactions. It was extended
and first applied to the study of low energy electron scattering by
complex atoms and ions by Burke et al (1971). The method has been
further refined to consider other atomic processes and there is now a
well developed suite of computer programs to perform R-matrix calculations
on atoms and ions for a great number of processes, Burke (1987).

Electron scattering by molecules includes the phenomena of exchange
and correlation already present in electron atom scattering but additional
problems arise from the multicentre nature of the interactions and the
presence of nuclear motion. The R-matrix method was further developed
and extensively applied, both by Schneider and Hay (1975, 1976) and
independently by Burke et al (1977), Buckley et al (1979) and Noble et al
(1982) to study electron scattering by diatomic molecules within the fixed
nuclei approximation, that is, it is assumed that the nuclei remain fixed
in space during the collision. Burke et al (1983) extended the theory to
represent the scattered electron by numerical, rather than analytic
functions, which enabled the calculation to be carried out over a wider
energy range, and Tennyson (1986) modified the theory to consider positron
scattering by diatomic molecules. The R-matrix method has also been
extended by Schneider et al (1979a), Morgan (1986) and Gillan et al (1987),
to include the nuclear motion thus enabling the processes of vibrational
excitation and dissociative attachment to be studied. The R-matrix method
is now an ab-initio technique comparable with the Linear Algebraic Method
of Collins and Schneider (1981) and the Schwinger Multichannel Method of
McKoy and co-workers (Watson and McKoy (1979), Huo et al (1987a)).

This article summarizes the salient features of the fixed nuclei
R-matrix theory before discussing the recent developments required to
include nuclear motion. A more detailed discussion can be found in
Gillan et al (1987). Results are presented from recent application of
the theory to electron scattering by N_2 and HF and finally some future
directions of research are anticipated.

THEORY

The basic philosophy behind the R-matrix approach is that the interaction of the electron with the target molecule has fundamentally different
physical characteristics in different regions of configuration space.
One takes advantage of this fact and formulates different equations
appropriate to the region in question joining the solutions at the
boundaries by an R-matrix.

In the fixed nuclei approximation, there are just two regions separated
by a sphere of radius r = a, where r is the distance of the scattering
electron from the centre of mass of the target.   A multi-centre configuration interaction type expansion of the scattering wavefunction is chosen
in the inner region, with the scattering wavefunction expanded in a basis
ψ_k

$$\psi_k = \mathcal{A} \sum_{ij} \bar{\Phi}_i \eta_j \alpha_{ijk} + \sum_j \chi_j \beta_{jk} \tag{1}$$

where $\mathcal{A}$ anti-symmetrises the co-ordinates of the scattering electron with
the coordinates of all the other electrons.   The functions $\bar{\Phi}_i$ are eigenfunctions of the total spin operator, S^2 and its z-component ^iS_z formed by
coupling the spin function of the continuum electron to the target elect-
ronic states Φ_i and pseudo-states Φ^{PS}_i which are included.   The η_j functions
are the continuum molecular orbitals, on the centre of gravity of the mole-
cule, which are non-zero on the boundary of the inner region.   The second
summation in equation (1) is over square integrable (L^2) functions χ_j where
all of the electrons are in bound orbitals.   These χ_j functions,   like
the $\bar{\Phi}_i$, are zero on the boundary of the inner region.   Their role is to
account for short range correlation effects between the target and projectile.   The constants α_{ijk}, β_{jk} are determined by diagonalizing H + L in the
basis giving

$$\langle \psi_k | H + L | \psi_t \rangle = E_t \delta_{kt} \tag{2}$$

where H is the electronic hamiltonian for the N + 1 electron system and L
is a surface projection operator introduced by Bloch (1957) which ensures
that H + L is hermitian.   It is defined by

$$L = \frac{1}{2} \sum_{i=1}^{N+1} \sum_j | \bar{\Phi}_j(r_i^{-1}) Y_{l_j m_j}(\hat{r}_i) > \delta(r_i - a) \left(\frac{d}{dr_i} - \frac{b - 1}{r_i} \right)$$

$$< \Phi_j(r_i^{-1}) Y_{l_j m_j}(\hat{r}_i) | \quad . \tag{3}$$

The diagonalization, which is similar to a molecular bound state problem,
is performed using a modified version of the IBM ALCHEMY quantum chemistry
program package, McLean (1971), Noble (1982).

Solving the Schrodinger equation for the scattering wavefunction Ψ_E
formally, yields

$$\Psi_E = \sum_k \frac{| \psi_k \rangle \langle \psi_k | L | \Psi_E \rangle}{E_k - E} \quad . \tag{4}$$

The R-matrix joining the inner and outer regions, is defined by projecting
(4) onto the channel function $\bar{\Phi}_i Y_{l_j m_j}$ and evaluating the expression at the
R-matrix boundary.

The collision problem can be solved at fixed internuclear separations
by adopting a single centre, no exchange, close coupling expansion of the

scattering wavefunction in the outer region and matching the two solutions at the boundary. T-matrices calculated for several internuclear separations can be adiabatically averaged using the theory of Chase (1956) to obtain vibrationally resolved T-matrices.

When nuclear motion is introduced into the theory, configuration space is again partitioned into different regions, but the inner region is now taken to be a hypersphere defined by: $0 < r_i < a$ for all i; $A_{in} < R < A_{out}$. Here r_i and a are defined as in the fixed nuclei approximation, R is the internuclear separation while A_{in} is chosen to exclude the singularity in the internuclear potential due to repulsion and A_{out} is chosen such that those target vibrational states, which are to be included in the model have negligible amplitude for $R > A_{out}$.   The different regions are represented schematically in figure 1.

Expanding the scattering wavefunction in a basis Θ_k, where the functions Θ_k are chosen to be eigenfunctions satisfying the following equation

$$< \Theta_k | H - \frac{1}{2M} \frac{d^2}{dR^2} + L_1 | \Theta_{k'} > = E_k^{NUC} \delta_{kk'}, \tag{5}$$

The Schrodinger equation may again be solved formally, in the absence of a dissociating channel, to give

$$\psi_E^{NUC} = \sum_k \frac{| \Theta_k > < \Theta |L_1| \psi_E >}{E_k^{NUC} - E} \tag{6}$$

where H is the hamiltonian of equation (2), $-\frac{1}{2M} \frac{d^2}{dR^2}$ is the nuclear kinetic energy operator and L_1 is again a Bloch surface projection operator. The generalised R-matrix is now obtained by projecting this equation onto the external region channel functions and evaluating at the R-matrix boundary.

To obtain vibrational excitation cross sections it is necessary to solve the collision problem in the external region.   Adopting the expansion

$$\psi_E^{NUC} = \sum_{i1_iv} \overline{\Phi}_i(x_1,\ldots,x_N,\sigma_{N+1}) Y_{1_im_i}(\hat{r}_{N+1}) \zeta_{iv}(R) \frac{1}{r} F_{i1_iv} \tag{7}$$

substituting into the Schrodinger equation and projecting out the channel functions $\Phi_i Y_{1_im_i} \zeta_{iv}$, where ζ_{iv} are vibrational wave functions, yields a set of coupled differential equations for the reduced radial wavefunctions F_{i1_iv} which are in fact the same as those obtained by Chandra and Temkin (1976) in the hybrid close coupling method. These are solved by R-matrix propagator, Morgan (1984), and accelerated Gailitis expansion techniques, Noble and Nesbet (1984).

APPLICATIONS TO e-N$_2$ SCATTERING

The low energy shape resonance in e-N$_2$ scattering now has an extensive literature and has become a favourite testing ground for new theories of electron molecule scattering since the mechanism responsible for its creation was unravelled by Schulz (1962,1966) and Herzenberg and Mandl (1962).   It is now established that the resonance is due to the trapping of the scattering electron in a d-wave centrifugal potential, corresponding to temporary capture into a π_g orbital to form a state of N_2^-.   The angular distribution of the scattering electron is expected to exhibit typical d-wave behaviour.   The lifetime of the resonance is approximately equal to the vibrational period of the compound state and this gives rise

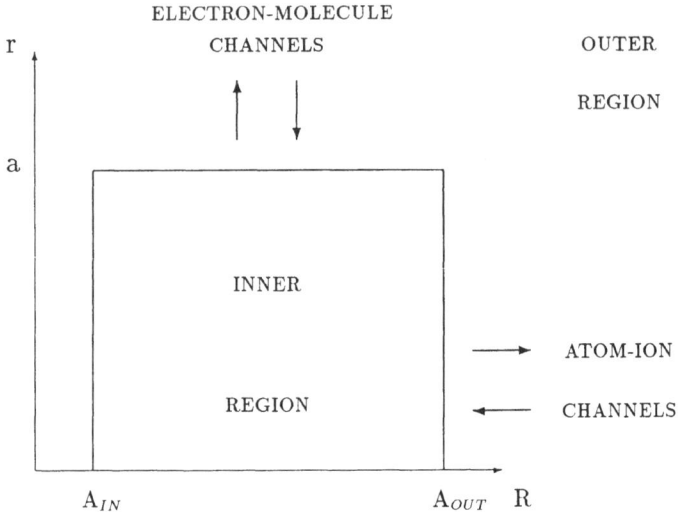

Fig 1.   Inner and outer regions for a nuclear motion calculation.

to sharp peaks in the total cross section.   The study of this scattering problem provided a stringent test of our theory in the absence of a dissociating channel and also allowed us to debug the associated part of the computer programs.

There have been a number of measurements of differential scattering cross sections for this vibrational excitation process;   however earlier calculations have so far been based on theories where either local exchange or model polarization potentials are adopted.   Our ab-initio approach is capable however, of providing information on the dynamics of the scattering process and may be compared with other recent ab-initio calculations Schneider et al, (1979b), Schneider and Collins (1984) and Huo et al (1987a,b).

We carried out our first calculations with only the ground state of the N_2 target included in the first expansion in equation (1).   Scattering calculations were carried out for the eight symmetries

$$^2\Sigma_g^+ \quad ^2\Sigma_u^+ \quad ^2\Pi_u \quad ^2\Pi_g \quad ^2\Delta_g \quad ^2\Delta_u \quad ^2\Phi_u \quad ^2\Phi_g$$

in the static exchange plus polarization (SEP) approximation.   Also, as well as the terms retained in the first expansion in (1) we also included 2p-1h configurations in the second expansion in (1).

Calculations on the $^2\Sigma_g^+$ symmetry were repeated by Gillan et al (1988) with two polarized pseudo-states included in addition to the ground state in equation (1).   The pseudo-states, one of symmetry $^1\Sigma_u^+$ and one of symmetry $^1\Pi_u$, were introduced to account for the two components of dipole polarizability, $\alpha_{||}$ and $\alpha_\perp$ respectively.   The dipole polarization inter-action between the scattering electron and the target, which is important at very low energies and in particular for the $l = 0$ partial wave, is not accounted for by the form of the χ_j functions that we used.

We present in figure 2 our total cross section results, obtained by summing $\sigma(0 \rightarrow v)$ over all vibrational levels, compared with the absolute measurements of Kennerly (1980).   There is satisfactory agreement in the magnitude of the cross section in the neighbourhood of the $^2\Pi_g$ resonance.

240

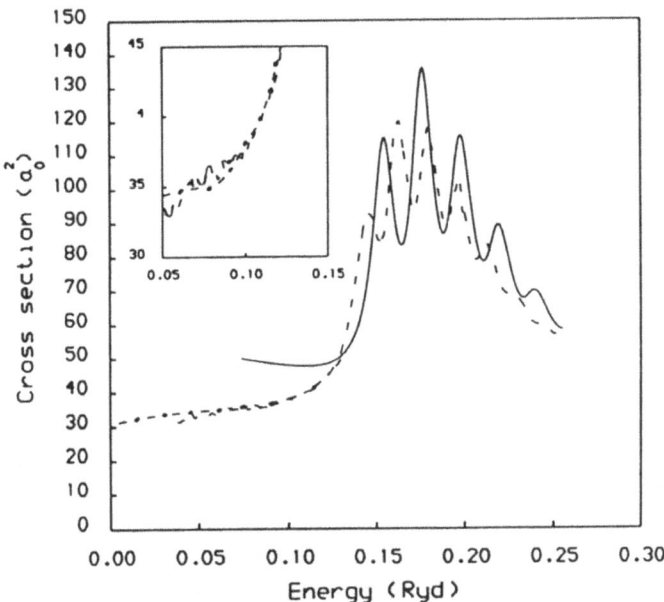

Fig. 2 Total cross section in a_0^2 for electron scattering from the ground
electronic state of N_2. $-\,-$, experiment of Kennerly (1980),
—— , calculated values Gillan et al (1987) without polarized
pseudo-states $\cdot-\cdot-\cdot-$ calculated values of Gillan et al (1988)
with pseudo-states.

The small differences in the spacing of the peaks is attributed to our
choice of an SCF N_2 potential curve. The vibrational spacing obtained
was significantly larger than the experimentally determined value of 0.294
eV. The effect of including the long range dipole polarization in our
calculations is clear. Without it, the cross section diverges from the
experimental results at very low energy. However by including the pseudo-
states we have brought our results into agreement with experiment. We
have not included target correlation effects in our model but this would
be the next step and would certainly be required to provide a very detailed
comparison of theory and experiment.

The inadequate representation of the long range dipole polarisation in
our work is also reflected in the elastic differential cross section. We
present our differential cross section for the elastic $v = 0 \rightarrow 0$ transition
at an incident energy of 5 eV in figure 3 compared with the experimental
measurements of Shyn and Carignan (1980) and the theoretical calculations
of Chandra and Temkin (1976) and of Truhlar et al (1976). Our results are in
good agreement with experiment for scattering angles greater than about
$40°$ but over estimate the cross section in the forward direction. The
hybrid theory of Chandra and Temkin (1976) gives better agreement with
experiment in the forward direction but this is because they included
polarisation effects by a model polarisation potential involving the
experimental polarizabilities and containing a cut-off parameter which
was tuned to give the experimentally observed energy for the $^2\Pi_g$ resonance.
The results of Truhlar et al (1976) who used a model interaction potential
with adjustable parameters are in poor agreement with experiment in both
forward and backward directions. This indicates that both the long range
and short range correlation effects are incompletely represented in their
work.

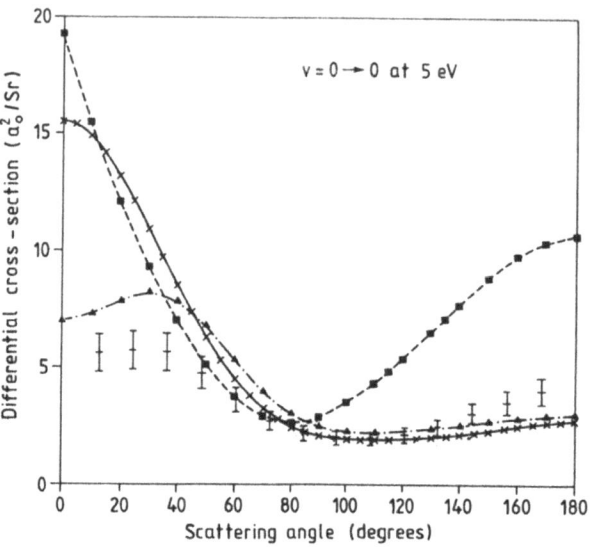

Fig. 3 Elastic $v = 0 \to 0$ differential cross section $a_o^2 sr^{-1}$ at 5 eV. $\cfrac{\ }{\ }$,
experiment of Shyn and Carignan (1980), ——X——, calculated
values Gillan et al (1987), $- \cdot \blacktriangle \cdot - \cdot$ Chandra and Temkin
(1976), $- - \blacksquare - -$ Truhlar et al (1976).

In contrast to elastic scattering, the vibrationally inelastic cross
sections are dominated by scattering in the $^2\Pi_g$ resonance state.   In this
case the short range correlation effects are more important than the long
range polarisation potential and hence we expected our results to be better.
This is confirmed by our differential cross sections for the $v = 0 \to 1$
transition at 5 eV which we compare in figure 4 with the experimental
measurements of Tanaka et al (1981) and the theoretical calculations of
Chandra and Temkin (1976) and of Truhlar et al (1976).   We see that the
shape of the cross section agrees well with experiment at all angles.

APPLICATION TO e - HF SCATTERING

Rohr and Linder (1976) have observed sharp peaks near threshold in the
vibrational excitation of HF and HCl molecules by electron impact.   They
also found that the differential cross sections were isotropic in the
vicinity of these peaks.   Recently, Knoth et al (1987), have confirmed the
existence of these peaks but measured differential cross sections which are
anisotropic.

There has been some controversy among theorists about the mechanism
responsible for the peak structure in the cross section.   On the one hand
Dube and Herzenberg (1977) have proposed the existence of a virtual state
but did not account for nuclear motion in their model while on the other
hand Gauyacq and Herzenberg (1982) and Domcke and Mundel (1985) explained
the peaks as nuclear excited Feshbach resonances.   Rescigno et al (1982),
performed a detailed ab-initio series of fixed nuclei calculations at the
static exchange level and then employed the energy modified adiabatic
approximation, Nesbet (1979), to obtain ro-vibrational cross sections.   They
obtained threshold peaks but no evidence of a resonance.

Morgan and Burke (1987) have performed a detailed ab-initio calculation,
on e - HF scattering using the nuclear motion R-matrix method.   A series of
fixed nuclei calculations were performed for R = 1.3 to 2.5 a.u. at the

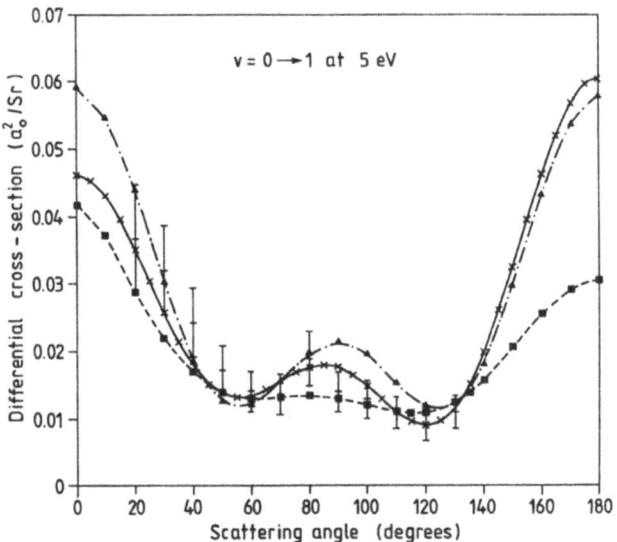

Fig. 4   Inelastic $v = 0 \to 1$ differential cross section in $a_o^2 sr^{-1}$ at 5 eV. $\mathbf{\mathbf{\Phi}}$, experiment of Tanaka et al (1981), ——X——, calculated values Gillan et al (1987), – · ▲ · – Chandra and Temkin (1976), – – ■ – – Truhlar et al (1976).

static exchange plus polarization level and then nuclear motion was accounted for as in e - N_2 scattering, for both the $^2\Sigma$ and $^2\Pi$ scattering symmetries.

A static exchange calculation at the equilibrium geometry was in excellent agreement with the results of Rescigno et al (1982) but short range polarization was found to make a significant contribution when introduced into the calculation.   The S-matrix poles were calculated in the complex energy plane but no resonance or virtual state poles were obtained in the vicinity of the real energy axis.   The HF⁻ curve obtained, which is shown in figure 5, does not appear to cross the HF SCF ground state and therefore plays the role of a bound state.   In addition no resonances were found in either scattering symmetry.

Integrated cross sections calculated for $v = 0 \to 1$ transition are shown in figure 6.   The differential cross section for this transition shows anisotropic behaviour.   The poles of the vibrationally resolved S-matrix were located in the complex plane.   The single bound state pole of the fixed nuclei calculations discussed above, was split into a sequence of poles, one lying slightly below each threshold.   The first, below the $v = 0$ threshold is still a bound state but the others lie off the physical axis and display all of the features of nuclear excited Feshbach resonances.

FUTURE DIRECTIONS

Ab-initio calculations on electron diatomic molecule scattering commenced only really in the last few years.   It is therefore a very young and dynamic field with methods, including the R-matrix, being constantly refined and improved.   The advance is directly linked to increases in computing power.   Currently, the R-matrix programs are being re-written to take full advantage of the vector processing and multi-tasking capabilities of the Cray XMP supercomputer.   Once this has been completed it will be possible to study more complicated problems, such as electronic excitation, with greater ease.

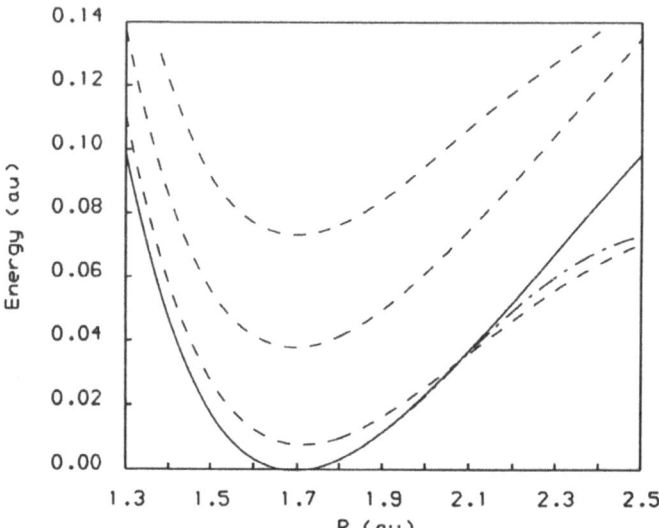

Fig. 5 Potential energy curves in a.u. relative to the energy of HF at
its equilibrium geometry. ———— SCF $^1\Sigma$ ground state of HF; –·–·,
$^2\Sigma$ bound state of HF$^-$; – – – – , R-matrix pole positions for
$^2\Sigma$ symmetry.

Preliminary calculations have already been performed on electronic
excitation of N_2 where the ground state and the eight lowest lying valence
excited states have been included in equation (1).   Encouraging results
have been obtained and we hope to pursue this work further.

A new method of performing R-matrix calculations which has been
discussed by Burke et al (1987), has been developed.   This will be particularly useful for studying intermediate energy scattering where there
are many open channels.   The method has been applied to electron scattering
by hydrogen atoms, Scholz et al (1988), and has yielded very accurate
results.

In addition to electronic excitation, it is anticipated that there will
be considerable interest in electron scattering from highly polar molecules.
This has already begun with the calculations on HF.   Currently work is in
progress on electron scattering from HCl.

The experience gained from the application of the R-matrix technique to
scattering from diatomic molecules will be of use in the development of
methods and programs to study scattering from polyatomic molecules.

REFERENCES

Bloch, C., 1957, Nucl. Phys. 4:503.
Buckley, B. D., Burke, P. G. and Vo Ky Lan, 1979, Comput. Phys. Commun.,
    17:175.
Burke, P. G., 1987, R-matrix method in Atomic Physics, in: Atomic Physics
    10, Publisher, Elsevier Science.
Burke, P. G., Hibbert, A. and Robb, W. D., 1971, J. Phys. B: At. Mol. Phys.
    4:153.
Burke, P. G., Mackey, I. and Shimamura, I., 1977, J. Phys. B: At. Mol. Phys.
    10:2497
Burke, P. G., Noble, C. J. and Salvini, S., 1983, J. Phys. B: At. Mol. Phys.
    16:L113.

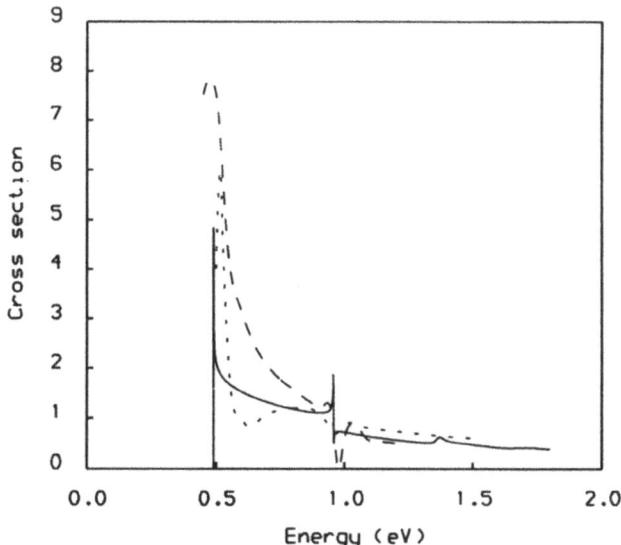

Fig. 6 Integrated cross sections for excitation of the v = 1 level.
———— , Morgan and Burke (1988) results; — — — — — — ,
Gauyacq (1985); ...... , Rohr and Linder (1976).

Burke, P. G., Noble, C. J. and Scott, P., 1987, Proc. Roy. Soc. (London),
    A410:289.
Chandra, N. and Temkin, A., 1976, Phys. Rev.,A13:188.
Chase, D. M., 1956, Phys. Rev., 104:838.
Collins, L. A. and Schneider, B. I., 1981, Phys. Rev., A24:2387.
Domcke, W. and Mundel, C., 1985, J. Phys. B: At. Mol. Phys., 18:4491.
Dube, L. and Herzenberg, A., 1977, Phys. Rev. Lett., 38:820.
Gauyacq, J. P. and Herzenberg, A., 1982, Phys. Rev.,A25:2959.
Gillan, C. J., Nagy, O., Burke, P. G., Morgan, L. A. and Noble, C. J.,
    1987, J. Phys. B: At. Mol. Phys., 20:4585.
Gillan, C. J., Noble, C. J. and Burke, P. G., 1988, J. Phys. B: At. Mol.
    Phys., 21:L53.
Herzenberg, A. and Mandl, F., 1962, Proc. Roy. Soc. (London), A270:48.
Huo, W. M., Gibson, T. L., Lima, M.A.P. and McKoy, V., 1987a, Phys. Rev.,
    A36:1632.
Huo, W. M., Lima, M.A.P., Gibson, T. L. and McKoy, V., 1987b, Phys. Rev.,
    A36:1642.
Kennerly, R. E., 1980, Phys. Rev.,A21:1876.
Knoth, G., Radle, M., Ehrhardt, H. and Jung, K., 1987 (see these
    proceedings).
McLean, A. D., 1971, Potential Energy Surfaces from Ab-initio Computation:
    Current and Projected Capabilities of the Alchemy Computer Program,
    in:Proc. Conf. on Potential Energy Surfaces in Chemistry, W. A. Lester
    Jr., ed., Publisher, IBM San Jose.
Morgan, L. A., 1986, J. Phys. B: At. Mol. Phys., 19:L439.
Morgan, L. A., 1984, Comput. Phys. Commun., 31:419.
Morgan, L. A. and Burke, P. G., 1988, submitted to J. Phys. B: At. Mol. Phys.
Nesbet, R. K., 1979, Phys. Rev., A19:551.
Noble, C. J., 1982, Daresbury Laboratory Report DL/SCI/TH33T.
Noble, C. J., Burke, P. G. and Salvini, 1982, J. Phys. B: At. Mol. Phys.,
    15:3779.
Noble, C. J. and Nesbet, R. K., 1984, Comput. Phys. Commun.,33:399.

Rescigno, T. N., Orel, A. E., Hazi, A. U. and McKoy, B. V., 1982, <u>Phys. Rev.</u>, A26:690.

Rohr, K. and Linder, F., 1976, <u>J. Phys. B: At. Mol. Phys.</u>, 9:2521.

Schneider, B. I., 1975, <u>Chem. Phys. Lett.</u>, 2:237.

Schneider, B. I. and Collins, L. A., 1984, <u>Phys. Rev.</u>, A30:95.

Schneider, B. I. and Hay, P. J., 1976, <u>Phys. Rev.</u>, A13:2049.

Schneider, B. I., LeDourneuf, M. and Burke, P. G., 1979a, <u>J. Phys. B: At. Mol. Phys.</u>, 12:L35.

Schneider, B. I., LeDourneuf, M. and Vo Ky Lan., 1979b, <u>Phys. Rev. Lett.</u>, 43:1926.

Scholz, T., Scott, M. P., Burke, P. G. and Noble, C. J., 1988, Electron Scattering at Intermediate Energies, in Invited Papers and Progress Reports XV ICPEAC, H. B. Gilbody et al. ed., Publisher North Holland.

Schulz, G. J., 1962, <u>Phys. Rev.</u>, 125:229.

Schulz, G. J. and Koons, H. C., 1966, <u>J. Chem. Phys.</u>, 44:1297.

Shyn, T. W. and Carignan, G. R., 1980, <u>Phys. Rev.</u>, A22:923.

Tanaka, H., Yamamoto, Y. and Okada, T., 1981, <u>J. Phys. B: At. Mol. Phys.</u>, 14:2081.

Tennyson, J. T., 1986, <u>J. Phys. B: At. Mol. Phys.</u>, 19:4025.

Truhlar, D. G., Brandt, M. A., Chutjian, A., Srivastava, S. K. and Trajmar, S., 1976, <u>J. Chem. Phys.</u>, 65:2962.

Watson, D. K. and McKoy, V., 1979, <u>Phys. Rev.</u>, A20:1474.

Wigner, E. P., 1946a, <u>Phys. Rev.</u>, 70:15.

Wigner, E. P., 1946b, <u>Phys. Rev.</u>, 70:606.

Wigner, E. P. and Eisenbud, L., 1947, <u>Phys. Rev.</u>, 72:29.

CALCULATED VIBRATIONALLY AND ROTATIONALLY RESOLVED PHOTOELECTRON SPECTRA
OF H_2

Jonathan Tennyson

Department of Physics and Astronomy
University College London
Gower Street, London WC1E 6BT, U.K.

INTRODUCTION

The photoionisation of molecular hydrogen is one of the most fundamental processes in molecular physics. Surprisingly, for such a seemingly
simple process, there are a number of questions which, despite much
experimental and theoretical work, remain unresolved. Some of these
problems have been the subject of a recent review by Itikawa (1987). In
the present work, ab initio calculations are presented for vibrationally
and rotationally resolved spectra below the possible resonance region
(Raseev 1985; Tennyson et al. 1986) at about 30eV.

THEORY

The fully ro-vibrationally resolved photoionisation process can be
written:

$$H_2(X\,^1\Sigma_g^+;\ v'',j'') + h\nu \rightarrow H_2^+(X\,^2\Sigma_g^+;\ v',j') + e^-. \tag{1}$$

For a particular photon energy and $(v'',j'') \rightarrow (v',j')$ the resolved cross-
section is usually parameterised

$$I(\theta) = \frac{1}{4\pi}\ \sigma\big[1 + \beta\,P_2(\cos\theta)\big] \tag{2}$$

where θ is measured from the direction of the light polarisation. The
parameter β represents the asymmetry of the distribution of photo electrons and, as will be seen, is generally the most difficult to measure or
calculate accurately.

The calculations presented here follow closely those of Tennyson
et al (1986) who performed photoionisation calculations using previous
$e^- - H_2^+$ scattering calculations (Tennyson et al, 1984; Tennyson and Noble,
1985). The scattering calculations employed a multicentre R-matrix
formalism and used the ground and first excited states of H_2^+ in the
close-coupled expansion. Coupling to the lowest three partial waves in
each symmetry was allowed for. Thus continuum states of $^1\Sigma_u^+$ and $^1\Pi_u$
symmetry included contributions from p,f and h waves coupling with
the $H_2^+(X\,^2\Sigma_g^+)$ state. Exchange and short-range polarisation effects
were also included in these calculations ab initio.

In the present calculations, the electronic ground state of H_2 was obtained from a scattering calculation performed with negative energy (Ohja and Burke, 1983). This method guarantees that the bound and continuum states are treated at the same level of approximation. Comparison with accurate electronic structure calculations shows that 90% of the H_2 correlation energy was recovered in this calculation.

The vibrational motion of the target and the ion was accounted for by adiabatically averaging a series of 13 fixed nuclei calculations with vibrational wavefunctions generated from accurate ab initio potentials. Similarly the rotational motion was treated within the adiabatic nuclei (AN) approximation (Itikawa, 1978) using the program of Tennyson and Chandra (1987).

This approximation leads to a marked simplification of the dependence of the β parameter on rotational state. As the process is dominated by $\Delta j = 0, \pm 2$, for a particular initial j'' only the parameters β_Q, β_S and β_O need be calculated. In the AN approximation β_S equals β_O and is only weakly dependent on j''. Furthermore, β_Q for $j'' = 0$ always equals 2.0, a value endorsed by analysis of experiment by Hara and Nakamura (1986).

For rotationally unresolved spectra the total cross section, $\sigma_{v'}$, is simply a sum over j' of the rotationally-resolved cross sections. The vibrationally-averaged asymmetry parameter, $\beta_{v'}$, is similarly given by a weighted sum of rotationally-resolved asymmetry parameters.

RESULTS

Recent calculations (Tennyson, 1987) on the photoionisation of vibrationally-excited H_2 and D_2 by 584Å photons showed structure in the total cross sections due to Franck-Condon factors in agreement with previous calculations (O'Neill and Reinhardt, 1978) and experiment (Van der Meer et al, 1985). A seemingly related structure was also predicted in the β parameters associated with photoionisation of vibrationally-excited targets. In particular, certain transitions, such as $v''=2 \rightarrow v'=2$ and $v''=1 \rightarrow v'=3$ for D_2, were predicted to have β values less than 1.0.

Tables 1 and 2 present results for ro-vibrationally resolved photoionisation of H_2 in its ground state. Comparison is made with previous theoretical and experimental results. More extensive comparisons can be found in Hara (1985) and Hara et al (1986).

DISCUSSION

The results in Tables 1 and 2 show a fair measure of agreement for the rotationally-averaged total cross sections, $\sigma_{v'}$, but are rather contradictory for the other parameters. In particular the experimental results are sensitive to the method used to interpret them: for example giving different values and error estimates according to whether $\beta_Q=2$ for $j''=0$ is assumed or not.

Theoretically the situation is no less clear. For example the present results give $\beta_{v'}$ and β_S values consistently lower than those obtained by Hara and co-workers. The parameter β_S is known to be highly sensitive to the details of the calculation, and in particular to the inclusion of higher partial waves, but the exact source of the discrepancy remains unclear.

Table 1

Vibrationally and rotationally resolved parameters for the photoionisation of $H_2(v''=0, j''=0)$ by 584Å radiation. (Absolute cross sections in atomic units are given in brackets).

v'	Ref.	σ_S/σ_Q	β_S	$\beta_{v'}$	$\sigma_{v'}/\sigma_{v'=2}$
0	a	0.087(3)	0.81(17)		
	b	0.131(2)	0.87(12)	1.868(10)	0.434
	c	0.095(4)	0.87(19)	1.862(13)	0.438
	e	0.241	0.749	1.758	0.4465
	f	0.103	0.831	1.891	0.4647
	g	0.154	0.568	1.807	0.4489
1	a	0.073(2)	0.71(16)		
	b	0.107(2)	0.75(11)	1.879(7)	0.839
	c				0.850
	e		0.720	1.793	0.8527
	f	0.085	0.782	1.904	0.8743
	g	0.124	0.501	1.833	0.8732
2	a	0.061(2)	0.73(15)		
	b	0.088(1)	0.75(11)	1.899(6)	
	c				(0.0366)
	e	0.159	0.687	1.820	(0.03124)
	f	0.071	0.746	1.917	(0.03399)
	g	0.101	0.459	1.859	(0.03859)
3	a	0.052(2)	0.80(16)		
	b	0.075(1)	0.83(11)	1.918(6)	0.937
	c				0.946
	f	0.059	0.723	1.929	0.9183
	g	0.082	0.422	1.880	0.9193
4	a	0.043(2)	0.66(19)		
	b	0.062(1)	0.69(3)	1.923(5)	0.778
	c				0.783
	f	0.050	0.702	1.938	0.7516
	g	0.068	0.392	1.897	0.7466
5	a	0.037(2)	0.65(21)		
	b	0.054(1)	0.68(14)	1.933(5)	0.593
	c				0.598
	f	0.044	0.680	1.945	0.5746
	g	0.058	0.369	1.911	0.5691
6	b				0.440
	c				0.453
	f	0.036	0.677	1.954	0.4237
	g	0.049	0.352	1.922	0.4176

Experiment
a. Ruf et al. (1983)
b. Deduced from Ruf et al. (1983) by Hara et al. (1986)
c. Pollard et al. (1982a,b)
Theory
e. Itikawa (1979), $\ell=1,3$
f. Hara et al. (1986), $\ell=1,3$
g. This work, $\ell=1,3,5$

Table 2

Vibrationally and rotationally resolved parameters for the photionisation of $H_2(v''=0, j''=0)$ by 736Å radiation. (Absolute cross section in atomic units are given in brackets).

v'	Ref.	σ_S/σ_Q	β_S	$\beta_{v'}$	$\sigma_{v'}/\sigma_{v'=2}$
0	a	0.064(1)	0.54(16)		
	b	0.093(2)	0.62(11)	1.882(9)	
	c	0.091(3)	0.09(13)	1.841(12)	
	e	0.212	0.348	1.711	
	f	0.087	0.643	1.891	0.465
	g	0.110	0.359	1.837	0.493
1	a	0.052(1)	0.47(14)		
	b	0.076(1)	0.52(9)	1.896(7)	
	d		0.83(48)	1.93(3)	
	e	0.167	0.330	1.761	
	f	0.069	0.598	1.910	0.877
	g	0.091	0.334	1.861	0.946
2	a	0.044(1)	0.59(13)		
	b	0.065(1)	0.67(9)	1.919(5)	
	f	0.055	0.575	1.926	(0.06108)
	g	0.075	0.310	1.882	(0.06518)

Experiment
  a. Ruf et al. (1982)
  b. Deduced from Ruf et al. (1983) by Hara (1985)
  c. Deduced from Pollard et al. (1982a,b) by Hara (1985)
  d. Niehaus and Ruf (1971) re-analysed by Ruf et al. (1983)

Theory
  e. Itikawa (1979), $\ell=1,3$
  f. Hara (1985), $\ell=1,3$
  g. This work, $\ell=1,3,5$

    Finally, mention should be made of a recent paper by Parr et al (1988). These workers performed vibrationally-resolved photoionisation experiments on H_2 at a range of photon frequencies using synchrotron radiation. Their results give vibrationally averaged asymmetry parameters, $\beta_{v'}$, in the region of 1.79. This is lower than most theoretical estimates but in fair agreement with the results presented here. Parr et al also tentatively observe structure in their $\beta_{v'}$ parameters above 30eV as predicted by theory (Raseev, 1985; Tennyson et al, 1986).

ACKNOWLEDGEMENT

    I would like to thank Dr. N. Chandra for stimulating my interest in this problem.

REFERENCES

Hara, S., 1985, J. Phys. B: At. Mol. Phys., 18:3759.

Hara, S., and Nakamura, M., 1986, J. Phys. B: At. Mol. Phys., 19:L467.

Hara, S., Sato, H., Ogata, S., and Tamba, N., 1986, J. Phys. B: At. Mol. Phys., 19:1177.

Itikawa, Y., 1978, Chem. Phys., 28:461.

Itikawa, Y., 1979, Chem. Phys., 37:401.

Itikawa, Y., 1987, Comments At. Mol. Phys., 20:51.

Niehaus, A., and Ruf, M.W., 1971, Chem. Phys. Lett., 11:55.

Ohja, P.C., and Burke, P.G., 1983, J. Phys. B: At. Mol. Phys., 16:3513.

O'Neill, S.V., and Reinhardt, W.P., 1978, J. Chem. Phys., 69:2126.

Parr, A.C., Hardis, J.E., Southworth, S.H., Feigerle, C.S., Ferrett, T.A., Holland, D.M.P., Quinn, F.M., Dobson, B.R., West, J.B., Marr, G.V., and Dehmer, J.L., 1988, Phys. Rev. A. (in press).

Pollard, J.E., Trevor, D.J., Reut, J.E., Lee, Y.T., Shirley, D.A., 1983a Chem. Phys. Lett., 88:434.

Pollard, J.E., Trevor, D.J., Reut, J.E., Lee, Y.T., Shirley, D.A., 1983b, J. Chem. Phys., 77:34.

Raseev, G., 1985, J. Phys. B: At. Mol. Phys., 18:423.

Tennyson, J., 1987, J. Phys. B: At. Mol. Phys., 20:L375.

Tennyson, J., and Chandra, N., 1987, Computer Phys. Commun., 46:99.

Tennyson, J., and Noble, C.J., 1985, J. Phys. B: At. Mol. Phys., 18:155.

Tennyson, J., Noble, C.J., and Burke, P.G., 1986, Int. J. Quantum Chem., 19:1033.

Tennyson, J., Noble, C.J., and Salvini, S., 1984, J. Phys. B: At. Mol. Phys., 17:905.

Van der Meer, W.J., Van Lonkhuyzen, H., Butsehaar, R.J., and De Lange, C.A., 1985, J. Chem. Phys., 83:6173.

THE EFFECT OF A CHANGE OF CHARGE: THE SCATTERING

OF LOW ENERGY POSITRONS BY HYDROGEN MOLECULES

E.A.G. Armour, D.J. Baker and M. Plummer

Mathematics Department
Nottingham University
Nottingham NG7 2RD
England

ABSTRACT

The lowest partial wave is calculated for low-energy e^+-H_2 scattering using the Kohn method. In order to obtain accurate results, it is necessary to use a trial function which includes basis functions containing the positron-electron distance as a linear factor, i.e. Hylleraas-type functions. The inclusion of such functions has a very significant effect on the low-energy total cross section, bringing it into agreement with experiment for incident energies less than about 2eV. It also brings the calculated value for the annihilation rate of thermal positrons much closer to the experimental value. Progress is also reported on the calculation of the lowest partial waves of Σ_u^+ and Π_u symmetry. As far as we are aware, this is the first time that Hylleraas-type functions have been used in a molecular scattering calculation.

INTRODUCTION

It has been well known since the work of Hylleraas[1] and James and Coolidge[2] that the inclusion of the interelectronic distance as a linear factor in the trial function brings about rapid convergence in variational calculations on bound states of two-electron atoms and molecules. However, trial functions involving only separable functions i.e. functions which can be expressed as a finite expansion of products of one-particle functions, are much easier to work with. The required matrix elements can be evaluated straightforwardly using Slater's rules for one and two electron operators. The speed of modern computers makes it possible to include a very large number of Slater determinants constructed from separable functions. This has made it possible to obtain satisfactory accuracy in variational calculations on atomic and molecular bound states (see, for example, Schaefer[3]). Trial functions of this type have also proved very successful in electron-atom and electron-molecule scattering calculations (see, for example, Burke et al[4,5], Schneider and Collins[6] and Lima et al[7]).

The advent of intense low-energy positron beams has made possible many positron-atom and molecule scattering experiments[8], and further experiments can be expected in the near future. The change of sign in positron as opposed to electron scattering has very important consequences. The positron is distinguishable from the target electrons and therefore exchange effects with the incident particle are absent. In contrast to the electron, the positron is attracted by the target electrons and repelled by the nuclei. In this situation an expansion for the wavefunction in terms of separable functions converges very slowly. The calculations by Armour[9], using a generalization of the Kohn method, and Tennyson[10], using the R-matrix method, show that it is possible to obtain a good qualitative description of the behaviour of the phase shift of the lowest partial wave in low-energy positron-hydrogen molecule scattering by including sufficient separable functions in the trial function.

However, such calculations are incomplete. As the positron is the antiparticle corresponding to the electron, it may annihilate with a target electron to form 2 or 3 gamma rays. The calculation of the annihilation rate provides a particularly stringent test of the accuracy of a wavefunction[11,12]. Only the region of configuration space where the positron is very close to a target electron is sampled and the error in the annihilation rate is first order in the error in the wavefunction[13]. Armour and Baker[14] have shown that the wavefunction obtained in the Kohn calculation[9] gives a value of the annihilation rate for thermal positrons which is much smaller than the experimental value[15].

The use of separable basis functions corresponds to including only even powers of the positron-electron distance in the trial function. Schwartz[16] showed in 1961 that the Kohn variational method with a basis set made up of functions containing linear and higher powers of the interparticle distance gave accurate results for low energy positron-hydrogen-atom scattering. Basis sets of this type have subsequently been employed in Kohn and other variational calculations on low energy positron-hydrogen-atom and positron-helium scattering. For details, see the review article by Humberston[13].

Calculations of this type are greatly facilitated by the spherical symmetry of atomic targets. Nevertheless, Clary[17] has succeeded in carrying out very accurate variational calculations of the energy of He_2^+ and He_2 for a given internuclear distance by using basis sets which included functions containing the interelectron distance as a linear factor.

In this paper we describe how, following a method similar to that used by Clary, we have carried out an accurate calculation of the lowest (Σ_g^+) partial wave in positron-hydrogen-molecule scattering by using Kohn trial functions which include basis functions containing the positron-electron distance as a linear factor. We also report on progress on the extension of the method to the lowest partial waves of Σ_u^+ and Π_u symmetry. As far as we are aware, this is the first time that Hylleraas-type functions have been used in a molecular scattering calculation.

CALCULATION

The basic form of the calculation was similar to earlier calculations using separable basis functions (Armour[9,18]). The lowest partial wave is of Σ_g^+ symmetry and the trial function was taken to be of the form

$$\Psi_T = \Omega(c,\lambda_3,\mu_3;\tau,a)\Psi_G + \sum_{i=1}^{M} g_i\chi_i\Psi_G \qquad (1)$$

where τ, a and $\{g_i\}$ are variable parameters and $c = \frac{1}{2}kR$, where k is the wave number of the incident positron and $R = 1.4a_0$ is the separation between the nuclei, which are fixed in their equilibrium position. λ_i, μ_i and ϕ_i are the prolate spheroidal (or confocal elliptical) coordinates of the i-th particle (see, for example, Flammer[19]), where particles 1 and 2 are the target electrons and particle 3 is the positron. $\Omega(c,\lambda_3,\mu_3;\tau,a)$ is an open-channel function appropriate to the lowest partial wave and the $\{\chi_i\}$ are short-range correlation functions used to take into account the interaction between the positron and the hydrogen molecule. Ψ_G is the 6-term approximate target wavefunction used earlier[9,18]. It takes into account 50% of the correlation energy of the hydrogen molecule.

Two different forms were used for the open-channel function $\Omega(c,\lambda_3,\mu_3;\tau,a)$. Firstly,

$$\Omega(c,\lambda_3,\mu_3;\tau,a) = \left[\frac{B}{(\lambda_3-1)}\right](\sin[c(\lambda_3-1)]\cos\tau$$

$$+ \cos[c(\lambda_3-1)]\{1-\exp[-\gamma(\lambda_3-1)]\}\sin\tau$$

$$+ a[\cos[c(\lambda_3-1)]\{1-\exp[-\gamma(\lambda_3-1)]\}\cos\tau$$

$$- \sin[c(\lambda_3-1)]\sin\tau] \tag{2}$$

where B and γ are constants. This is basically the form of open-channel function introduced by Massey and Ridley[20] and was used in earlier calculations[9,18]. The parameter τ has been introduced to permit variation of the open-channel function to avoid the anomalous singularities that can arise in applications of the Kohn method[21,18].

Secondly,

$$\Omega(c,\lambda_3,\mu_3;\tau,a) = D(R_{00}^{(1)}(c,\lambda_3)\cos\tau$$

$$- R_{00}^{(2)}(c,\lambda_3)\{1-\exp[-\delta(\lambda_3-1)]\}^2\sin\tau$$

$$+ a[R_{00}^{(2)}(c,\lambda_3)\{1-\exp[-\delta(\lambda_3-1)]\}^2\cos\tau$$

$$+ R_{00}^{(1)}(c,\lambda_3)\sin\tau])S_{00}(c,\mu_3) \tag{3}$$

where D and δ are constants. $R_{00}^{(1)}(c,\lambda_3)$ and $R_{00}^{(2)}(c,\lambda_3)$ are the radial solutions to the free-particle equation in prolate spheroidal coordinates which are regular and irregular, respectively, at $\lambda_3 = 1$ and are associated with $S_{00}(c,\mu_3)$, the lowest spheroidal μ_3 function (see Flammer[19] and Takagi and Nakamura[22]). This form of open-channel function is exact if mixing of spheroidal partial waves is neglected[23].

In earlier calculations[9,18] the short-range correlation functions $\{\chi_i\}$ were taken to be of the form

$$\chi_i = \{\lambda_1^{a_i}\lambda_2^{b_i}\mu_1^{c_i}\mu_2^{d_i}[M_{13}\cos(\phi_1-\phi_3)]^{p_i}$$

$$+ \lambda_1^{b_i}\lambda_2^{a_i}\mu_1^{d_i}\mu_2^{c_i}[M_{23}\cos(\phi_2-\phi_3)]^{p_i}\}\exp[-\beta(\lambda_1+\lambda_2)]$$

$$\times B\lambda_3^{r_i}\mu_3^{s_i}\exp(-\alpha\lambda_3) \tag{4}$$

where

$$M_{13} = [(\lambda_1^2-1)(\lambda_3^2-1)(1-\mu_1^2)(1-\mu_3^2)]^{\frac{1}{2}} \tag{5}$$

$$= \frac{4}{R^2}(x_1x_3+y_1y_3), \tag{6}$$

a_i, b_i, c_i, d_i, p_i, r_i and s_i are non-negative integers and α and β are constants. The Cartesian coordinates are with respect to the nuclear centre of mass as origin and the z-axis is directed along the internuclear axis. $c_i + d_i + s_i$ must be even so that χ_i is of overall Σ_g^+ symmetry. p_i was taken to have the values 0 and 1. Note that if $p_i = 0$ the electronic part of χ_i is of Σ symmetry whereas if $p_i = 1$, it is of Π symmetry.

The crucial change in the present calculation is the inclusion of additional short-range correlation functions of the form

$$\chi_i = (\lambda_1^{a_i} \lambda_2^{b_i} \mu_1^{c_i} \mu_2^{d_i} r_{13} + \lambda_1^{b_i} \lambda_2^{a_i} \mu_1^{d_i} \mu_2^{c_i} r_{23})$$

$$\times \exp[-\beta(\lambda_1 + \lambda_2)]B\lambda_3^{r_i} \mu_3^{s_i} \exp(-\alpha\lambda_3) \tag{7}$$

where $\quad r_{13} = |\underline{x}_3 - \underline{x}_1|$. $\tag{8}$

These are the Hylleraas-type functions containing the positron-electron distance as a linear factor.

The inclusion of such functions means that to calculate the matrix elements in the Kohn equations, integrals involving up to three particles have to be evaluated, rather than just two particles as in the case of the separable functions used in the earlier calculation. However, as explained by Armour and Baker[24,25], use of the method of models[26,13], together with the fact that the absence of exchange makes it possible to include Ψ_G as a factor in Ψ_T, made it unnecessary to evaluate integrals involving the very complicated factor, $\dfrac{r_{13}r_{23}}{r_{12}}$. It was necessary to evaluate integrals involving the factors $r_{13}r_{23}$ and $\dfrac{r_{23}}{r_{13}}$. These were evaluated using the Neumann expansion for $\dfrac{1}{r_{13}}$ and the 'boundary derivative reduction' method of Handy and Boys[27,17,24]. Full details of the evaluation procedure will be given in a later publication[28]. Apart from the inclusion of the Hylleraas-type functions, the calculation was carried out in the same way as in earlier calculations[9,18].

RESULTS AND DISCUSSION

It is of interest to begin by examining the effect the inclusion of Hylleraas-type has on the calculated value of the annihilation rate. The determining parameter in this case is $Z_{eff}(k)$, the number of electrons per molecule available to the positron for annihilation[13,14]. The results obtained for $Z_{eff}(k)$ for k in the range $0.1 - 1.0a_0^{-1}$ using various basis sets are given in Table 1. It can be seen that the inclusion of just a small number of Hylleraas-type functions, in addition to the functions of Σ and Π electronic symmetry used in the earlier calculation[9], brings about a large increase in the contribution to $Z_{eff}(k)$ at the low energy values[29]. This was to be expected on theoretical grounds[14,25]. Investigation shows, however, that the omission from the basis set of the functions of Σ or Π electronic symmetry very much reduces the increase in the contribution to $Z_{eff}(k)$ at the low energy values[29,25].

For thermal positrons at 293K, where $kT = 25meV$, the results obtained with $\alpha = 0.575$ and the first open-channel function (equation (2)) predict a value[29] of 10.2 for $\bar{Z}_{eff}$, the Boltzmann average value of $Z_{eff}(k)$. This is much closer to the experimental value[15] of 14.8 than has been obtained in any previous calculation. Note that the other calculations using Hylleraas-type functions will give values of $\bar{Z}_{eff}$ grouped around this value.

Table 1. Contributions to $Z_{eff}(k)$ from the lowest partial wave in the range $k = 0.1$ to $1.0a_0^{-1}$

$k(a_0^{-1})$	E(eV)	Contribution to $Z_{eff}(k)$							
		1	2	3	4	5	6	7	8
0.1	0.14	3.28	9.42	9.62	9.77	8.46	10.45	8.64	11.50
0.2	0.54	2.78	7.31	7.47	6.85	6.22	7.44	5.38	7.74
0.3	1.2	2.02	5.28	5.40	5.26	4.96	5.57	4.23	5.14
0.4	2.2	1.74	4.43	4.54	4.47	3.78	4.68	3.58	5.05
0.5	3.4	1.62	4.13	4.22	4.29	3.41	3.88	3.08	4.04
0.6	4.9	1.60	3.57	3.64	3.19	2.80	3.12	2.70	3.20
0.7	6.7	1.29	2.74	2.79	2.78	2.37	2.71	2.35	2.73
0.85	9.8	1.15	2.33	2.37	2.62	2.07	2.31	2.06	2.30
1.0	13.6	1.15	2.49	2.49	2.06	3.08	2.00	1.95	1.99

Basis sets

First open-channel function with $\gamma = 0.75$

1 The 64 separable functions used in the earlier calculation[9] with $\alpha = 0.575$.

2 As in 1 plus functions 1-3 in Table 2 with $\alpha = 0.575$.

3 As in 1 plus functions 1-3 and 9 in Table 2 with $\alpha = 0.575$.

4 As in 1 plus functions 1-8 in Table 2 with $\alpha = 0.575$.

5 As in 1 plus functions 1-6 in Table 2 with $\alpha = 0.3$.

Second open-channel function with $\delta = 1.0$

6 As in 4, but without χ_1^{24}.

7 As in 5, but without χ_1.

Second open-channel function with $\delta = 0.75$

8 As in 1, but without χ_1, plus functions 1-6 in Table 2 with $\alpha = 0.575$.

In all cases, $\beta = 0.2$.

Table 2. Hylleraas-type basis functions of the form given in equation (7)

i	a_i	b_i	c_i	d_i	r_i	s_i
1	0	0	0	0	0	0
2	0	0	0	0	1	0
3	1	0	0	0	0	0
4	1	0	0	0	1	0
5	0	0	1	1	0	0
6	0	0	1	1	1	0
7	1	1	0	0	0	0
8	1	1	0	0	1	0
9	0	0	1	0	0	1

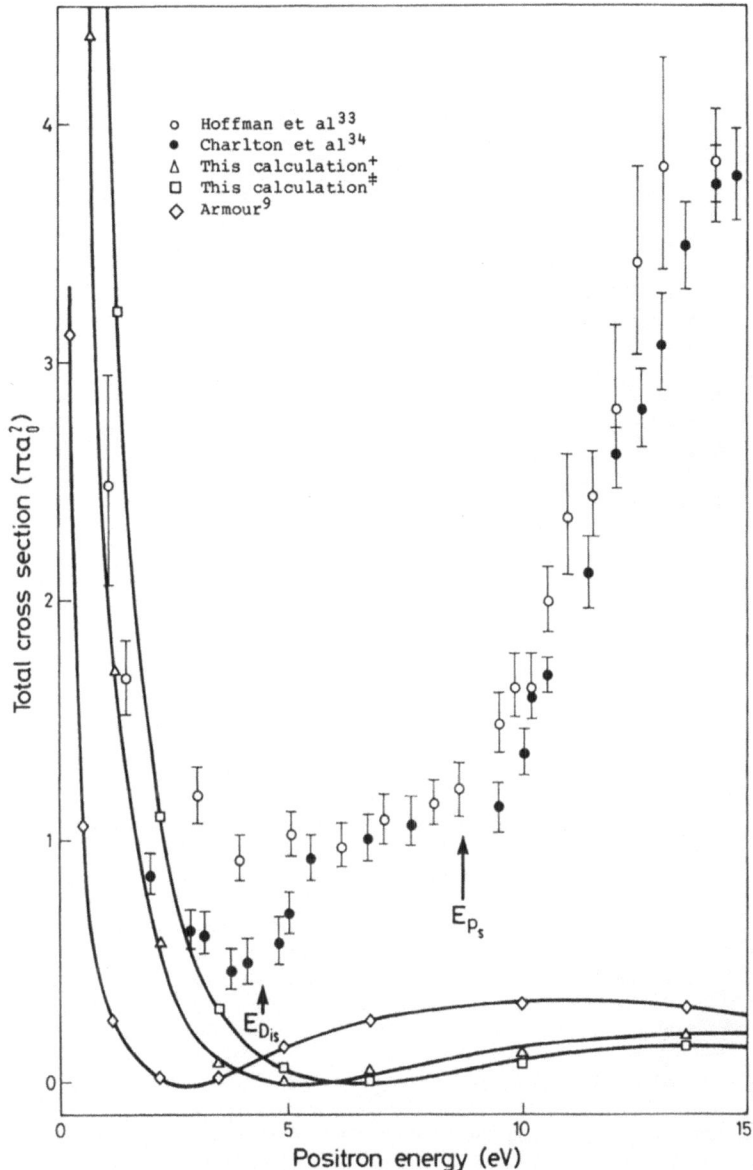

Figure 1. Comparison of experimental and theoretical total cross sections.

+First type of open-channel function and basis set 3 in Table 3.
‡Second type of open-channel function and basis set 6 in Table 3.

Table 3. Phase shifts in the range $k = 0.1$ to $1.0 a_0^{-1}$ obtained using both types of open-channel function

$k(a_0^{-1})$	$E(\text{eV})$	1	2	3	4	5	6	7	8	9
0.1	0.14	0.090	0.157	0.157	0.174	0.088	0.229	0.223	0.294	0.209
0.2	0.54	0.104	0.200	0.213	0.219	0.114	0.293	0.281	0.303	0.285
0.3	1.2	0.077	0.188	0.199	0.183	0.084	0.277	0.253	0.286	0.267
0.4	2.2	0.026	0.142	0.155	0.128	0.028	0.215	0.186	0.219	0.208
0.5	3.4	-0.034	0.081	0.066	0.060	-0.038	0.140	0.105	0.141	0.134
0.6	4.9	-0.113	-0.002	-0.002	-0.016	-0.104	0.064	0.022	0.063	0.061
0.7	6.7	-0.180	-0.066	-0.058	-0.076	-0.168	-0.010	-0.055	-0.012	-0.012
0.85	9.8	-0.243	-0.138	-0.137	-0.140	-0.249	-0.111	-0:151	-0.112	-0.112
1.0	13.6	-0.278	-0.210	-0.217	0.055	-0.308	-0.193	-0.221	-0.191	-0.194

Basis sets

First type of open-channel function with $\gamma = 0.75$

1 The 64 separable functions used in the earlier calculation[9] with $\alpha = 0.575$.

2 As in 1 plus functions 1-3 in Table 2 with $\alpha = 0.575$.

3 As in 1 plus functions 1-8 in Table 2 with $\alpha = 0.575$.

4 As in 1 plus functions 1-6 in Table 2 with $\alpha = 0.3$.

Second type of open-channel function with $\delta = 1.0$

5 As in 1, but without χ_1[24].

6 As in 3, but without χ_1.

7 As in 4, but without χ_1.

Second type of open-channel function with $\delta = 0.75$.

8 As in 1, but without χ_1, plus functions 1-6 in Table 2 with $\alpha = 0.575$.

Second type of open-channel function with $\delta = 1.25$.

9 As in 8.

In all cases, $\beta = 0.2$.

The results obtained for the phase shift for k in the range $0.1 - 1.0a_0^{-1}$ using both types of open-channel function and various basis sets are given in Table 3. The results obtained with only separable functions, i.e. basis sets 1 and 5, are in good overall agreement with the results obtained by Tennyson[10]. This has already been discussed by Tennyson (see also, Armour[23]).

It can be seen that the inclusion of just a small number of Hylleraas-type functions, in addition to the separable functions, brings about a large increase in the phase shift over the entire k range. The results obtained with $\alpha = 0.575$ and $\alpha = 0.3$ are similar except in the case of basis set 4 and $k = 1.0a_0^{-1}$, where there is evidence of instability in the Kohn calculation. Note that the increase in the phase shift is much greater in the case of the calculation using the second-type of open-channel function. This has been discussed by Armour and Baker[24,25].

The contribution to the total cross section from the lowest partial wave obtained using both types of open-channel function are compared with experiment in Figure 1. It can be seen that the inclusion of the 8 Hylleraas-type functions very much improves the agreement with experiment at very low energies. The contribution to the total cross section is now sufficient to account for the experimentally observed total cross section up to about 2eV. The first significant rearrangement threshold is the positronium formation threshold at 8.63eV[18]. No attempt has been made to take this into account in this calculation.

Improvements to the calculation have been discussed by Armour and Baker[24,25]. However, it is not expected that they will have as big an effect as the introduction of Hylleraas-type functions into the basis set.

EXTENSION OF THE CALCULATION TO HIGHER PARTIAL WAVES

It is straightforward to modify the above calculation to apply it to the lowest Σ_u^+ wave. The open-channel function in equation (3) is replaced by the function appropriate to this partial wave[30,23] and $c_i + d_i + s_i$ in the short-range correlation functions in equations (4) and (7) is required to be odd rather than even.

However, unlike the lowest Σ_g^+ wave, the Σ_u^+ wave experiences a centrifugal barrier. Thus at sufficiently low energies the phase shift is determined by the asymptotic form of the positron-molecule interaction potential,

$$V(r_3,\theta_3) \underset{r_3 \to \infty}{\sim} \frac{QP_2(\cos\theta_3)}{r_3^3} - \frac{\alpha_0}{2r_3^4} - \frac{\alpha_2 P_2(\cos\theta_3)}{2r_3^4} \tag{9}$$

where Q is the permanent quadrupole moment of the hydrogen molecule and

$$\alpha_0 = \tfrac{1}{3}(\alpha_\parallel + 2\alpha_\perp) \tag{10}$$

$$\alpha_2 = \tfrac{2}{3}(\alpha_\parallel - \alpha_\perp). \tag{11}$$

$\alpha_\parallel$ and $\alpha_\perp$ are the parallel and perpendicular polarizabilities, respectively, of the hydrogen molecule. r_3, θ_3, ϕ_3 are the spherical polar coordinates of the positron with respect to the Cartesian coordinates referred to earlier.

The first term is the tail of the static potential and can be straightforwardly taken into account[23] in a Kohn calculation. The second and third terms are the tail of the polarization potential. They are second order terms arising from virtual excitations involving target states of Σ_u^+ and Π_u symmetry[23]. To take these terms into account in a Kohn calculation requires the inclusion of further basis functions[23], some of Σ_u^+ and some of Π_u electronic symmetry, which give accurate values[31] for $\alpha_{\parallel}$ and $\alpha_{\perp}$ and behave asymptotically like $\left.\begin{array}{c}\cos c\lambda_3\\ \sin c\lambda_3\end{array}\right\} \times \lambda_3^{-3}$.

Work is in progress on the inclusion of oscillatory basis functions of this type in the Kohn trial function. Preliminary calculations have been carried out including both these basis functions and the separable basis functions of the form given in equation (4). The further extension of the basis set to include Hylleraas-type functions, as well as oscillatory functions and separable functions, should present no problems.

Preliminary calculations of the lowest Π_u wave have also been carried out using separable short-range correlation functions[32,23] similar to those in equation (4). This calculation needs further extension to include Hylleraas-type functions and also oscillatory functions. This will be more complicated than for the Σ_u^+ wave but, once again, should present no problems.

CONCLUSION

The change in the charge when positron rather than electron scattering is considered means that the interaction between the projectile and the bound but mobile target electrons is attractive rather than repulsive. To take it into account adequately at low energies requires the inclusion of Hylleraas-type functions in the basis set. Hitherto this has only been possible for simple atoms. This calculation shows that the inclusion of such functions is also possible in the case of hydrogen molecules. Refinement of the method used in this calculation should make it possible to obtain as detailed an understanding of positron-hydrogen molecule scattering as has already been obtained for positron-helium scattering[13].

ACKNOWLEDGEMENTS

We are grateful to the SERC(UK) for financial support for this research.

REFERENCES

1. E.A. Hylleraas, Z. für Phys. 54: 347 (1929).
2. H.M. James and A.S. Coolidge, J. Chem. Phys. 1: 825 (1933).
3. H.F. Schaefer, "Quantum Chemistry" (Oxford University Press, Oxford, 1984).
4. P.G. Burke, A. Hibbert and W.D. Robb, J. Phys. B 4: 153 (1971).
5. P.G. Burke, I. Mackey and I. Shimamura, J. Phys. B 10: 2497 (1977).
6. B.I. Schneider and L.A. Collins, J. Phys. B 18: L857 (1985).
7. M.A.P. Lima, T.L. Gibson, W.M. Huo and V. McKoy, J. Phys. B 18: L865 (1985).
8. "Proceedings of the NATO Advanced Research Workshop on Atomic Physics with Positrons, University College London, 1987", J.W. Humberston and E.A.G. Armour, eds, Plenum, New York (1988).
9. E.A.G. Armour, J. Phys. B 18: 3361 (1985) (Corrigendum Ibid. 20: 5255 (1987)).
10. J. Tennyson, J. Phys. B 19: 4255 (1986).
11. P.A. Fraser, B.H. Bransden, P.G. Coleman and W. Raith, Can. J. Phys. 60: 565 (1982).

12. J.W. Humberston, Adv. At. Mol. Phys. 22: 1 (1986).
13. J.W. Humberston, Adv. At. Mol. Phys. 15: 101 (1979).
14. E.A.G. Armour and D.J. Baker, J. Phys. B 18: L845 (1985).
15. J.D. McNutt, S.C. Sharma and R.D. Brisbon, Phys. Rev. A 20: 347 (1979).
16. C. Schwartz, Phys. Rev. 124: 1468 (1961).
17. D.C. Clary, Mol. Phys. 34: 793 (1977).
18. E.A.G. Armour, J. Phys. B 17: L375 (1984).
19. C. Flammer, "Spheroidal Wavefunctions", Stanford University Press, Stanford, CA (1957).
20. H.S.W. Massey and R.O. Ridley, Proc. Phys. Soc. (London) A69: 659 (1956).
21. C. Schwartz, Ann. Phys. NY 16: 36 (1961).
22. H. Takagi and H. Nakamura, J. Phys. B 13: 2619 (1980).
23. E.A.G. Armour, submitted to Physics Reports.
24. E.A.G. Armour and D.J. Baker, J. Phys. B 20: 6105 (1987).
25. E.A.G. Armour, "Proceedings of the NATO Advanced Research Workshop on Atomic Physics with Positrons, University College London, 1987", J.W. Humberston and E.A.G. Armour, eds, Plenum, New York (1988) p.95.
26. R.J. Drachman, J. Phys. B 5: L30 (1972).
27. N.C. Handy and S.F. Boys, Theo. Chim. Acta 31: 195 (1973).
28. E.A.G. Armour and J.W. Humberston, Computer Physics Reports, in preparation.
29. E.A.G. Armour and D.J. Baker, J. Phys. B 19: L871 (1986).
30. E.A.G. Armour, "Proc. 3rd Int. Workshop on Positron(Electron)-Gas Scattering, Detroit, 1985", W.E. Kauppila, T.S. Stein and J.M. Wadehra, eds, World Scientific, Singapore (1986) p.85.
31. W. Kołos and L. Wolniewicz, J. Chem. Phys. 46: 1426 (1967).
32. E.A.G. Armour, D.J. Baker and M. Plummer, "Abstracts of the XV ICPEAC, Brighton 1987", J. Geddes, H.B. Gilbody, A.E. Kingston, C.J. Latimer and H.J.R. Walters, eds, p.408.
33. K.R. Hoffman, M.S. Dababnek, Y.-F. Hsieh, W.E. Kauppila, V. Pol, J.H. Smart and T.S. Stein, Phys. Rev. A25: 1393 (1982).
34. M. Charlton, T.C. Griffith, G.R. Heyland and G.L. Wright, J. Phys. B 16: 323 (1983).

LIST OF DELEGATES

R. ABOUAF, Laboratoire des Collisions Atomiques et Moleculaires, Bâtiment 351, Université Paris-Sud, 91405 Orsay Cedex, France.

T. AJIRO, Institute of Laser Science, University of Electro-Communications, Chofugaoka, Chofushi, Tokyo 182, Japan.

M. ALLAN, Institut de Chimie Physique, University of Fribourg, Perolles, CH-1700 Fribourg, Switzerland.

J.M. BENSON, Department of Physics, King's College London, Strand, London WC2R 2LS.

C.E. BIELSHOWSKY, Instituto de Quimica da UFRJ, Department de Fisico Quimica, CT Bloco A s/406, Cidade Universitaria, Ilha do Fundao, 21910 Rio De Janeiro, Brazil.

N. BOWERING, Fakultat für Physik, Universität Bielefeld, Pf 8640, D-4800 Bielefeld, Fed. Rep. Germany.

C. BRION, Department of Chemistry, University of British Columbia, 2036 Main Mall, Vancouver, BC, Canada V6T 1Y6.

S.J. BUCKMAN, Institute of Advanced Studies, School of Physical Sciences, Australian National University, P.O. Box 4, Canberra ACT 2600, Australia.

P.G. BURKE, Department of Applied Mathematics, Queen's University Belfast, Belfast BT7 1NN, N. Ireland.

A.A. CAFOLLA, Department of Physics, University of Manchester, Manchester M13 9PL.

L.A. COLLINS, Group T-4 MS B212, Los Alamos National Laboratory, Los Alamos, NM 87545, USA.

J. COMER, Department of Physics, University of Manchester, Manchester M13 9PL.

G. DANBY, Department of Physics & Astronomy, University College London, Gower Street, London WC1E 6BT.

P. DECLEVA, Dipartimento di Scienze Chimiche, Universita di Trieste, Piazzale Europa 1, I-34127 Trieste, Italy.

J. DEHMER, Radiological & Environmental Research Division, Building 203, Argonne National Laboratory, Argonne, IL 60439, USA.

P.J. DRALLOS, Department of Physics & Astronomy, Wayne State University, 666 W. Hancock, Detroit, Michigan 48202, USA.

H. EHRHARDT, Fachbereich Physik der Universität, Erwin-Schrödinger Strasse 46, Universität Kaiserslautern, D-6750 Kaiserslautern, Fed. Rep. Germany.

J.H.D. ELAND, Physical Chemistry Laboratory, University of Oxford, South Parks Road, Oxford OX1 3QZ.

D. FIELD, School of Chemistry, University of Bristol, Cantocks Close, Bristol BS8 1TS.

V. GALASSO, Dipartimento di Scienze Chimiche, Universita degli Studi di Trieste, I-34127 Trieste, Italy.

J.-P. GAUYACQ, LCAM, Bâtiment 351, Université Paris-Sud, 91405 Orsay Cedex, France.

F.A. GIANTURCO, Universita degli Studi di Roma 'La Sapienza', Dipartimento di Chimica Nuovo Edificio Chimico, Citta Universitaria, 00185 Roma AD, Italy.

C.J. GILLAN, Department of Applied Mathematics and Theoretical Physics, David Bates Building, College Park, Queen's University of Belfast, Belfast BT7 1NN.

A.-M. GRISOGONO, Institute of Atomic Studies, School of Physical Sciences, Flinders University of South Australia, Bedford Park, 5042 South Australia.

R. HALL, Groupe de Spectroscopie par Impact Electronique et Ionique, Université de Pierre et Marie Curie, 4 Place Jussieu T-12 E-5, 75252 Paris Cedex 05, France.

J.E. HANSEN, Zeeman-Laboratorium, Universität van Amsterdam, Plantage Muidergracht 4, 1018 TV Amsterdam, The Netherlands.

S. HARA, Institute of Physics, University of Tsukuba, Ibaraki 305, Japan.

P.A. HATHERLY, J.J. Thomson Physical Laboratory, University of Reading, Whiteknights, Reading RG6 2AF.

T. HAYAISHI, Institute of Applied Physics, University of Tsukuba, Sakura, Ibaraki 305, Japan.

L. HELLNER, Laboratoire de Photophysique Moleculaire, CNRS, Bâtiment 213, Université Paris-Sud, 91405 Orsay Cedex, France.

A. HERZENBERG, Department of Applied Physics, Yale University, P.O. Box 2157, New Haven, CT 06520, USA.

A. PEET HICKMAN, Molecular Physics Department, SRI International, 333 Ravenswood Avenue, Menlo Park, CA 94025, USA.

T. HIRAYAMA, c/o Professor Hirosi Suzuki, Department of Physics, Sophia University, Kioicho, Chiyodaku, Tokyo 102, Japan.

J.R. HISKES, Lawrence Livermore National Laboratory, University of California, P.O. Box 5511 L-630, Livermore, CA 94550, USA.

D. HOLLAND, SERC Daresbury Laboratory, Warrington WA4 4AD, U.K.

X.-M. HU, Department of Physics, University of Aberdeen,
    Aberdeen AB9 2UE.

M. JUNGEN, Institute of Physical Chemistry, University of Basle,
    Klingelbergstrasse 80, CH 4056, Basle, Switzerland.

A. KATASE, Department of Nuclear Engineering, Faculty of Engineering,
    Kyushu University, Fukuoka 812, Japan.

K. KAUFMANN, Institute of Physical Chemistry, University of Basle,
    Klingelbergstrasse 80, CH 4056, Basle, Switzerland.

G.C. KING, Department of Physics, Schuster Laboratory, University of
    Manchester, Manchester M13 9PL.

D.W. KNIGHT, Department of Science, Bristol Polytechnic, Coldharbour
    Lane, Frenchay, Bristol BS16 1QY.

T. KOIZUMI, Department of Physics, Faculty of Science, Rikkyo University,
    34-1 Nishi-Ikebukuro 3, Toshima-ku, Tokyo 171, Japan.

Y.-W.J. KOO, Department of Physics, Natural Philosophy Building,
    University of Aberdeen, Aberdeen AB9 2UE.

I. KOYANO, Institute for Molecular Science, Myodaiji, Okazaki 444, Japan.

M. LE DOURNEUF, Observatoire de Paris, Section d'Astrophysique, 5 Place
    Jules Janssen, 92195 Meudon Principal Cedex, France.

B.G. LINDSAY, Department of Pure & Applied Physics, Queen's University of
    Belfast, Belfast BT7 1NN.

W.C. LINEBERGER, JILA CB 440, University of Colorado, Boulder,
    Colorado 80309-0440, USA.

A. LISINI, Dipartimento di Scienze, Universita di Trieste, Piazzale
    Europa 1, I-34127 Trieste, Italy.

R. LUCCHESE, Department of Chemistry, Texas A & M University, College
    Station, Texas 77843, USA.

D.L. LYNCH, MS J569, Los Alamos National Laboratory, Los Alamos,
    New Mexico 87545, USA.

H. MAKOTO, Nagoya Institute of Technology, Gokiso-cho, Showa-ku,
    Nagoya 466, Japan.

G.V. MARR, Department of Physics, Natural Philosophy Building, University
    of Aberdeen, Aberdeen AB9 2UE.

N.J. MASON, Department of Physics & Astronomy, University College London,
    Gower Street, London WC1E 6BT.

H.-D. MEYER, Theoretische Chemie, Physikalisch-Chemisches Institut,
    Universität Heidelberg, Im Neuenheimer Feld 253, D-6900 Heidelberg,
    W. Germany.

H.-H. MICHELS, Department of Applied Physics, United Technologies
    Research Centre, East Hartford, Connecticut 06108, USA.

L. MORGAN, University of London Computer Centre, Royal Holloway & Bedford New College, Egham Hill, Egham, Surrey TW20 0EX.

T. MORI, c/o Professor H. Suzuki, Department of Physics, Sophia University, Kioicho, Chiyodaku, Tokyo 102, Japan.

A.Z. MSEZANE, Department of Physics, Atlanta University, 223 James P. Brawley Drive SW, Atlanta, Georgia 30314, USA.

M. MULLER, Fritz-Haber-Institut der Max-Planck-Gesellschaft, Faradayweg 4-6, D1000 Berlin 33, Fed. Rep. Germany.

K. NAKASHIMA, Department of Molecular Science, Kyushu University, Kasuga-shi, Fukuoka 816, Japan.

I. NENNER, Department de Physico-Chimie, CEN Saclay, 91191 Gif-Sur-Yvette, France.

R.K. NESBET, IBM Almaden Research Centre, K31/802, 650 Harry Road, San Jose, CA 95120-6099, USA.

C.J. NOBLE, SERC Daresbury Laboratory, Warrington WA4 4AD, U.K.

D.W. NORCROSS, JILA, Box 440, University of Colorado, Boulder, Colorado 80309-0440, USA.

T. OGAWA, Department of Molecular Science & Technology, Kyushu University, Kasuga-shi, Fukuoka 816, Japan.

T.F. O'MALLEY, 2276 Lucretia No. 4, San Jose, CA 95122, USA.

R. PALMER, Cavendish Laboratory, University of Cambridge, Madingley Road, Cambridge CB3 0HE.

A.V. PHELPS, JILA, Box 440, University of Colorado, Boulder, Colorado 80309-0440, USA.

M. PLUMMER, Mathematics Department, University of Nottingham, Nottingham NG7 2RD.

E.D. POLIAKOFF, Department of Chemistry, University of Boston, 590 Commonwealth Avenue, Boston, MA 02215, USA.

A.W. POTTS, Department of Physics, King's College London, University of London, Strand, London WC2R 2LS.

S.D. PRICE, Physical Chemistry Laboratory, University of Oxford, South Parks Road, Oxford OX1 3QZ.

G. RASEEV, Laboratoire de Photophysique Moleculaire, CNRS, Bâtiment 213, Université Paris-Sud, 91405 Orsay Cedex, France.

T.J. REDDISH, Department of Physics, University of Manchester, Oxford Road, Manchester M13 9PL.

B.C. SAHA, Department of Physics and Astronomy, University of Oklahoma, Norman, OK 73019, USA.

T. SAKAE, Department of Nuclear Engineering, Faculty of Engineering, Kyushu University, Fukuoka 812, Japan.

Y. SAKAI, c/o Professor H. Suzuki, Department of Physics, Sophia
University, Kioicho, Chiyodaku, Tokyo 102, Japan.

J.A.R. SAMSON, Behlen Laboratory of Physics, University of Nebraska,
Lincoln, Nebraska 68588-0111, USA.

H. SATO, Department of Physics, Faculty of Science, Ochanomizu
University, 1-1 Otsuku 2, Bunkyo-ku, Tokyo, Japan.

T. SCHOLZ, DAMTP, Queen's University of Belfast, Belfast BT7 1NN,
N. Ireland.

S. SCIALLA, Department of Applied Mathematics, Queen's University of
Belfast, Belfast BT7 1NN, N. Ireland.

I. SHIMAMURA, RIKEN, Institute of Physics & Chemistry Research, Hirosawa,
Wako Saitama 351, Japan.

D. SPENCE, Argonne National Laboratory, 9700 South Cass Avenue,
Building 203, Argonne, IL 60439, USA.

S. SRIVASTAVA, Jet Propulsion Laboratory, MS 183-601, 4800 Oak Grove
Drive, Pasadena, CA 91109, USA.

G. STEFANI, Consiglio Nazionale delle Ricerche Istituto di Met. Avanzate,
Inorganiche Area della Ricerca di Roma, Via Salaria Km 29 300, 00016
Monterotondo Scalo (Roma), Italy.

M.B. SUTCLIFFE, Department of Chemistry, Imperial College, London.

A. SVENSSON, SERC Daresbury Laboratory, Warrington WA4 4AD, U.K.

H. TAKAGI, Kitasato University, Physics Laboratory, School of Medicine,
Kitasato 1-15-1, Sagimihara, Kanagawa 228, Japan.

D. TEILLET-BILLY, Laboratoire des Collisions Atomiques et Moleculaires,
Bâtiment 351, Université Paris-Sud, 91405 Orsay Cedex, France.

A. TEMKIN, Goddard Space Flight Centre, Greenbelt, Maryland 20771, USA.

J. TENNYSON, Department of Physics & Astronomy, University College
London, Gower Street, London WC1E 6BT.

P. TEUBNER, School of Physical Sciences, The Flinders University of South
Australia, Bedford Park, South Australia 5042, Australia.

D.G. THOMPSON, Department of Applied Mathematics & Theoretical Physics,
Queen's University of Belfast, Belfast BT7 1NN, N. Ireland.

J. TULLY, Observatoire de Nice, BP 139, 06003 Nice Cedex, France.

M. UKAI, Department of Chemistry, Tokyo Institute of Technology,
Meguro-ku, Tokyo 152, Japan.

R.J. VAN BRUNT, National Bureau of Standards, Building 220, Room B344,
Gaithersburg, MD 20899, USA.

A. WAGUE, Department of Physics, Université de Dakar, Dakar-Fann,
Senegal.

E. WEIGOLD, School of Physical Sciences, Flinders University of South
    Australia, Sturt Road, Bedford Park 5042, Australia.

J.B. WEST, SERC Daresbury Laboratory, Warrington WA4 4AD, U.K.

A. OREL WOODIN, M5/754 The Aerospace Corporation, P.O. Box 92957,
    Los Angeles, CA 90009, USA.

A. YAGISHITA, c/o Professor B. Sonntag, DESY, Notkestr. 85, 2000
    Hamburg 52, Fed. Rep. Germany.

M. ZUBEK, Department of Physics, Technical University of Gdansk,
    Majakowskiego 11/12, 80-952 Gdansk-Wrzeszcz, Poland.

INDEX